Complex Webs

Complex Webs synthesizes modern mathematical developments with a broad range of complex network applications of interest to the engineer and system scientist, presenting the common principles, algorithms, and tools governing network behavior, dynamics, and complexity. The authors investigate multiple mathematical approaches to inverse power laws and expose the myth of normal statistics to describe natural and man-made networks. Richly illustrated throughout with real-world examples including cell phone use, accessing the Internet, failure of power grids, measures of health and disease, distribution of wealth, and many other familiar phenomena from physiology, bioengineering, biophysics, and informational and social networks, this book makes thought-provoking reading. With explanations of phenomena, diagrams, end-of-chapter problems, and worked examples, it is ideal for advanced undergraduate and graduate students in engineering and the life, social, and physical sciences. It is also a perfect introduction for researchers who are interested in this exciting new way of viewing dynamic networks.

BRUCE J. WEST is Chief Scientist Mathematics with the Information Science Directorate at the Army Research Office, a position he has held for the last 10 years. After receiving his Ph.D. in Physics from the University of Rochester in 1970, he was Associate Director of a small private research institute (La Jolla Institute) for almost 20 years and a Professor at the University of North Texas for a decade. His research interests are in the nonlinear dynamics of complex networks. He has over 350 scientific publications, including 11 books and 8,500 citations, and he has received multiple academic and government awards for his research and publications.

PAOLO GRIGOLINI is currently a Professor in the Physics Department and the Center for Nonlinear Science at the University of North Texas. He is an internationally recognized theorist interested in the foundations of quantum mechanics, including wave-function collapse and the influence of classical chaos on quantum systems. His other research interests include the foundations of statistical physics, biophysical problems such as DNA sequencing, and the network science of human decision making and cognition.

Complex Webs

Anticipating the Improbable

BRUCE J. WEST
Army Research Office, USA

PAOLO GRIGOLINI
University of North Texas

CAMBRIDGE
UNIVERSITY PRESS

CAMBRIDGE UNIVERSITY PRESS
Cambridge, New York, Melbourne, Madrid, Cape Town, Singapore,
São Paulo, Delhi, Dubai, Tokyo, Mexico City

Cambridge University Press
The Edinburgh Building, Cambridge CB2 8RU, UK

Published in the United States of America by Cambridge University Press, New York

www.cambridge.org
Information on this title: www.cambridge.org/9780521113663

First published 2011

Printed in the United Kingdom at the University Press, Cambridge

A catalog record for this publication is available from the British Library

ISBN 978-0-521-11366-3 Hardback

Contents

Preface

The Italian engineer turned social scientist Vilfredo Pareto was the first investigator to determine that the income in western society followed a law that was fundamentally unfair. He was not making a value judgement about the poor and uneducated or about the rich and pampered; rather, he was interpreting the empirical finding that in 1894 the distribution of income in western societies was not "normal," but instead the number of people with a given income decreased as a power of the level of income. On bi-logarithmic graph paper this income distribution graphs as a straight-line segment of negative slope and is called an inverse power law. He interpreted his findings as meaning that a stable society has an intrinsic imbalance resulting from its complex nature, with the wealthy having a disproportionate fraction of the available wealth. Since then staggeringly many phenomena from biology, botany, economics, medicine, physics, physiology, psychology, in short every traditional discipline, have been found to involve complex phenomena that manifest inverse power-law behavior. These empirical laws were explained in the last half of the twentieth century as resulting from the complexity of the underlying phenomena.

As the twentieth century closed and the twenty-first century opened, a new understanding of the empirical inverse power laws emerged. This new understanding was based on the connectedness of the elements within the underlying phenomena and the supporting web structure. The idea of networks became pervasive as attention was drawn to society's reliance on sewers and the electric grid, cell phones and the Internet, banks and global stock markets, roads and rail lines, and the multitude of other human-engineered webbings that interconnect and support society. In parallel with the studies of social phenomena came new insight into the distribution in size and frequency of earthquakes and volcanic eruptions, global temperature anomalies and solar flares, river tributaries and a variety of other natural phenomena that have eluded exact description by the physical sciences. Moreover, the inverse power laws cropped up in unexpected places such as in heart rates, stride intervals and breathing, letter writing and emails, cities and wars, heart attacks and strokes; the inverse power law is apparently ubiquitous.

The synthesis of complexity and networks emphasizes the need for a new kind of scientific understanding, namely a grasp of how things work that exceeds the traditional mechanistic approach taken by science ever since Newton introduced gravity to explain planetary orbits and why things fall. The historical scientific approach reveals the workings of the two-body problem, but when three or more bodies interact in this way the strategy breaks down; chaos typically takes over and a different kind of thinking is required. This book is about how this new way of thinking has struggled to overcome the shackles of the "normal" distribution and the domination of the mean and standard deviation within the traditional disciplines.

The final result of our studies, ruminations and collaborations is not a standard textbook, although there are extended explanations of phenomena, diagrams, problems and worked-out examples. Instead this book has many characteristics associated with more idiosyncratic monographs; including well-labeled speculations, arguments that are meant to provoke response and stimulate thinking, and connections made to modern research discoveries that are usually denied to elementary texts. We have labored to get the mathematics right while not disregarding the language and the spirit of scientific discovery that is usually scrubbed from more traditional texts.

A number of people assisted us in preparing the present book. Gerhard Werner contributed directly by reading a draft of the manuscript and providing extensive feedback on how to improve its readability, suggestions that we took seriously. Our past students and present collaborators have contributed less directly to the manuscript and more directly to the research on which various sections of the book are based. The contributions of P. Allegrini, G. Aquino, F. Barbi, M. Bianucci, M. Bologna, M. Ignaccolo, M. Latka, A. Rocco and N. Scafetta are extensively referenced throughout, indicating their contributions to our fundamental theoretical understanding of the ubiquity of empirical inverse power-law distributions in complex webs.

1   Webs

The science of complex webs, also known as network science, is an exciting area of contemporary research, overarching the traditional scientific disciplines of biology, economics, physics, sociology and the other compartments of knowledge found in any college catalog. The transportation grids of planes, highways and railroads, the economic meshes of global finance and stock markets, the social webs of terrorism, governments, and businesses as well as churches, mosques, and synagogs, the physical lattices of telephones, the Internet, earthquakes, and global warming, in addition to the biological networks of gene regulation, the human body, clusters of neurons and food webs, share a number of apparently universal properties as the webs become increasingly complex. This conclusion is shared by the recent report *Network Science* [23] published under the auspices of the National Academy of Science. The terms networks and network science have become popular tags for these various areas of investigation, but we prefer the image of a web rather than the abstraction of a network, so we use the term web more often than the synonyms network, mesh, net, lattice, grille or fret. Colloquially, the term web entails the notion of entanglement that the name network does not share. Perhaps it is just the idea of the spider ensnaring its prey that appeals to our darker sides.

Whatever the intellectual material is called, this book is not about the research that has been done to understand complex webs, at least not in the sense of a monograph. We have attempted to put selected portions of that research into a pedagogic and often informal context, one that highlights the limitations of the more traditional descriptions of these areas. In this regard we are obligated to discuss the state of the art regarding a broad sweep of complex phenomena from a variety of academic disciplines. Sometimes properly setting the stage requires a historical approach and other times the historical view is replaced with personal perspectives, but with either approach we do not leave the reader alone to make sense of what can be difficult material. So we begin by illuminating the basic assumptions that often go unexamined in science.

1.1   The myth of normalcy

Natural philosophers metamorphosed into modern scientists in part by developing a passion for the quantifiable. But this belief in the virtue of numbers did not come about easily. In the time of Galileo Galilei (1564–1642), who was followed on his death by

the birth of Sir Isaac Newton (1642–1727), experiments did not yield the kind of reproducible results we are accustomed to accepting today. Every freshman physics course
has a laboratory experiment on measurement error that is intended to make students
familiar with the fact that experiments are never exactly reproducible; there is always
experimental error. But this pedagogic exercise is quickly forgotten, even when well
learned. What is lost in the student's adjustment to college culture is that this is probably the most important experiment done during that first year. The implications of the
uncertainty in scientific investigation extend far beyond the physics laboratory and are
worthy of comments regarding their significance.

Most people recognize that they do not completely control their lives; whether it is
the uncertainty in the economy and how it will affect a job, the unexpected illness that
disrupts the planned vacation, or the death of one near and dear, all these things are
beyond one's control. But there is solace in the belief, born of the industrial revolution,
that we can control our destiny if only we had enough money, or sufficient prestige
and, most recently, if we had an adequate amount of information. This is the legacy of
science from the nineteenth and twentieth centuries, that the world can be controlled if
only we more completely understood and could activate the levers of power. But is this
true? Can we transfer the ideas of predictability and controllability from science to our
everyday lives? Do the human sciences of sociology and psychology have laws in the
same way that physics does?

In order to answer these and other similar questions it is necessary to understand how
scientists have traditionally treated variability and uncertainty in the physical sciences.
We begin with a focus on the physical sciences because physics was historically the first
to develop the notion of quantification of physical laws and to construct the underlying
mathematical infrastructure that enabled the physicist to view one physical phenomenon
after another through the same lens and thereby achieve a fundamental level of understanding. One example of this unity of perspective is the atomic theory of matter, which
enables us to explain much of what we see in the world, from the sun on our face to the
rain wetting our clothes or the rainbow on the horizon, all from the same point of view.
But the details are not so readily available.

We cannot predict exactly when a rain shower will begin, how long it will last, or how
much ground it will cover. We might understand the basic phenomenon at the microscopic level and predict exactly the size and properties of molecules, but that does not
establish the same level of certainty at the macroscopic level where water molecules
fall as rain. The smallest seemingly unimportant microscopic variation is amplified by
means of nonlinear interactions into a macroscopic uncertainty that is completely unpredictable. Therefore it appears to us that in order to develop defenses against the vagaries
of life it is necessary to understand how science treats uncertainty and explore the limitations of those treatments. Most importantly, it is necessary to understand what science
got right and what it got wrong. For that we go back to the freshman physics laboratory
experiment on measurements.

Each time a measurement is made a certain amount of estimation is required, whether
it is estimating the markings on a ruler or the alignment of a pointer on some gauge. If
one measures a quantity q a given number of times, N say, then instead of having a

single quantity Q the measurements yield a collection of quantities $Q_1, Q_2, \ldots, Q_N$. Such a collection is often called an ensemble and the challenge is to establish the best representation of the ensemble of measurements. Simpson, of "Simpson's rule" fame in the calculus, was the first scientist to recommend in print [30] that all the measurements taken in an experiment ought to be utilized in the determination of a quantity, not just those considered to be the most reliable, as was the custom in the seventeenth century. He was the first to recognize that the observed discrepancies between successively measured events follow a pattern that is characteristic of the ensemble of measurements. His observations were the forerunner to the *law of frequency of errors*, which asserts that there exists a relationship between the magnitude of an error and how many times it occurs in an ensemble of experimental results. Of course, the notion of an error implies that there is an exact value that the measurement is attempting to discern and that the variability in the data is a consequence of mistakes being made, resulting in deviations from the exact value, that is, in errors.

This notion of a correct value is an intriguing one in that it makes an implicit assumption about the nature of the world. Judges do not allow questions of the form "Have you stopped beating your wife?" because implicit in the question is the idea that the person had beaten his wife in the past. Therefore either answer, yes or no, confirms that the prisoner has beaten his wife in the past, which is, presumably, the question to be determined. Such leading questions are disallowed from the courtroom but are the bread and butter of science. Scientists are clever people and consequently they have raised the leading question to the level of hypothesis and turned the tables on their critics by asking "Have you measured the best value of this experimentally observed phenomenon?" Of course, either answer reinforces the idea of a best value. So what is this mysterious best value?

To answer this question we need to distinguish between statistics and probability; statistics has to do with measurements and data, whereas probability has to do with the mathematical theory of those measurements. Statistics arise because on the one hand individual results of experiments change in unpredictable ways and, on the other hand, the average values of long data sequences show remarkable stability. It is this statistical regularity that suggests the existence of a best value and hints at a mathematical model of the body of empirical data [8]. We point this out because it is not difficult to become confused over meaning in a discussion on the probability associated with a statistical process. The probability is a mathematical construct intended to represent the manner in which the fluctuating data are distributed over the range of possible values. Statistics represent the real world; probability represents one possible abstraction of that world that attempts to make quantitative deductions from the statistics. The novice should take note that the definition of probability is not universally accepted by the mathematical community. One camp interprets probability theory as a theory of *degrees of reasonable belief* and is completely disassociated from statistics in that a probability can be associated with any proposition, even one that is not reproducible. The second camp, with various degrees of subtlety, interprets probability theory in terms of the relative frequency of the occurrence of an event out of the universe of possible events. This second definition of probability is the one used throughout science and is adopted below.

Consider the relative number of times N_j a measurement error of a given magnitude occurs in a population of a given (large) size N; the relative frequency of occurrence of any particular error in this population is

$$p_j = \frac{N_j}{N}.$$
(1.1)

Here j indexes the above measurements into M different bins, where typically $N \gg M$. The relative number of occurrences provides an estimate of the probability that a measurement of this size will occur in further experiments. From this ensemble of N independent measurements we can define an average value,

$$\overline{Q} = \sum_{j=1}^{M} Q_j\, p_j,$$
(1.2)

with the average of a variable being denoted by the overbar and the N measurements put into M bins of equal size. The mean value $\overline{Q}$ is often thought to be an adequate characterization of the measurement and thus an operational definition of the experimental variable is associated with $\overline{Q}$. Simpson was the first to suggest that the mean value be accepted as the *best* value for the measured quantity. He further proposed that an isosceles triangle be used to represent the theoretical distribution in the measurements around the mean value. Of course, we now know that using the isosceles triangle as a measure of variability is wrong, but don't judge Simpson too harshly, after all, he was willing to put his reputation on the line and speculate on the possible solution to a very difficult scientific problem in his time and he got the principle right even if he got the equation wrong.

Subsequently, it was accepted that to be more quantitative one should examine the degree of variation of the measured value away from its average or "true" value. The magnitude of this variation is defined not by Simpson's isosceles triangles but by the standard deviation σ or the variance σ^2 of the measurements,

$$\sigma^2 \equiv \sum_{j=1}^{M} \left(Q_j - \overline{Q} \right)^2 p_j,$$
(1.3)

which, using the definition of the average and the normalization condition for the probability

$$\sum_{j=1}^{M} p_j = 1,$$
(1.4)

reduces to

$$\sigma^2 = \overline{Q^2} - \overline{Q}^2.$$
(1.5)

These equations are probably the most famous in statistics and form the basis of virtually every empirical theory of the physical, social and life sciences that uses discrete data sets.

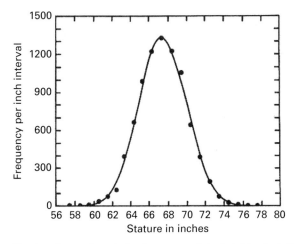

Figure 1.1.   The dots denote the relative frequency of the heights of adult males in the British Isles [41]. The solid curve is the normal distribution with the same mean and variance as that of the data points.

In the continuum limit, that is, the limit in which the number of independent observations of a web variable approaches infinity, the characteristics of any measured quantity are specified by means of a distribution function. From this perspective any particular measurement has little or no meaning in itself; only the collection of measurements, the ensemble, has a scientific interpretation that is manifest through the distribution function. The distribution function is also called the probability density and it associates a probability with the occurrence of an event in the neighborhood of a measurement of a given size. For example in Figure 1.1 is depicted the frequency of occurrence of adult males of a given height in the general population of the British Isles. From this distribution it is clear that the likelihood of encountering a male six feet in height on your trip to Britain is substantial and the probability of meeting someone more than ten feet tall is zero.

Quantitatively, the probability of meeting someone with a height Q in the interval $(q, q + \Delta q)$ is given by the product of the distribution function and the size of the interval $P(q)\Delta q$. The solid curve in Figure 1.1 is given by a mathematical expression for the functional form of $P(q)$. Such a bell-shaped curve, whether from measurements of heights or from errors, is described by the well-known distribution of Gauss, and is also known as the normal distribution.

Half a century after Simpson's work the polymath Johann Carl Friedrich Gauss (1777–1855) [12] systematically investigated the properties of measurement errors and in so doing set the course of experimental science for the next two centuries. Gauss postulated that if each observation in a sequence of measurements $Q_1, Q_2, \ldots, Q_j, \ldots,$ Q_N was truly independent of all of the others then the deviation from the average value is a random variable

$$\xi_j = Q_j - \overline{Q}, \tag{1.6}$$

so that ξ_j and ξ_k are statistically independent of one another if $j \neq k$. This definition has the virtue of defining the average error in the measurement process to be zero,

$$\overline{\xi} = \sum_{j=1}^{N} \xi_j p_j = 0, \tag{1.7}$$

implying that the average value is the best representation of the data. Gauss determined that the variance defined by (1.3) in terms of the error (1.6) takes the form

$$\sigma^2 \equiv \sum_{j=1}^{N} \xi_j^2 p_j \tag{1.8}$$

and can be used to measure how well the average characterizes the ensemble of measurements. Note that it is not necessary to introduce p_j for the following argument and, although its introduction would not change the presentation in any substantial way, the discussion is somewhat simpler without it.

Gauss used the statistical independence of the measured quantities to prove that the average value gave their best representation and that, with a couple of physically reasonable assumptions, the associated statistical distribution was normal, an unfortunate name that had not been introduced at that time. We present a modified version of his arguments here to lay bare the requirements of normalcy. The probability I of obtaining a value in the interval $(Q, Q + \Delta Q)$ in any measurement is given by

$$I = P(Q)\Delta Q \tag{1.9}$$

and in a sequence of N measurements the data are replaced with the deviations from the average, that is, by the errors $\xi_1, \ldots, \xi_N$, allowing us to segment the range of values into N intervals,

$$P(\xi_j)\Delta\xi_j \equiv \text{probability of observing the deviation } \xi_j. \tag{1.10}$$

In (1.10) the probability of making the N independent measurements in the ensemble together with the property that the probability of the occurrence of any two independent events is given by the product of their individual probabilities, and assuming $\Delta\xi_j = \Delta\xi$ for all j, is

$$I = \prod_{j=1}^{N} P(\xi_j)\Delta\xi_j = P(\xi_1)P(\xi_2)\ldots P(\xi_N)\Delta\xi^N. \tag{1.11}$$

According to Gauss the estimation of the value for Q appears *plausible* if the ensemble of measurements resulting in $\overline{Q}$ is the most *probable*. Thus, Q is determined in such a way that the probability I is a maximum for $Q = \overline{Q}$. To determine this form of the probability density we impose the condition

$$\frac{d \ln I}{d\overline{Q}} = 0 \tag{1.12}$$

and use (1.11) to obtain

$$\frac{d \ln I}{d\overline{Q}} = \sum_{j=1}^{N} \frac{d\xi_j}{d\overline{Q}} \frac{\partial \ln I}{\partial \xi_j} = -\sum_{j=1}^{N} \frac{\partial \ln P(\xi_j)}{\partial \xi_j}, \tag{1.13}$$

where we have used the fact that

$$\frac{d\xi_j}{d\overline{Q}} = -1$$

for all j. The constraint (1.12) applied to (1.13) is the mathematical rendition of the desirability of having the average value as the most probable value of the measured variable.

We now solve (1.13) subject to the constraint

$$\sum_{j=1}^{N} \xi_j = 0 \tag{1.14}$$

by assuming that the jth derivative of the logarithm of the probability density can be expanded as a polynomial in the random error

$$\frac{\partial \ln P\left(\xi_j\right)}{\partial \xi_j} = \sum_{k=0}^{\infty} C_k \xi_j^k, \tag{1.15}$$

where the set of constants $\{C_k\}$ is determined by the equation of constraint

$$-\sum_{j=1}^{N}\sum_{k=0}^{\infty} C_k \xi_j^k = 0. \tag{1.16}$$

All the coefficients in (1.16) vanish except $k = 1$, since by definition the fluctuations satisfy the constraint equation (1.14) so the coefficient $C_1 \neq 0$ satisfies the constraint. Thus, we obtain the equation for the probability density

$$\frac{\partial \ln P(\xi_j)}{\partial \xi_j} = C_1 \xi_j, \tag{1.17}$$

which integrates to

$$P(\xi_j) \propto \exp\left(\frac{C_1}{2}\xi_j^2\right). \tag{1.18}$$

The first thing to notice about this solution is that its extreme value occurs at $\xi_j = 0$, that is, at $Q_j = \overline{Q}$ as required. For this to be a maximum as Gauss required and Simpson speculated, the constant must be negative, $C_1 < 0$, so that the second derivative of P at the extremum is positive. With a negative constant the function decreases symmetrically to zero on either side, allowing the function to be normalized,

$$\int_{-\infty}^{\infty} P(\xi_j)d\xi_j = 1, \tag{1.19}$$

and because of this normalization the function can be interpreted as a probability density. Moreover, we can calculate the variance to be

$$\sigma_j^2 = \int_{-\infty}^{\infty} \xi_j^2 P(\xi_j)d\xi_j, \tag{1.20}$$

allowing us to express the normalized probability density as

$$P(\xi_j) = \frac{1}{\sqrt{2\pi}\sigma_j} \exp\left(-\frac{\xi_j^2}{2\sigma_j^2}\right).$$

(1.21)

Equation (1.21) is the normal or Gauss distribution of individual errors. A property of this distribution is the multiple convolution relation, that being

$$\frac{1}{\sqrt{2\pi}\sigma_z}e^{-\frac{z^2}{2\sigma_z^2}} = \frac{1}{2\pi\sigma_x\sigma_y}\int e^{-\frac{(z-y)^2}{2\sigma_x^2}}e^{-\frac{y^2}{2\sigma_y^2}}\,dy,$$

(1.22)

where the sum variable is

$$z = x + y,$$

(1.23)

with the variance

$$\sigma_z^2 = \sigma_x^2 + \sigma_y^2.$$

(1.24)

Consequently we obtain by applying the convolution N times that the ensemble distribution function for the sum of errors becomes

$$P(\xi) = \frac{1}{\sqrt{2\pi}\sigma}e^{-\frac{\xi^2}{2\sigma^2}}$$

(1.25)

with the total variance given by

$$\sigma^2 = \sum_{j=1}^{N}\sigma_j^2.$$

(1.26)

So this is how things stood at the beginning of the nineteenth century. No experimental results were exactly reproducible, but the errors made in any given set of experiments obeyed the law of frequency of error given by the normal distribution. In this environment scientists performed large numbers of experiments giving credence to the average value being the best representation of the data, provided that the standard deviation was sufficiently small relative to the average value $\sigma \ll \overline{Q}$. Indeed, physics did very well with this interpretation of experimental error.

1.1.1   Knowledge, information and uncertainty

We have taken the first step on the long journey towards understanding complex webs and their implications; that first step was identifying the origin of the normal distribution. But we should point out that there are other, independent, origins of the distribution, some of which predate Gauss. So we have pocketed one piece of traditional wisdom that has been transferred from the domain of science to the world at large: no phenomenon is *exactly* predictable, but measurements of it fluctuate and the statistics of these fluctuations obey a normal distribution. Of course, this observation is accompanied by the notion that these fluctuations (errors) can be made arbitrarily small, and achieving that perfect end state of no variability has been the pursuit of much of experimental science and all of manufacturing for over 200 years.

The next logical question concerns how we explain the nearly universal acceptance of the normal distribution by the scientific community. Answering this question constitutes the second step of our journey and, like the first, requires some historical review. We need to understand how scientists first began to learn about the sources of complexity that the fluctuations in measurements often manifest. The arguments of the last section explained the formal properties of the experimental results but not the reasons for those properties. Without the reasons we have a mathematical model, but with the reasons the mathematical model can become a scientific theory. In the area of mathematics these concerns spawned the disciplines of statistics and probability theory; the difference between the two will be taken up in due course. In the physical sciences these concerns led to investigations of the influence of the microscopic world on the macroscopic world and to the laws of thermodynamics.

Historically thermodynamics was the first quantitative discipline to systematically investigate the order and randomness of complex webs, since it was here that the natural tendency of phenomena to become disordered was first articulated. As remarked by Schrödinger in his ground-breaking work *What Is Life?* [28]:

The non-physicist finds it hard to believe that really the ordinary laws of physics, which he regards as prototypes of inviolable precision, should be based on the statistical tendency of matter to go over into disorder.

In the context of physical networks the quantitative measure of disorder, which has proven its value, is entropy and the idea of thermodynamic equilibrium is the state of maximum entropy. Of course, since entropy has been used as a measure of disorder, it can also be used as a measure of complexity. If living matter is considered to be among the most complex of webs then it is useful to understand how the enigmatic state of being alive is related to entropy. Schrödinger maintained that a living organism can hold off the state of maximum entropy, namely death, only by giving up entropy or by absorbing negative entropy, or negentropy, from the environment. He pointed out that the essential thing in metabolism is that the organism succeeds in freeing itself from all the entropy it cannot help producing by being alive and sheds it into the environment.

Here it is probably reasonable to point out that we have entered a conceptual thicket associated with the concepts of knowability, disorder, complexity and entropy, as well as other related notions. Various mathematical theories have developed over the years to illuminate these difficult concepts, including generalized systems analysis, cybernetics, complexity science, information theory, graph theory and many more. Each theory has its advocates and detractors, each has its own domain of application, and each has its non-overlapping sets of limitations. We have elected to present from this plethora of intellectual activity only those concepts that contribute to our vision of what constitutes a simple or a complex web. Consequently, we are not always overly precise in our definitions, intending to be inclusive rather than exclusive, hoping that the reader will not be too harsh in having to wait while the big picture is being laid out. Key among the choices made is probably the decision to use the term web rather than network or system in order to be selective regarding the properties we choose to associate with a web, some of which overlap with what other investigators would call a network or system. We do this because there are fewer preconceptions associated with a web than with the other more familiar terms. A complex web in some cases would be the same as a "complex adaptive system" and in other cases it would be indistinguishable from a "complex graphical network" and we intend to use it for either, both or neither depending on the context.

The term web spans a spectrum from complex at one end of a continuum to simple at the other. It is possible to associate simplicity with absolute knowability, so we can know everything about a simple web, even one that is dynamic. On the other hand, we associate complexity with disorder, which is to say with limited knowability. So it is probably not possible to know everything about a complex web. This rather comfortable separation of the world of webs into the complex and the simple, or the knowable and the unknowable, breaks down under scrutiny, but is useful for some purposes, which we intend to explore.

As emphasized by Schrödinger, all the laws in the physical and life sciences are statistical in nature. This is especially true when dealing with webs. Most physical laws have imagined that the number of web elements is very large. Consider the example of a gas in a container under pressure at a given temperature. If the container confines N molecules of gas then at any moment in time the relations among the thermodynamic variables such as pressure and volume, given by Boyle's law, could be tested and would be found to be inaccurate by departures of order $\sqrt{N}$ so that the average of any intensive macroscopic variable has fluctuations of order $1/\sqrt{N}$, on the basis of the ideas of Gauss. According to Schrödinger [28] the $\sqrt{N}$ rule concerns the degree of inaccuracy to be expected in any physical law. The simple fact is that the descriptive paradigm of natural order, the empirical laws of physics, is inaccurate within a probable error of order $1/\sqrt{N}$. The application of these ideas to physical phenomena had led to the view that the macroscopic laws we observe in natural phenomena, such as Boyle's law, Ohm's law, Fick's law, and so on, are all consequences of the interaction of a large number of particles, and therefore can be described by means of statistical physics.

Much has changed in the sixty years since Schrödinger first passed along his insights regarding the application of physical law to biology [28]. In particular, the way we understand the statistics of complex phenomena in the context of the natural and social sciences as well as in life science is quite different from what it was then. Herein we examine evidence for the deviation from his $\sqrt{N}$ rule, which suggests that the statistical basis for Schrödinger's understanding of biology, or complexity in general, has changed. The normal statistics resulting from the central limit theorem (CLT) have been replaced by apparently more bizarre statistics indicating that the CLT is no longer applicable when the phenomena become complex. What has not changed in this time is the intuition that webs in the natural, social and life sciences are complex, or, stated differently, complex webs can be used to categorize phenomena from all these disciplines.

The words complex and complicated are often used to mean equivalent things. It is only recently that science has recognized that a process need not be complicated to generate complexity; that is, complex phenomena need not have complex or complicated dynamical descriptions. In such cases the complicated nature of the phenomena resides in the way the various scales contributing to the process are interconnected. This interconnectedness often cannot be described by analytic functions, as one might expect from such traditional disciplines as celestial mechanics. In these areas the tying together of different scales prohibits the taking of derivatives for the definition of functions and, without any derivatives, it is not possible to construct constitutive equations for describing the behavior of the phenomenon. This suggests the need for a new way of thinking about how to represent complex webs.

The entropy measure of complexity can be given a dynamical interpretation, but in order to do this one has to interpret the dynamics using a probability density. This is traditionally done by discarding the notion of a closed web and recognizing that every web has an environment with which it necessarily interacts. By explicitly eliminating the environmental variables from the description of the web dynamics one obtains a description that is statistical in nature. The absolute predictability which was apparently present in the deterministic nature of Newton's equations is abandoned for a more tractable description of the web having many fewer variables, but the price is high and that price is the loss of absolute predictability. In a general social context the notion of predicting an outcome is replaced by *anticipating* an outcome.

Implicit in the concept of entropy is the idea of uncertainty. That idea is reasonable only in a context where there is a conscious being that is extracting information from the web, and is therefore subjective. Uncertainty means that not all the information one needs for a complete description of the behavior of a web is available. Even the term "needs" is in this sense subjective, because it depends on the questions the observer poses, which in turn depend on the "purpose" of the observer [26]. This is where all subjectivity enters, and we do not go further into the philosophical implications of having an observer with a purpose conducting the experiment. We wish merely to be clear that a web containing conscious individuals cannot be treated in a deterministic way since the objectivity stemming from determinism conflicts with the subjectivity of the individual (free will).

However, we can safely say that entropy is a measure of uncertainty as it relates to order in a physical network, and, like uncertainty, entropy is a non-decreasing function of the amount of information available to the observer. This connection between information and statistical physics is quite important, and at this stage of our discussion we can say that it is the uncertainty that allows us to describe dynamical webs in thermodynamic terms. Szilard [31] was the first to identify information with entropy in his explanation of why Maxwell's demon does not violate the second law of thermodynamics.

Perhaps we should say a few words here about phase space and what it is. We understand physical space to be what we measure with a ruler and a clock, the space-time continuum. However, historically, in trying to understand the behavior of complex systems physicists realized that it was necessary to know the position and momentum of every particle of that system, say of a gas. Therefore, consider a space in which the coordinate axes are the positions and momenta of all the particles, and consequently the "phase" or state of the complex system is specified by a single point in this space. In this way the dynamics of the gas could be described by a single trajectory in this high-dimensional space as the gas diffuses through the room or undergoes any other dynamics. A collection of initial states for the system is a bunch of points in phase space, which are described by a phase-space distribution function that occupies a certain volume. Imagine that the phase space for an isolated network can be partitioned into a large number of cells and that each cell is statistically equivalent to each of the other cells. There is therefore an equal probability of a particle occupying any one of the cells in phase space. The definition of entropy in this phase space is fairly abstract, and depends on the volume of the phase space occupied by the particles.

Boltzmann's expression for entropy is

$$S = k_B \ln \Omega, \tag{1.27}$$

where Ω is the volume of phase space occupied by the web of interest and the proportionality constant k_B is the Boltzmann constant. If one considers two independent networks B_1 and B_2 with entropies S_1 and S_2, respectively, then the entropy of the combined network is just the arithmetical sum $S_1 + S_2$, as it would be for the energy. The entropy is consequently extensive through the logarithmic assumption in (1.27), which means that the measure of disorder Ω for the combined system Ω_{com} is given by the product of the individual volumes, that is, $\Omega_{com} = \Omega_1 \Omega_2$. The quantity Ω indicates disorder that in classical physics is due to thermal motion of the particles in the environment.

Entropy can also be expressed in more physical terms through the use of the continuous phase-space distribution function, $\rho(\Gamma, t)$, where Γ represents the phase-space variables, the displacements and momenta describing the dynamics of the N-particle system. The phase-space function keeps track of where all the particles in the system are as a function of time and what they are doing. Boltzmann (1844–1906) was able to show that the entropy could be defined in terms of the phase-space distribution function as

$$S(t) = -k_B \int_\Omega \rho(\Gamma, t) \ln \rho(\Gamma, t) d\Gamma, \tag{1.28}$$

which described the fluid-like motion of a gas, and is a non-decreasing function of time. This definition of entropy attains its maximum value when the web achieves thermodynamic equilibrium, in which case the phase-space distribution function becomes independent of time.

We refer to the definition of entropy as given by Boltzmann (1.28) as the statistical entropy. This development reached maturity in the hands of Gibbs, who attempted to provide the mechanical basis of the description of thermodynamic phenomena through the formulation of statistical mechanics. Gibbs gave a probability interpretation to the phase-space distribution function, and introduced the notion of ensembles into the interpretation of physical experiments. The above statistical definition of entropy is very general and is one of the possible measures of complexity that we seek. In fact, if the network is simple and we are able to measure the coordinates and the momenta of all the particles with extreme precision, we have from (1.28) that this entropy is a minimum. A simple web, namely one that is closed and whose equations of motion are integrable, does not have any growth of entropy, due to the time-reversibility of the dynamical equations. Here the terms "integrable" and "time-reversible" dynamics are a consequence of Newton's laws of motion.

Wiener [39] and subsequently Shannon [29] determined how to construct a formal measure of the amount of information contained by a web and the problems associated with the transmission of a message within a web and between webs. Shannon expressed information in terms of bits, or the number of binary digits in a sequence. He proved that a web with N possible outputs, where the output i has the probability of occurring p_i, can be described by a function

$$H = -\sum_{i=1}^{N} p_i \log_2 p_i,$$ (1.29)

which is the information entropy, with the logarithm to the base two and Boltzmann's constant set to unity. The information entropy H attains its maximum value at the extremum where its variation denoted by the operator δ vanishes subject to the normalization condition

$$\delta \left[H + \lambda \left(\sum_{i=1}^{N} p_i - 1 \right) \right] = 0$$ (1.30)

and λ is a Lagrange multiplier. The solution to (1.30) yields the maximum information entropy when each of the possible states of the web has the same probability of occurrence, that is when maximal randomness (maximum uncertainty) occurs,

$$p_i = 1/N.$$ (1.31)

In this case of maximal randomness the entropy

$$H_{\max} = \log_2 N$$ (1.32)

is also maximum.

The information entropy is the discrete equivalent of Gibbs' treatment of Boltzmann's entropy, as Shannon discusses at the end of his 1948 article. The analytic expressions for the entropy, disregarding the unit of measure and therefore the base of the logarithm, are exactly the same, but this informational interpretation offers possibilities of extending the definition of entropy to situations using conditional probabilities, resulting in conditional entropies, mutual entropies, and so on. This means that it is possible to recognize two equivalent pieces of information, and to disregard the "copy" because nothing new is learned from it. It is possible to extract the new pieces of information from a message of which the majority of the content is already known, and therefore it is useful for separating the knowable but unknown from the known. New pieces of information decrease the level of uncertainty, and the more complex the web the more information it can carry.

It appears that an advanced version of this information-selection process evolved in psychological networks as is manifest in the phenomenon of habituation. Habituation is the decremental response to a new repetitive stimulus without the strength of the stimulus changing. After a short time humans no longer hear the background conversation at a party; a person soon forgets the initial chill experienced upon entering a room; and the evil smell of the newly opened cheese fades from consciousness with the passage of time. Humans and other animals become unresponsive to repeated stimulation and this enables them to attend to new and, arguably, more important sources of information. How information is transported in complex networks and how this is related to the phenomenon of habituation will be discussed later.

1.1.2   Maximum entropy

In paraphrasing Gauss' argument in an earlier section we introduced the idea of maximizing a function, that is, determining where the variation of a function with respect to an independent variable vanishes. Recall that the derivative of a function vanishes at an extremum. This trick has a long pedigree in the physical sciences, and is not without precedent in the social and life sciences as well. We use the vanishing of the variation here to demonstrate another of the many methods that have been devised over the years to derive the normal distribution, to gain insight into its meaning, and to realize the kind of phenomena that it can be used to explain. The variational idea was most ambitiously applied to entropy by Jaynes [17] as a basis for a formal derivation of thermodynamics. Here we are much less ambitious and just use Jaynes' methods to emphasize the distinction between phenomena that can be described by normal statistics and those that cannot.

In the previous subsection we mentioned the idea of entropy being a measure of order, the more disordered the web the greater the entropy. This idea has been used to maximize the entropy subject to experimental constraints. The way this is done is by determining the least-biased probability density to describe the network that is consistent with observations. This is an extension of Gauss' approach, where he used the vanishing of the average value and the maximization to determine the normal distribution. The parameters for the distribution were then fixed by normalization and the variance.

Here we use the experimental observation of the second moment for a zero-centered variable and define the constant given by (1.8) and the normalization of the probability density to maximize

$$I = -\int p(q)\ln p(q)dq - \alpha \left[\int p(q)dq - 1\right] - \beta \left[\int q^2 p(q)dq - \sigma^2\right]. \quad (1.33)$$

To maximize the distribution function the variation in expression (1.33) must vanish,

$$\delta I = 0, \quad (1.34)$$

and the parameters α and β, called Lagrange multipliers, are adjusted to satisfy the constraints enclosed in their respective brackets. Taking the variation with respect to the probability density yields

$$\int dq \left[-1 - \ln p(q) - \alpha - \beta q^2\right]\delta p(q) = 0 \quad (1.35)$$

and, since the variation in the probability density $\delta p(q)$ is arbitrary, the term within the coefficient brackets must vanish. Consequently, we obtain the probability density

$$p(q) = \exp[-1 - \alpha - \beta q^2]; \tag{1.36}$$

imposing the normalization integral relates the two parameters by

$$e^{-1-\alpha} = \sqrt{\frac{\beta}{\pi}}, \tag{1.37}$$

which, when inserted back into (1.36), yields

$$p(q) = \sqrt{\frac{\beta}{\pi}} e^{-\beta q^2}. \tag{1.38}$$

The second constraint, that of the second moment, then yields the parameter value

$$\beta = \frac{1}{2\sigma^2}, \tag{1.39}$$

so the distribution which maximizes the entropy subject to a finite variance is normal with the same parameter values as given in (1.25). Using kinetic theory, it is readily shown that the parameter β is the inverse of the thermodynamic temperature.

Obtaining the normal distribution from the maximum-entropy formalism is a remarkable result. This distribution, so commonly found in the analyses of physical, social and life phenomena, can now be traced to the idea that normal statistics are a consequence of maximum disorder in the universe that is consistent with measurements made in the experiment under consideration. This is similar to Laplace's notion of unbiased estimates of probabilities. Marquis Pierre-Simon de Laplace (1740–1827) believed that in the absence of any information to the contrary one should consider alternative possible outcomes as equally probable, which is what we do when we assert that heads or tails is equally likely in a fair coin toss. Of course, Laplace did not provide a prescription for the general probability consistent with a given set of statistically independent experimental data points. He did, however, offer the probability calculus as the proper foundation for social science.

But this is not the end of the story. The existence of these multiple derivations of the normal distribution is reassuring since normal statistics permeates the statistics taught in every elementary data-processing course on every college campus throughout the world. The general impression is that the normal distribution is ubiquitous because everyone says that it is appropriate for the discipline in which they happen to be working. But what is the empirical evidence for such claims? In Figure 1.1 the distribution of heights was fit with the normal distribution, that is true. But one swallow does not make a summer.

Every college student in the United States is introduced to the bell-shaped curve of Gauss when they hear that the large class in sociology (economics, psychology and, yes, even physics and biology) in which they have enrolled is graded on a curve. That curve invariably involves the normal distribution, where, as shown in Figure 1.2 between $+1$ and -1 lie the grades of most of the students. This is the C range, between plus and minus one standard deviation of the class average, which includes 68% of the students.

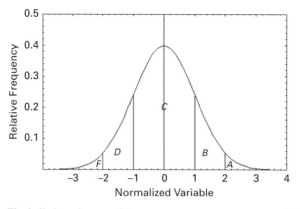

Figure 1.2.   The bell-shaped curve of Gauss concerns errors; consequently, the experimental value has the average value subtracted so that the curve is centered on zero, and the new variable is divided by the standard deviation (width of the distribution) of the old variable, leaving the new variable dimensionless. The units are therefore in terms of standard deviations so that 68% of the student grades are in the interval $[-1, 1]$ the C range; 95% are in the interval $[-2, 2]$ with the D range $[-2, -1]$ and the B range $[1, 2]$; and the remaining 5% are divided between the Fs and As.

The next range is the equally wide B and D intervals from one to two standard deviations in the positive and negative directions, respectively. These two intervals capture another 27% of the student body. Finally, the top and bottom of the class split the remaining 5% equally between A and F (or is it E?).

One author remembers wondering how such a curve was constructed when he took his first large class, which was freshman chemistry. It appeared odd that the same curve was used in the upper-division courses even though the students from the bottom of the curve had failed to advance to the upper-division classes. Such thoughts were put on hold until graduate school, when he was grading freshman physics exams, when it occurred to him again that he had never seen any empirical evidence for the bell curve in the distribution of grades; the education web merely imposed it on classes in the belief that it was the right thing to do.

Recently we ran across a paper in which the authors analyzed the achievement tests of over 65,000 students graduating from high school and taking the university entrance examination of the Universidade Estadual Paulista (UNESP) in the state of São Paulo, Brazil [13]. In Figure 1.3 the humanities part of the entrance exam is recorded for different forms of social parsing of the students. The upper panel partitions the students into those that attended high school during the day and those that attended at night; the middle panel groups the students into those that went to public and those that went to private schools; finally, the bottom panel segments the students into high-income and low-income students. The peaks of the distributions move in these comparisons, as do the widths of the distributions, but in each and every case it is clear that the bell-shaped curve gives a remarkably good fit to the data.

Figure 1.3 appears to support the conjecture that the normal distribution is appropriate for describing the distribution of grades in a large population of students under a variety of social conditions. However, we have yet to hear from the sciences. In Figure 1.4 the

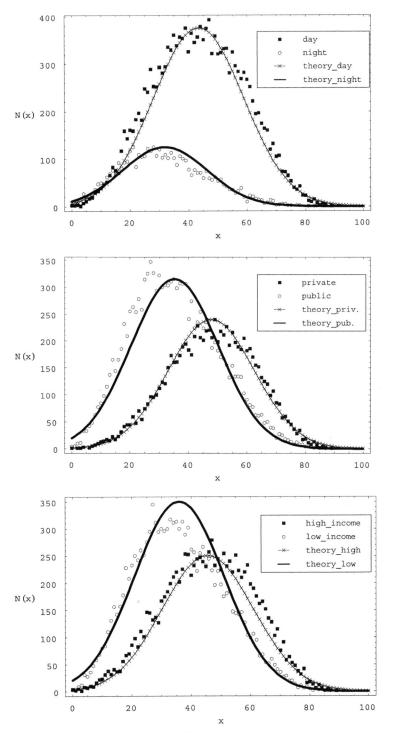

Figure 1.3. The data for the approximately 65,000 students taking the university entrance examination of the Universidade Estadual Paulista (UNESP) in São Paulo, Brazil [13] graphed for the humanities (reproduced with permission). The students were separated into various comparative groups: top, day and night students; middle, public and private schools; and bottom, high and low income.

data from the physical sciences are graphed under the same groupings as those of the humanities in Figure 1.3. One thing that is clear from the three panels in Figure 1.4 is that the distribution of grades is remarkably different from the bell-shaped curve. But are these the only data for which the grade distribution is not normal?

Figure 1.5 depicts the distribution of grades under the same set of groupings for the biological sciences. The first thing to notice is that the distributions of grades in the biological sciences are more like those of the physical sciences than they are like those in the humanities, although they are not exactly the same. In fact, the distribution of grades in the sciences is nothing like that in the humanities. There is no peak, no aggregation of grades around a characteristic average value. The grades seem to dominate the region of the origin and spread throughout all values with diminishing values.

So why does normalcy apparently apply to the humanities but not to the sciences?

One possible explanation for this difference in the grade distributions between the humanities and the sciences has to do with the structural difference between the two categories. Under the heading of the humanities are collected a disjoint group of disciplines including language, philosophy, sociology, economics, and a number of other relatively independent areas of study. We use the term independent because what is learned in sociology is not dependent on what is learned in economics, but may weakly depend on what is learned in language. Consequently, the grades obtained in each of these separate disciplines are relatively independent of one another, thereby satisfying the conditions of Gauss' argument. In meeting Gauss' conditions the distribution of grades in the humanities takes on normality.

On the other hand, science builds on previous knowledge. Elementary physics cannot be understood without algebra and the more advanced physics cannot be understood without the calculus, which also requires an understanding of algebra. Similarly, understanding biology requires a mastery of some chemistry and some physics. The disciplines of science form an interconnecting web, starting from the most basic and building upward, a situation that violates Gauss' assumptions of independence and the idea that the average value provides the best description of the process. The empirical distribution of grades in science clearly shows extensions out into the tail of the distribution and consequently there exists no characteristic scale like the average, with which to characterize the data.

The distinction between the distribution in grades in the humanities and that for the sciences is clear evidence that the normal distribution does not describe the usual or normal situation. The bell-shaped curve of grades is imposed through educational orthodoxy by our preconceptions and is not indicative of the process by which students master material. So can we use one of the arguments developed for deriving the normal distribution to obtain the empirical distribution with the long tails shown in Figures 1.4 and 1.5? Let us reconsider maximizing the entropy, but from a slightly different perspective.

Recall that the entropy-maximization argument had three components: (1) the definition of entropy; (2) the moments of the empirical data to restrict the distribution; and (3) determining the Lagrange multipliers from the data in (2) through maximization. Here let us replace step (2) with the more general observation that the grade distributions

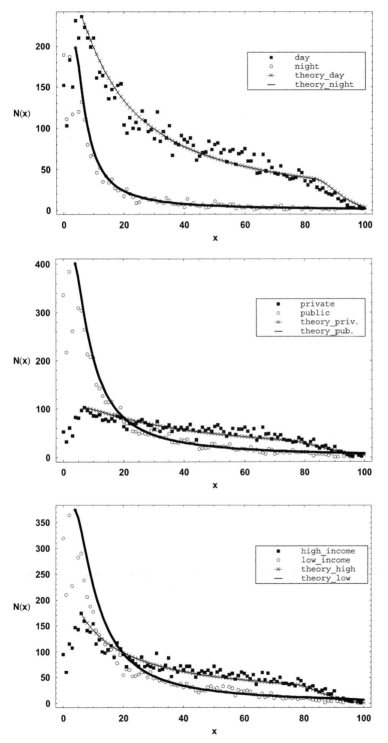

Figure 1.4. The data for the approximately 65,000 students taking the university entrance examination of the Universidade Estadual Paulista (UNESP) in São Paulo, Brazil [13] graphed for physics (reproduced with permission). The students were separated into various comparative groups: top, day and night students; middle, public and private schools; and bottom, high and low income.

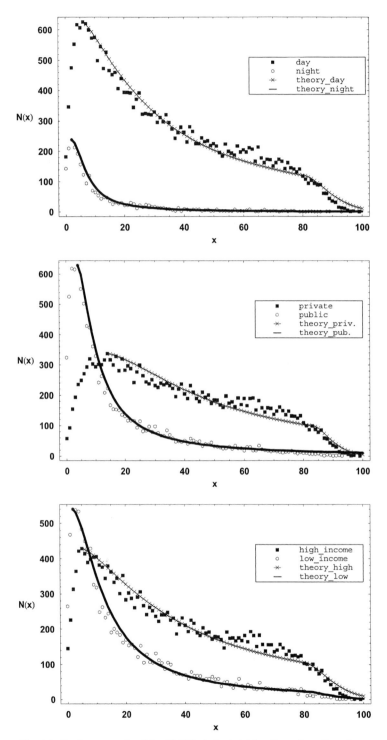

Figure 1.5. The data for the approximately 65,000 students taking the university entrance examination of the Universidade Estadual Paulista (UNESP) in São Paulo, Brazil [13] graphed for biology (reproduced with permission). The students were separated into various comparative groups: top, day and night students; middle, public and private schools; and bottom, high and low income.

in science do not have a characteristic scale. We introduce that condition through a function that scales, that is, the logarithm. The quantity to be varied is then

$$I = -\int p(q)\ln p(q)dq - \alpha\left[\int p(q)dq - 1\right] - \beta\left[\int p(q)\ln(q+T)dq - \text{constant}\right],$$

$$(1.40)$$

where we have included the normalization and the average logarithm $\ln(q+T)$ as constraints. The variation of I being set to zero yields the relation

$$\ln p(q) = \alpha - 1 - \beta\ln(q+T),$$

which through the normalization condition yields the properly normalized distribution on the domain $(0, \infty)$:

$$p(q) = \frac{(\beta - 1)\, T^{\beta-1}}{(q+T)^\beta}.$$

$$(1.41)$$

The entropy-maximization argument therefore determines that a distribution that has no characteristic scale, is finite at the origin, and is maximally random with respect to the rest of the universe is given by a hyperbolic distribution (1.41), which asymptotically becomes an inverse power-law distribution $p(q) \propto q^{-\beta}$.

This hyperbolic distribution has the kind of heavy tail that is observed qualitatively in the distribution of grades in the sciences. Consequently, it would be prudent to explore the differences between the normal and hyperbolic distributions in order to understand the peculiarities of the various phenomena they represent. But it is not just the grade distribution that is described by a distribution with an inverse power-law tail but literally hundreds of complex webs in the physical, social and life sciences. This statistical process is shown to be representative of many more phenomena than are described by the distribution of Gauss. In fact, most of the situations where the normal distribution has been applied historically are found to be the result of simplifying assumptions and more often than not the use of the bell-shaped curve is not supported by data. These are the phenomena we go after in this book, to show that when it comes to describing complex webs the normal distribution and nomalcy are myths.

1.2 Empirical laws

The first investigator to recognize the existence of inverse power-law tails in complex phenomena by looking at empirical data was the social economist Marquis Vilfredo Frederico Damaso Pareto (1848–1923). Unlike most of his nineteenth-century contemporaries, he believed that the answers to questions about the nature of society could be found in data and that social theory should be developed to explain information extracted from such observations, rather than being founded on abstract moral principles. As Montroll and Badger [22] explain, Pareto collected statistics on individual income and wealth in many countries at various times in history and his analysis convinced him of the following [25]:

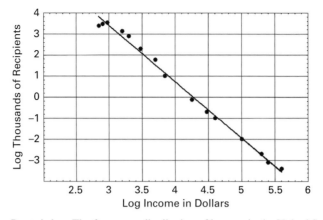

Figure 1.6.   Pareto's law. The frequency distribution of income in the United States in 1918 is plotted on log–log graph paper. The slope of the down-going straight line gives the power-law index. Adapted from Montroll and Badger [22].

In all places and at all times the distribution of income in a stable economy, when the origin of measurement is at a sufficiently high income level, will be given approximately by the empirical formula $y = ax^{-\nu}$, where y is the number of people having income x or greater and ν is approximately 1.5.

   In Figure 1.6 the income distribution of people in the United States in 1918 is depicted on graph paper, where both axes are logarithmic. On the vertical axis is the logarithm of the number of individuals with a given level of income and on the horizontal axis is the logarithm of the income level. The closed circles are the data showing the number of people with a given level of income and the solid-line segment with a negative slope indicates Pareto's law of income [25] fit to the data. It is from such curves as the one shown in Figure 1.6 that Pareto sought to understand how wealth was distributed in western societies, and, from this understanding, determine how such societies operate.

   Unlike the normal distribution, for which the measurements cluster in the vicinity of the average value, with 95% of the population residing between +2 and −2 standard deviations of the average; income data are broadly spread out with some individuals at either extreme. The region of the distribution where the large values of income exist is the tail, and is called the tail because the income levels are far from the central region, where one would expect the mean income of ordinary wage earners to reside. Such observations lead Pareto to draw the following conclusion [25]:

These results are very remarkable ... The form of this curve seems to depend only tenuously upon different economic conditions of the countries considered, since the effects are very nearly the same for the countries whose economic conditions are as different as those of England, of Ireland, of Germany, of the Italian cities, and even of Peru.

   Another conclusion that Pareto drew from his insight into the data structure is that the world is not fair. If the world were fair then everyone would have almost the same income, which would be the mean. In the world of Gauss that would be how income is distributed, with some individuals making more than average and some making less than

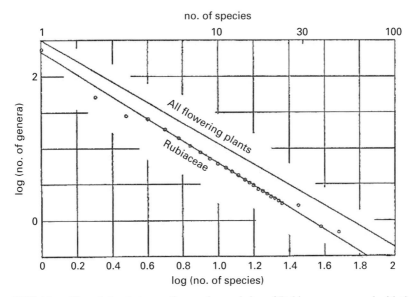

no. of species

Figure 1.7.   Willis' law. The relation between the number and size of Rubiaceae compared with the slope of the fit to all flowering plants [18].

average, but the majority making a fair living. Pareto observed that in stable societies that is not what occurs; there is a fundamental imbalance in how income is distributed. He came to believe that the inverse power-law distribution of income, which subsequently became known as Pareto's law, was the expression of a social law. This was the first glimpse into how complex webs organize their complexity or, alternatively, are organized by their complexity.

Over the last century a substantial number of books addressing the complexity of webs has been written, and we shall draw from as many of them as we need in our discussions. Lotka's book [18] on biological webs focused on the macroscopic level using statistical physics of near-equilibrium phenomena as a modeling strategy, but he was very sensitive to the data available to test various biophysical hypotheses. One data set had been compiled by J. C. Willis, who had published a book on evolutionary biology [40] in which he (Willis) argued that the spatial area occupied by a biological species is directly proportional to the age of the species, so that the area occupied becomes a measure of evolutionary age. The details of Willis' 1922 theory do not concern us here, in part because the critique of it contained in Lotka's book was not particularly positive, but Willis' data are still interesting. Willis collected data on a variety of natural families of plants and animals and graphed the number of genera as the ordinate (x) and the number of species in each genus (y) as the abscissa. Figure 1.7 depicts the number and size of Rubiaceae[1] on log–log graph paper. The data are fit by a line segment satisfying the linear equation in the logarithms of the variables

---

[1] The Rubiaceae is a family of flowering plants consisting of trees, shrubs, and occasionally herbs comprising about 650 genera and 13,000 species.

$$\log y + \beta \log x - b = 0 \qquad (1.42)$$

or, in more compact form,

$$yx^\beta = \text{constant}. \qquad (1.43)$$

Equation (1.43) is a generalized hyperbolic relation between genera and number of species and covers a wide variety of cases including both plants and animals.

As noted by Lotka, the monotypic genera, with one species each, are always the most numerous; the ditypics, with two species each, are next in rank; genera with higher numbers of species become increasingly fewer with increasing number. In Figure 1.7 it is clear that the fitting curve is an inverse power law with slope $-\beta$. The pure inverse power law was found to be true for all flowering plants and for certain families of beetles as well as for many other groupings of plants and animals. Note that the pure inverse power law is the asymptotic form of the distribution resulting from entropy maximization when the logarithm is used as the constraint to impose scaling on the random variable. Moreover, what is meant by asymptotic is determined by the size of the constant T in (1.41).

Lotka did not restrict his observations to biodiversity; his complexity interests extended to the spreading of humans over the Earth's surface in the form of urban concentration. The build up of urban concentration had been investigated earlier by F. Auerbach [2], who recognized that, on ordering the cities of a given country in order of decreasing size from the largest to the smallest, the product of the city's size and the city's rank order in the sequence is approximately constant, as given by (1.43). Lotka applied Auerbach's reasoning to the cities in the United States, as shown in Figure 1.8, which is a log–log graph of population versus rank order of the 100 largest cities in the United States in a succession of years from 1790 to 1930. It is apparent from this figure that, for the straight-line segment given by (1.42), where y is the city population and x is the city rank, the empirical parameter b changes from year to year, but the power-law index $\beta = 0.92$ remains fairly constant. Lotka leaves open the question of the significance of this empirical relation, although he does give some suggestive discussion of the properties of certain thermodynamic relations that behave in similar ways. In the end Lotka maintains that he does not know how to interpret these empirical relations, but subsequently he constructed an empirical relation of his own in a very different context.

As an academic Lotka was also curious about various measures of scientific productivity and whether or not they revealed any underlying patterns in the social activity of scientists. Whatever his reasons, he was the first to examine the relation between the number of papers published and the number of scientists publishing that number of papers. This relationship provides a distribution of scientific productivity that is not very different from biodiversity and the urban-concentration context just discussed. Lotka's law, shown in Figure 1.9, is also an empirical relation that connects the fraction of the total number of scientists publishing a given number of papers with the number of papers published in order of least frequent to most frequent. Here again, as in the earlier examples, the distribution is an inverse power law in the number of papers published but with a power-law index $\beta \approx 2$. From Figure 1.9 we see that for every 100 authors

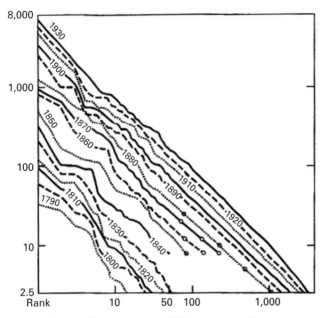

Figure 1.8. Auerbach's law. The populations of United States cities are graphed as the ordinate and on the abscissa is the corresponding city rank in order of magnitude, both on logarithmic scales [42].

who publish a single paper in a given period of time there are 25 authors who publish two papers, 11 who publish three, and so on. Do not be concerned about the details of this interpretation for the moment; how the interpretation is determined will be laid bare subsequently.

These last two examples involve social webs. The former has to do with how people interact with one another to form communities and how these communities aggregate to form towns, which in turn merge to form cities. It seems that most individuals are more strongly attracted to regions where others have already settled, but it is the nature of the inverse power law that the hermit who wants no part of society is also accommodated by the distribution. It might even be possible to construct a theory in which the attraction to these regions depends on the local concentration density of people, but developing such a theory would be getting ahead of ourselves. It is worth mentioning now, however, that sociologists refer to this in a modern context as *Matthew's effect*, for reasons that will eventually be clear. We return to this question in later chapters. The second social web discussed has to do with the requirements to publish a paper and these requirements will be discussed subsequently. However, these are not the only, or even the dominant, phenomena that are described by inverse power laws. Such distributions also occur in many different kinds of biological phenomena.

The complex interrelation in biological development between form and function was addressed at the turn of the last century in the seminal work of D'Arcy Thompson [33]. He argued that the elements of biological webs are physical and therefore subject to physical laws, but the fact that they are animate implies rules that may be distinct from those rules that govern inanimate webs. The growth of biological webs is therefore

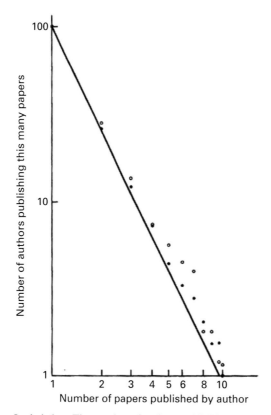

Number of authors publishing this many papers (vertical axis)

Number of papers published by author (horizontal axis)

Figure 1.9.   Lotka's law. The number of authors publishing exactly n papers as a function of n. The open circles represent data taken from the first index volume of the abridged *Philosophical Transactions of the Royal Society of London* (seventeenth and early eighteenth centuries); the filled circles are those from the 1907–18 decennial index of *Chemical Abstracts*. The straight-line segment shows the exact inverse-square law of Lotka. All data are reduced to a basis of exactly 100 authors publishing only one paper [19].

subject to physical constraints and these constraints lead to the formulation of important scaling relations in biology. For example, the growth of one organ in a body may be related to the growth of the body as a whole, called allometric growth by Huxley [16]. If W_b is the weight of a deer's body and W_a the weight of the antlers, then for certain species of deer the two are given by the allometric scaling relation

$$W_a = \alpha W_b^\beta, \qquad (1.44)$$

where α and β are constants.

The term allometric means "by a different measure," so this kind of relation implies that the phenomenon being described has more than one kind of measure. More importantly, the measures tie the different scales of the phenomenon together. For example, take the derivative of (1.44) and divide the resulting equation by the original equation to obtain

$$\beta \frac{dW_b}{W_b} = \frac{dW_a}{W_a}, \tag{1.45}$$

indicating that the percentage change in the body weight, its growth, is directly proportional to the percentage change in the weight of the organ. The constant of proportionality is dimensionless and indicates how the relative growth of the entire organism is related to the relative growth of the part. This is the scaling index sought in a wide variety of empirical studies.

Perhaps the most famous allometric relation does not concern the relative growth within a given animal, but concerns a property interrelating a variety of species. This is the allometric relation between the metabolic rate and the body mass of multiple mammalian species. The metabolic rate R refers to the total utilization of chemical energy for the generation of heat by the body of an animal. In Figure 1.10 the "mouse-to-elephant" curve depicts the metabolic rate for mammals and birds plotted versus body weight on log–log graph paper. The straight-line segment on this graph indicates that the metabolic rate is of the power-law form

$$R = \alpha W^{\beta}, \tag{1.46}$$

where again α and β are empirical constants. The value of the dimensionless power-law index in (1.46) is given by the slope of the curve in Figure 1.10 to be $\beta \approx 0.75$; however, the exact value of the power-law index is still the subject of some controversy.

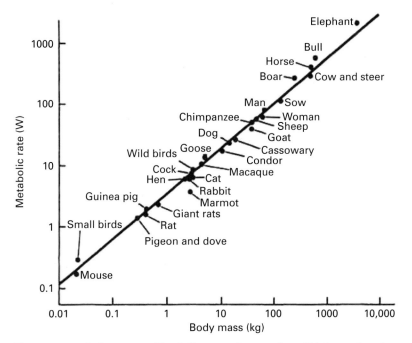

Figure 1.10. The mouse-to-elephant curve. Metabolic rates of mammals and birds are plotted versus the body weight (mass) on log–log graph paper. The solid-line segment is the best linear regression to the data from Schmidt-Neilson [27]. Reproduced with permission.

Allometric scaling indicates that the size of animals is clearly important, but, perhaps more significantly, a clue to understanding the web of life may be found in how the sizes of different animals are interconnected. In 1929 Haldane made the following observation on the importance of scale [14]:

You can drop a mouse down a thousand-yard mine shaft and, on arriving at the bottom, it gets a slight shock and walks away. A rat would probably be killed, though it can fall safely from the eleventh story of a building, a man is broken, a horse splashes.

Of course, the recognition of the importance of scaling did not begin with Haldane, or even with Thompson, but it did reach a new level of refinement with their studies. Consider the observations of Galileo Galilei, who in 1638 reasoned that animals cannot grow in size indefinitely, contrary to the writings of Swift. It was recognized by the Lilliputians that since Gulliver was 12 times their stature his volume should exceed theirs by a factor of 1,728 (12^3) and he must therefore be given a proportionate amount of food. Galileo observed that the strength of a bone increases in direct proportion to the bone's cross-sectional area (the square of its linear dimension), but its weight increases in proportional to its volume (the cube of its linear dimension). Consequently, there is a size at which the bone is not strong enough to support its own weight. Galileo captured all of this in one prodigious sentence [11]:

From what has already been demonstrated, you can plainly see the impossibility of increasing the size of structures to vast dimensions either in art or in nature; likewise the impossibility of building ships, palaces, or temples of enormous size in such a way that their oars, yards, beams, iron bolts and, in short, all their other parts will hold together; nor can nature produce trees of extraordinary size because the branches would break down under their own weight; so also it would be impossible to build up the bony structures of men, horses, or other animals so as to hold together and perform their normal functions if these animals were to be increased enormously in height; for this increase in height can be accomplished only by employing a material which is harder and stronger than usual, or by enlarging the size of the bones, thus changing their shape until the form and appearance of the animals suggest a monstrosity.

A variation of Galileo's bone-crushing example has to do with life and size. The weight of a body increases as the cube of its linear dimension and its surface area increases as the square of the linear dimension. This principle of geometry tells us that, if one species is twice as tall as another, it is likely to be eight times heavier, but to have only four times as much surface area. This raises the question of how larger plants and animals compensate for their bulk. Experimentally we know that respiration depends on surface area for the exchange of gases, as does cooling by evaporation from the skin and nutrition by absorption through membranes. Consequently the stress on animate bodies produced by increasing weight must be compensated for by making the exterior more irregular for a given volume, as Nature has done with the branches and leaves on trees. The human lung, with 300 million air sacs, has the kind of favorable surface-to-volume ratio enjoyed by trees. We discuss this in some detail later.

A half century after Lotka, the biologist MacDonald [20] and zoologist Calder [6] both adopted a different theoretical approach to the understanding of natural history, shying away from the dynamic equations of physics and explicitly considering the scaling of webs within organisms. Scaling and allometric growth laws are inextricably intertwined, even though the latter were introduced into the biology of the nineteenth century and the former gained ascendancy in engineering in the latter part of the twentieth century. There is no uniform agreement regarding the relation between the two, however. Calder pointed out that it would be "a bit absurd" to require the coefficient α in (1.44) to incorporate all the dimensions necessary for dimensional consistency of the allometric equation. He asserts that

The allometric equations ... are empirical descriptions and no more, and therefore qualify for the exemption from dimensional consistency in that they relate some quantitative aspect of an animal's physiology, form, or natural history to its body mass, usually in the absence of theory or prior knowledge of causation.

Allometric growth laws are analytic expressions relating the growth of an organ to the growth of an organism of which the organ is a part, but historically they do not have a dynamical foundation. A contemporary of Lotka was Julian Huxley, who did attempt to put allometric relations on a dynamical basis by interrelating two distinct equations of growth. In the simplest form we write two growth laws for parts of the same organism as

$$\frac{dW_b}{dt} = bW_b,$$
$$\frac{dW_a}{dt} = aW_a,$$

and, on taking the ratios of the rate equations, obtain

$$\frac{dW_b}{W_b} = \frac{b}{a}\frac{dW_a}{W_a}. \tag{1.47}$$

This is just the equation obtained from the allometric relation (1.45), from which we see that the parameters can be related by

$$\beta = \frac{a}{b} \tag{1.48}$$

so that the allometric index is given by the ratio of the growth rate of the entire body to that of the member organ. In his book *Problems of Relative Growth* [16] Huxley records data ranging from the claws of fiddler crabs to the heat of combustion relative to body weight in larval mealworms, all of which satisfy allometric growth laws.

Some forty years after Lotka, the physicist de Solla Price [9] picked up on the former's study of scientific publications, extended it, and founded a new area of social investigation through examination of the number of citations of scientific papers during varying intervals of time. Figure 1.11 depicts the number of papers published having a given number of citations. Let us stop a moment and interpret this distribution. It is evident that the first 35% of all scientific papers published have no citations, the next 49% have one citation, there are two citations for the next 9% and so on. By the time the average number of 3.2 citations per year is reached, 96% of the distribution has been exhausted. It should be clear from this that the average number of citations does not characterize this social web.

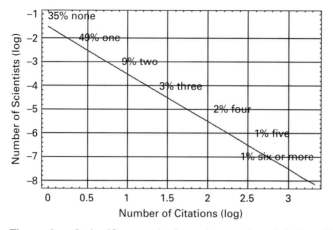

Figure 1.11. The number of scientific papers having a given number of citations. This is averaged over all scientific publications for one year and the average number of citations of a paper is 3.2 per year. Adapted from [9].

A frequent criticism of a paper in science is that it has only a few citations. In the world of Gauss this would be devastating because the paper would be average and we all know the value of being exceptional. The decision regarding whether to award tenure to a junior faculty member often rests on such tenuous evidence of scientific ability as the number of citations to published papers. In the real world, however, having the average number of citations means that such a paper is in a category shared by only the top 4% of the world's scientific publications. So the average paper is produced by the exceptional scientist.

The first monograph to recognize the importance of inverse power-law distributions in the natural, social and life sciences and to attempt their systematic explanation was George Kingsley Zipf's *Human Behavior and the Principle of Least Effort* [42]. This remarkable book had the subtitle *An Introduction to Human Ecology*, which indicates that the author put human behavior in the same category as that in which scientists traditionally put other naturally occurring phenomena, that is, natural science. In other words he was attempting to uncover the natural laws governing human behavior and articulate their underlying principles by examining and interpreting data. In his preface Zipf offers a word of caution that is as salient today as it was in 1949:

Nor are these principles particularly out of place at the present time, when we seem to be faced with an impending planned economy in which a few persons will tell many others how they *should behave* – often perhaps without regard to how people *do* behave.

As the title of his book implies, Zipf offered the principle of least effort as the primary principle that governs individual and collective behavior of all kinds and his book was devoted to presenting the empirical data supporting his hypothesis, clearly explaining the implications of his hypothesis and making the underlying theory logically consistent. Zipf saw people as being in continuous motion and, in selecting the path of their activity, viewing their decisions against a perceived background of alternatives. He hypothesized that they decided on their paths in such a way as to minimize the

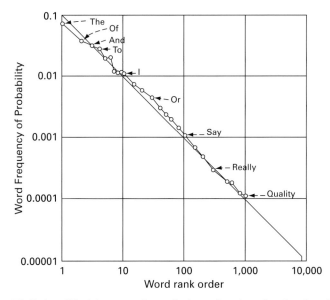

Figure 1.12. Zipf's law. Word frequency is graphed as a function of rank order of frequency usage for the 8,728 most frequently used words in the English language. The inverse power law is that of Zipf [42].

probable average rate of work expended. In this way the effort involved in making a decision to solve a problem, that is, in traversing a path, is minimized.

Unfortunately, most people who quote Zipf's scientific contributions refer only to his law regarding the relative frequency of use of words. Zipf's law is of the form (1.43) and is shown in Figure 1.12, with y being the relative frequency of word usage in the English language and x the rank of that word. In this example Zipf would argue that long words are used less frequently by the general population than are short words because they require more effort, so that ordering the words of a language by their frequency of usage results in a decreasing function. The length of a word is proportional to the time needed to produce the word, whether by writing, speaking, or otherwise, so that we may associate correlation times with word length. If we assume there is no characteristic time scale for language, or correspondingly no typical word length, then we may conclude that language ought to follow an inverse power law. We present the mathematical basis for this conclusion later. Zipf found that many languages have inverse power laws for the frequency of word usage, but that the slopes vary slightly from language to language.

The second monograph to "explain" the importance of inverse power laws in the natural, social and life sciences was Mandelbrot's now classic book [21], in which he introduced the concept of fractals. In our analyses it will become apparent that many of Mandelbrot's ground-breaking concepts have found application even beyond the expansive range he envisioned. A more recent book, but one focused on the morphogenesis of physiologic organs, is due to Weibel [36]. In Weibel's book the fractal concept is hypothesized to be a design principle in physiology, a concept that had previously been put forward by West and Goldberger [37]. In addition there are tracts on information networks, exploring the interface between man and machine as did Wiener, who initiated the modern theory of communication through his 1948 book *Cybernetics* [39].

Each of the above webs and others are generated by mechanisms specific to the complex phenomenon being considered, the psychology of human interaction in the social domain; the connectivity across scales in the natural sciences, such as across neurons in the brain; and the explosive growth of information in nonlinear dynamical networks through the generation of chaos. On the other hand, there are properties common to all these webs that enable us to identify them as such. These common features, if sufficiently robust, may form the foundation of a science of networks, to revert back to the accepted vocabulary. In this book we focus on the fact that to understand how these complex webs change over time the traditional analysis of information flow through a web, involving the exponential distributions of message statistics and consequently Poisson statistics of traffic volume, must be abandoned and replaced with less familiar statistical techniques. The techniques we introduce and discuss have appeared in the diaspore of the physical-sciences literature on networks and complexity and we shall call on them when needed.

Some excellent review articles on complex webs have been published recently, each with its own philosophical slant, and each of which is different from the perspective developed herein. Albert and Barabási [1] review complex webs from the perspective of non-equilibrium statistical physics, giving a rigorous foundation for the small-world models discussed in the lay literature [4, 5, 34, 35]. In some ways this approach is a natural extension of Lotka's approach, but it concentrates on what has been learned in the intervening century. Newman [24] organized his review around the various applications that have been made over the past decade, including social, information, technology and biology webs and the mathematical rendition of the common properties that are observed among them. The more recent survey by Costa *et al.* [7] views complex network research as being at the intersection of graph theory and statistical physics and focuses on existing measurements that support the principal models found in the literature.

The approach taken herein is a hybrid of the above views and the ideas were originally put into perspective by us with the able assistance of the then student Elvis Geneston and may be found in condensed form in West *et al.* [38]. We took non-equilibrium statistical physics and its techniques as our starting point, used recent experiments done on complex webs to highlight the inadequacies of the traditional approaches, and applied these data to motivate the development of new mathematical modeling techniques necessary to understand the underlying web structure. Much of that work has been folded into the present pages, but including all the background, gaps and details not contained in the original journal publication. This was done having the student in mind.

For clarity it is useful to list the properties associated with complex webs, because we are seeking a quantitative measure that may include an ordinal relation for web complexity. We note, however, that in everyday usage phenomena with complicated and intricate features having both the characteristics of randomness and order are called complex. Furthermore, there is no consensus on what constitutes a good quantitative measure of complexity, or what distinguishes a simple from a complex web. Any list of characteristics of complex webs is therefore idiosyncratic. So, with all these caveats, here is our list of properties that a complex web possesses.

(i) Diversity and non-extensivity. A complex web typically contains many elements or nodes and the complexity expands in a nonlinear way with the number of elements; it is not monotonic.

(ii) Structure. A complex web typically contains a large number of relations among its elements. These nodal relations usually constitute the number of independent dynamical equations that determine the dynamics of the web.

(iii) Nonlinearity. The relations among the elements are generally nonlinear, occasionally being of a threshold or saturation character, or more simply of a coupled, deterministic, nonlinear form. The web often uses these relations to evolve in self-conscious ways.

(iv) Non-stationarity. The relations among the elements of the web are constrained by the environment and often take the form of being externally driven or having time-dependent coupling. This coupling is often random and is a way for the web to probe the environment and adapt its evolution for maximal survival.

(v) Memory. A complex web typically remembers its evolution for a long time and is therefore able to adapt its future behavior to past changes in internal and environmental conditions.

(vi) Aging. Complex webs often have non-stationary properties and exhibit behaviors that depend separately on the time of preparation and on the time of measurement.

(vii) Scaling or scale heterogeneity. Complex webs often exhibit scaling behavior over a wide range of time and/or length scales, indicating that no one scale is able to characterize their dynamics.

These are among the most common properties modelers select to characterize complex webs and that they choose to incorporate into sets of dynamical equations. These properties can often be theoretically kept under control by one or more parameters, whose values can sometimes be taken as measures of the complexity of the web. This way of proceeding sometimes allows comparisons between the complexities of distinctly different phenomena, or more precisely between distinctly different models of phenomena. The above properties, however, which involve both topological relations among elements (properties (i) and (ii)), and nonlinear dynamical equations (properties (iii)–(vi)), are typically encountered together; properties (vi) and (vii) reflect the fact that complexity in structure and complexity in time correlations are connected by emerging properties that couple dynamical variables across time.

The present book is about presenting a coherent description of web complexity that incorporates most if not all of the above properties in a way that does not result in page after lifeless page of turgid equations. However, mathematics is not to be shunned, but welcomed when it is developed in a context that illuminates and makes understandable otherwise formidable phenomena. Therefore we proceed using multiple data sets extracted from complex webs drawn from a variety of phenomena in order to motivate the various mathematical models presented.

1.3   Why people forecasts are usually wrong

Social connections are different from physical and biological connections. In biological and physical webs the connections involve passing something tangible between the nodes; be it electrons, phonons, momentum, or energy, in each case a physical quantity is transferred. In a social web connections are made through telephone calls, Internet messages, conversations in a coffee shop, letters, a meaningful glance at a cocktail party and hundreds of other ways that involve the transmission of that ephemeral quantity information. Consequently the prediction of the behavior of people involves knowing how people create, send, receive and interpret information. It is therefore worthwhile to take a little time to examine, at least qualitatively, what some of the limitations of present-day information theory are when the web involves people.

Information transmission is considered classically to be the bundling of data into a message, followed by propagating that message in space and time from the sender to the receiver. The condition of the sender is not taken into account in considering the creation and transmission of information, but in a social web the sending of a message may markedly change the sender. This is the basis of psychoanalysis, for example, in which telling the story of psychological trauma can change a person's experience of trauma and modify subsequent behavior.

Historically information theory has a similar disregard for the receiver of information. Receiving a message is not expected to change the receiver's ability to receive or interpret subsequent messages and yet bad news can often result in a person's blocking subsequent messages, at least in the short term. A spouse, on hearing of the psychological trauma of his mate, can be overcome with pity, grief, remorse, pain (take your pick) and disconnect from the sharing. What the receiver hears in the message determines whether they continue to listen or how they interpret what they hear if they do continue to listen.

Only the messages measured in bits, information throughput in a channel, and error rates are considered in traditional information theory, with the sender and receiver being ignored. This failure to account for what some might consider to be the more important parts of an information web may be related to the mistaken idea that meaning and understanding are not part of information. Meaning can be included in the theory only through the characteristics of the sender, whereas understanding can be included in the theory only through the properties of the receiver. Meaning and understanding are psychological properties of the humans involved in the web and are not merely properties of a configuration of abstract symbols that constitute the information; they are meta-properties of a message and they are not unique.

But how is such a glaring omission as that of meaning and understanding made in such an important area of science as information theory? The answer is that it was a conscious choice made by Shannon in his 1948 paper and was, for better or worse, accepted by subsequent generations of scientists and engineers. Shannon restricted the application of information theory in the following way:

Frequently the messages have meaning; that is they refer to or are correlated according to some system with certain physical or conceptual entities. These semantic aspects of communication are irrelevant to the engineering problem.

Today we might say that this dismissal of semantics was a working hypothesis for the engineer. A working hypothesis is an assumption that one makes to simplify analysis, but which is subsequently relaxed in order to analyze the complete problem. It is unfortunate that the validity of this hypothesis was not more closely examined in subsequent discussions of information in order to take the properties of the sender and the receiver into account.

Meaning emanates outward from the message created by the sender, whereas understanding flows inward with the message interpreted by the receiver. However, neither understanding nor meaning is a simple concept. One message may be interpreted very differently by different people. Telling a co-worker that "they did a good job" may elicit a smile and nod from a subordinate; but it may be interpreted as patronizing by a competitor, and even as presumptuous by a superior. Consequently the meaning of this apparently positive phrase is not fixed, but depends on the context in which it is used. To the subordinate it is acknowledgment, to the peer it is condescension, and to management it is arrogance. Such confusion is at the heart of "politically correct" speaking.

Let us leave meaning for the moment and turn to understanding. Of course, in one sense understanding is determining the intended meaning of a message. So when the sender's intended meaning is the same as that apprehended by the receiver, the receiver is said to understand the message. Of course, the equivalence of the meaning and understanding of a message can be determined only by the sender and receiver getting together after reception of the transmission and comparing what was said with what was heard. But, since this happens all too rarely, perhaps we should look elsewhere for understanding.

From a different perspective, understanding is the completion of a mental map of a phenomenon. In the physical world an event occurs, say lightning flashes from the clouds connecting the sky and the earth with an erratic line of bright light, such as that shown in Figure 1.13. A second or so later there is a deafening sound of thunder. Primitive man on the Serengeti knows the thunder to be the voice of god and the lightning to be god's hand. Man therefore understands the nature of thunder and lightning and his mental map connects him to the elements. He is part of the awesome display of power that god puts on for him.

Man's understanding of such natural phenomena as thunder and lightning also gave those events meaning. The events were neither random nor capricious, but were part of the interaction between man and his god. It often fell to the shaman of the tribe to interpret the specific meaning of an interaction. For example, how the lightning, which caused a fire in the village, was punishment for something they had done, or for something they had failed to do. The shaman also knew how dancing might produce rain to end the drought, or what sacrifice might placate a god's anger. To some this "wisdom"

Figure 1.13.   A typical bolt of lightning as seen from the shore coupling the sky to the water. It takes little imagination to realize how awesome this spectacle must have been to primitive man.

of the shaman is exploitation of the masses by religion, but to others this is faith-based modeling of the unexplained.

Meaning and understanding are intertwined, inextricably tied together. The world of the primitive was no less understood than is the technological society of today. The map is not the reality, but the map is what we humans have. The map changes over time, sometimes in systematic ways and other times in jumps, with the future being disconnected from the past. For example, Benjamin Franklin changed the map by showing that lightning is the discharge of electricity from the clouds to the earth and thunder is the subsequent slapping together of the air masses that had been so dramatically cleaved by the discharge. This scientific understanding of the phenomenon is part of a mental map that today captures physical phenomena in a web of interleaving theories.

This change in understanding of the physical world has brought with it an associated change in meaning. In the case of lightning the change is a loss of teleologic meaning, in the sense of phenomena having no more purpose than the unfolding of physical law. But to some this in itself, the manifestation of scientific law, is equally a manifestation of god; a god that is as pervasive as, if less personal than, the god of the primitive. Others completely disconnect the scientific laws from any idea of a god and are left only with the map.

This map of the physical world seems to work quite well. It provides the engineer with the basic understanding to design and build things that had not previously existed. The understanding of the physical scientist in the hands of the skilled engineer is rendered into airplanes, computers, mobile phones, computer games, new material for clothing, sandals, and on and on. Wherever you look, whatever you see, if it was not built by human hands, such hands modified it. This history of being able to modify the physical world is empirical evidence that the scientist's mental map of these things is very good, which is to say that the map is a faithful rendition of physical phenomena.

The success of science does not mean that the map of the physical world is the same as the physical world, or even that the map stands in one-to-one correspondence to the physical world. What it means is that scientists have succeeded in developing codes that enable them to manipulate symbols and predict the outcome of certain actions done within the world. The codes are the mathematical representation of natural laws at the heart of physics and chemistry, and to a somewhat lesser extent the life sciences, and the actions they perform are experiments. The more physical the phenomenon the more closely the mental calculations correspond to what occurs in the real world. These manipulations result in the formulation of predictions, anticipating in quantitative detail the outcome of experiments.

When one establishes the conditions for an experiment and uses natural law to predict the outcome of that experiment, does that lead to understanding? In the physical sciences it is often stated that we understand a phenomenon only when we can predict its behavior and verify that prediction by experiment. A prediction enables us to transcribe one set of numbers into another set of numbers and vindicate that both sets of numbers exist within our mental map. The map must be sufficiently complex that the transcription can be manifested by means of a well-designed experiment. What we need to determine is whether or not this ability to anticipate or reproduce patterns in data is the same as understanding.

For some there is no difference between predicting and understanding. For example, if one could predict the change in the prices of stocks over time one could make a great deal of money. In the real world making money is a true indicator of understanding. The truth of this lies in the expression "If you're so smart, why ain't you rich?," for which we never found a satisfactory response, or at least not to a person who would think such a charge had merit. But, in any event, for some people anticipating the consequences of an action is indistinguishable from understanding.

We are not born with a mental map of the world, we construct them from our experience. Sometimes the construction is conscious, as in the study of science, but more often than not the map building goes on when we are least aware. In any event we spend the first two decades of life constructing the map and the next two decades trying to understand what we have built and deconstruct it, so that the last two decades can be enjoyed. Very often the most prominent features of our mental map have little to do with the world in which we live and a great deal to do with our reaction to that world. How is it that siblings raised in a single household can turn out so differently, one a philanthropist and another a serial killer? Well, maybe not usually that extreme, but on the whole very different, even when both parents are present throughout the children's formative years. Assuming that two siblings are presented with the same external stimuli, they apparently choose to experience those stimuli very differently. The sarcastic humor of the father one son adopts as his own and there is regular father–son bantering that draws the two of them closer together. The other son, who perhaps views the world in more humanist terms, interprets the sarcasm as a rejection of him and his world view, eventually becoming estranged from the father, but not necessarily from his brother.

One son's map is filled with conflict, contradiction and humor. The tension of paradox is resolved with humor, not with intellectual synthesis. Whereas the other son's map seeks to reward compatibility and to isolate paradox, and the intellectual synthesis is much stronger, but the humor is almost completely absent. The mental map of one draws him to interacting with other people, whereas the map of the other leads to introspection and understanding rather than sharing.

Understanding the psyche of an individual is not like understanding a physical process. There are no laws to guide the exploration of the psychological domain. But wait; it is not entirely true that there are no laws. In the middle of the nineteenth century, physics was considered the paradigm of objective science. In their rush to understand individuals and the world in which they live, scientists invented *psychophysics* and *sociophysics* in attempts to apply the methods of physics to the understanding of social and psychological phenomena. On the whole, psychophysics was the more successful of the two since it involved direct experimental verification of theory. Much of the theory was concerned with *psychometrics*, which is to say, the construction of metrics to enable the quantification of psychological phenomena. The theory of measurement, which is now a rather sophisticated branch of mathematics, grew from the efforts to understand these first experiments of their kind. Much of today's experimental design is concerned with the formulation of experimentally accessible measures, which, more often than not, is the most difficult part of the design of an experiment.

1.3.1   Power laws, prediction and failure

So let us return to the question of the utility of predictions. In the social and life sciences the predictions do not have the quantitative precision of the physical sciences. It might be argued that the precision is lacking because the natural laws in biology are empirical, such as the allometric laws. But that would overlook similar empirical laws in physics, for example, Boyle's law according to which the product of the pressure and volume of a gas is constant; this is the same type of relation that Auerbach found for urban concentration. When the temperature is kept constant Boyle's law can be used predictively to interpret experimental results. However, Boyle's law does not stand in isolation, but can be derived from the general theory of thermodynamics or from kinetic theory. Consequently its predictions are constrained by a web of physical observations and theory. This is usually not the case in the life sciences, where the allometric relation does stand alone and consequently its predictive value is limited.

Complex webs described by hyperbolic or inverse power-law distributions are not amenable to the predictions presented for simple physical phenomena. We shall find that the more complex a phenomenon becomes the less predictable it becomes; less predictable in that those events which are the most influential are the most difficult to predict. The complexity of these webs is manifest through the domination of the phenomena by the extreme values of the dynamics. Taleb calls this the *Black Swan* [32] because it is the deviant, the black swan, that ultimately determines the outcome of an experiment, not the average behavior. Recall our discussion of the citations of scientific articles; the average number of citations cannot be used to characterize the citation process because of its inverse power-law form. The black swan is unpredictable and yet its existence very often determines what can and cannot be known about a complex web.

Perhaps an example will serve to clarify this point. The black swan was set in an economic context and requires a certain amount of background to understand the influence of extrema, but there is another, more direct phenomenon with which we are familiar; a blackout. Hines [15] defines a blackout as being any power-network event that results in an involuntary interruption of power to customers and lasts longer than 5 minutes. When the power grid fails its influence is geographically dispersed, potentially affecting literally millions of people and crippling entire regions of the country. The data available from the North American Electrical Reliability Council (NERC) for 1984–2006 indicate that the frequency of blackouts follows an inverse power law and the frequency of large blackouts in the United States is not decreasing, in spite of the safety measures put into place to mitigate such failure over that time period; see, for example, Figure 1.14.

Part of what makes this example so interesting is that, in order to allocate resources and anticipate the event's effect, policy makers must know the patterns that appear in the data, namely the relationship between the size of an event (a blackout in terms

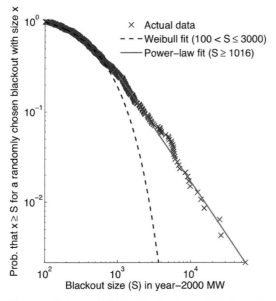

Figure 1.14. The cumulative probability distributions of blackout sizes for events greater than S in MW. The crosses are the raw data; the solid line segments show the inverse power-law fits to the data and the dashed curve is the fit to the Weibull distribution. Reproduced from [15] with permission.

of number of people affected or MW lost) and its probability of occurrence. Recall hurricane Katrina, whose initial damage could not be avoided but whose subsequent destructive impact, many argue, resulted from a lack of preparedness for its magnitude, duration and secondary destructive influence. It is important to design the infrastructure to be robust against large failures regardless of whether one is talking about hurricanes, blackouts, or, indeed, economic recessions. Moreover, it is not just the United States that has an inverse power-law size–frequency relationship in event size in either megawatts or customers, or both, but the international blackout data show the same pattern [10]. This universality of the blackout data indicates that the inverse power-law relationship is a reflection of the fundamental structure of the electrical power grid.

It is useful to make some formal observations here. The first is that the probability of failure in simple statistical phenomena is exponential, as one finds in uncomplicated manufacturing processes, say in the making of light bulbs. Empirically it is observed that for a large number of light bulbs there is a constant rate of failure per unit of time, say λ. This constant rate of failure implies that the probability a given fraction of this large number of light bulbs fails in a time t is given by

$$F(t) = e^{-\lambda t}, \tag{1.49}$$

so that, although one cannot say whether this bulb or that bulb will fail after a time t, it is possible to discuss fairly accurately what will happen to a large number of bulbs in that time. Consequently, it is useful for the factory owner to know this number in order to anticipate what the manufacturing cost of the bulbs will be over the next year and to design the production line to have an acceptable rate of failure. The assumption is that, although the specifics of the process cannot be predicted or controlled, an overall parameter such as the average failure rate λ can be changed to make the production process more efficient.

However, this exponential functional form is not observed in blackouts, or, indeed, in the failure of any other complex network.

It is web complexity that invalidates this exponential distribution of failure and its simple one-parameter characterization. Quite generally complexity is manifest in the loss of web characterization by a small number of parameters. In this chapter we have seen that normal statistics, with its mean and variance, is replaced by inverse power-law or Pareto statistics in which one or both of these quantities diverge. In subsequent chapters we show that these complex webs have failure probabilities that are also given by inverse power laws, just like the blackouts. Large blackouts typically involve complicated and cascading chains of events leading to failure that propagate through a power network by means of a variety of processes. It is useful to point out that each particular blackout can be explained as a causal chain of events after the fact, but that explanation cannot be constructed predictively prior to the blackout [3]. The same condition exists for most complex networks, so we do not spend much time on discussing particular failure mechanisms. We leave that to the subject-matter experts.

But we are getting ahead of ourselves. We need to see a great deal more data before the ubiquity of the inverse power-law structure of complex webs becomes evident. Moreover, we need to see the various ways complexity blurs and distorts our traditional understanding of phenomena, before the multitude of ways inverse power-law behavior can be mathematically modeled is appreciated.

1.4   Recap

In this first chapter we argued that normal statistics do not describe complex webs. Phenomena described by normal statistics can be discussed in terms of averages and in a relative sense are simple, not complex. An example of a simple process is manufacturing. The distribution of errors in the manufacture of products must be kept within narrow limits or the manufacturer will soon be out of business. Thus, the industrial revolution was poised to embrace the world according to Gauss and the mechanistic society of the last two centuries flourished. But as the connectivity of the various webs within society became more complex the normal distribution receded further and further into the background until it completely disappeared from the data, if not from our attempted understanding.

The world of Gauss is attractive, in part, because it leads to consensus. A process that has typical outcomes, real or imagined, enables people to focus attention on one or a few quantities and through discussion decide what is important about the process. Physicians do this in deciding whether or not we are healthy. They measure our heart rate and blood pressure, and perhaps observe us walking; all to determine whether these quantities fall within the "normal" range. The natural variabilities of these physiologic quantities have in the past only caused confusion and so their variation was ignored, by and large. A similar situation occurs when we try to get a salary increase and the boss brings out the charts to show that we already have the average salary for our position and maybe even a little more.

The normal distribution is based on one view of knowledge, namely a view embracing the notion that there is a best outcome from an experiment, an outcome that is predictable and given by the average value. Natural uncertainty is explained by means of the maximum-entropy argument, where the average provides the best representation of the data and the variation is produced by the environment having the maximum entropy consistent with the average. But this interesting justification for normalcy turned out to be irrelevant, because normal statistics do not appear in any interesting data (complex webs), not even in the distribution of grades in school. We did look at the distributions from some typical complex phenomena: the distribution in the number of scientific papers published, the frequency of words used in languages, the populations of cities and the metabolic rates of mammals and birds. The distributions for these complex webs were all inverse power laws, which from a maximum-entropy argument showed that the relevant property was the scale-free nature of the dynamic variable.

The replacement of the normal distribution with the hyperbolic distribution implies that it is the extremes of the data, rather than the central tendency, which dominate complex webs. Consequently, our focus was shifted from the average value and the standard deviation to the variability of the process being investigated, a variability whereby the standard deviation diverges. Of particular significance was the difference in the predictability of the properties of such networks, in particular their failure characteristics. No longer does the survival probability of an element or a network have the relatively rapid decline of an exponential, but instead failure has a heavy tail that extends far from the central region, making it both rarer and less reliably predictable. The why of this observation comes later.

We can pocket another piece of wisdom from the discussion in this chapter. While it is true that no experimental output is exactly predictable, it is not true that the variability in the experimental data is necessarily normal, or that the average is the best representation of those data. It is safe to conclude that, rather than normal statistics, it is hyperbolic statistics that apply to complex webs, supporting the observation that normalcy is a myth.

1.5   Problems

1.1 The normal distribution

Verify that the convolution of two Gaussian distributions is also a Gaussian distribution and thereby verify Equations (1.22)–(1.24).

1.2 Maximum information

Consider one state with the probability of being occupied p and a second state with a probability of being occupied $1 - p$. Consider the information contained in the two states and show geometrically that the information obtained is maximum when $p = 1/2$.

1.3 Resolving a contradiction

The argument of Huxley implicitly addresses the criticism of Calder since the overall coefficient in the allometric relation becomes a constant of integration. Express the coefficient of the allometric equation α in terms of the initial values $W_a(0)$ and $W_b(0)$ and use the constant of integration to refute Calder's objection.

1.4 The divine proportion

Another treatment of scaling dates back to the Greeks and is a consequence of a simple geometric construction. Draw a line of unit length and divide it into two segments of length α and β such that the ratio of the original line $(\alpha + \beta)$ to the longer segment α is the same as the ratio of α to the shorter segment β,

$$\frac{\alpha + \beta}{\alpha} = \frac{\alpha}{\beta},$$

and define the ratio of the two lengths as $\phi = \alpha/\beta$. What is the value of ϕ? This scaling proportion was called the golden section or golden mean by the ancient Greeks and Kepler called it the "divine proportion." Search the web to find structures that scale according to this ratio.

References

[1] R. Albert and A.-L. Barabási,"Statistical mechanics of complex networks," *Rev. Mod. Phys.* **74**, 48 (2002).

[2] F. Auerbach, "Das Gesetz der Bevölkerungskonzentration," *Petermanns Mitteilungen* **59**, 74 (1913).

[3] R. Baldick, B. Chowdhury, I. Dobson *et al.*, "Initial review of methods for cascading failure analysis in electric power transmission systems," *IEEE Power Engineering Society General Meeting,* Pittsburgh, PA (2008).

[4] A.-L. Barabási, *Linked*, New York: Plume (2003).

[5] M. Buchanan, *Nexus*, New York: W. W. Norton & Co. (2002).

[6] W. A. Calder III, *Size, Function and Life History*, Cambridge, MA: Harvard University Press (1984).

[7] L. da F. Costa, F. A. Rodrigues, G. Travieso and P. R. Villas Boas, *Adv. Phys.* **56**, 167 (2007).

[8] H. Cramér, *Mathematical Methods of Statistics*, Princeton, NJ: Princeton University Press (1946).

[9] D. de Sola Price, *Little Science, Big Science*, New York: Columbia University Press (1963).

[10] I. Dobson, B. A. Carreras, V. E. Lynch and D. E. Newman, "Complex systems analysis of series of blackouts: cascading failures, critical points and self-organization," *Chaos* **17**, 026103 (2007).

[11] G. Galilei, 1638, *Two New Sciences*, trans. H. Crew and A. de Salvio in 1914, New York: Dover (1954).

[12] F. Gauss, *Theoria motus corporum coelestrium*, Hamburg (1809).

[13] H. M. Gupta, J. R. Campanha and F. R. Chavorette, "Power-law distribution in high school education: effect of economical, teaching and study conditions," arXiv.0301523v1 (2003).

[14] J. B. S. Haldane, *On Being the Right Size and Other Essays*, ed. J. M. Smith, New York: Oxford University Press (1985).

[15] P. Hines, J. Apt and S. Talukdar, "Trends in the history of large blackouts in the United States," *Energy Policy* **37**, 5249–5259 (2009).

[16] J. S. Huxley, *Problems of Relative Growth*, New York: The Dial Press (1931).

[17] E. T. Jaynes, "Information theory and statistical mechanics," *Phys. Rev.* **106**, 620 (1957); "Information theory and statistical mechanics II," *Phys. Rev.* **108**, 171 (1957).

[18] A. J. Lotka, *Elements of Mathematical Biology*, New York: Dover (1956) (first published by The Williams and Wilkins Co. in 1924).

[19] A. J. Lotka, "The frequency distribution of scientific productivity," *J. Wash. Acad. Sci.* **16**, 317 (1926).

[20] N. MacDonald, *Trees and Networks in Biological Models*, New York: John Wiley & Sons (1983).

[21] B. B. Mandelbrot, *Fractals, Form and Chance*, San Francisco, CA: W. F. Freeman (1977).

[22] E. W. Montroll and W. W. Badger, *Introduction to the Quantitative Aspects of Social Phenomena*, New York: Gordon and Breach (1974).

[23] *Network Science*, National Research Council of the National Academies, Washington DC, www.nap.edu (2005).

[24] M. E. J. Newman, "The structure and function of complex networks," *SIAM Rev.* **45**, 167 (2003).

[25] V. Pareto, *Cours d'Economie Politique*, Lausanne and Paris (1897).

[26] D. Ruelle, *Chance and Chaos*, Princeton, NJ: Princeton University Press (1991).

[27] K. Schmidt-Nielsen, *Scaling; Why Is Animal Size so Important?*, London: Cambridge University Press (1984).

[28] E. Schrödinger, *What Is Life?*, New York: Cambridge University Press (1995) (first published in 1944).

[29] C. E. Shannon, "A mathematical theory of communication," *Bell Syst. J.* **27**, 379–423 and 623–656 (1948).

[30] T. Simpson, *Phil. Trans. Roy. Soc. Lond.* **49**, Part 1, 82 (1755).

[31] L. Szilard, "On the decrease of entropy in a thermodynamic system by the intervention of intelligent beings," *Behav. Sci.* **9**, 301 (1964), translated from *Z. Phys.* **53**, 840 (1929).

[32] N. N. Taleb, *The Black Swan; The Impact of the Highly Improbable*, New York: Random House (2007).

[33] D. W. Thompson, *On Growth and Form*, Cambridge: Cambridge University Press (1917); 2nd edn. (1963).

[34] D. J. Watts, *Small Worlds*, Princeton, NJ: Princeton University Press (1999).

[35] D. J. Watts, *Six Degrees*, New York: W. W. Norton (2003).

[36] E. R. Weibel, *Symmorphosis*, Cambridge, MA: Harvard University Press (2000).

[37] B. J. West and A. Goldberger, "Physiology in fractal dimensions," *American Scientist* **75**, 354 (1987).

[38] B. J. West, E. L. Geneston and P. Grigolini, "Maximizing information exchange between complex networks," *Physics Reports* **468**, 1–99 (2008).

[39] N. Wiener, *Cybernetics*, Cambridge, MA: MIT Press (1948).

[40] J. C. Willis, *Age and Area: A Study in Geographical Distribution and Origin of Species*, Cambridge: Cambridge University Press (1922).

[41] W. G. Yule and M. G. Kendal, *Introduction to the Theory of Statistics*, 12th edn., London: C. Griffin and Co. (1940).

[42] G. K. Zipf, *Human Behavior and the Principle of Least Effort: An Introduction to Human Ecology*, Cambridge, MA: Addison-Wesley (1949).

2   Webs, trees and branches

In the previous chapter we examined examples of hyperbolic and inverse power-law distributions dating back to the beginning of the last century, addressing biodiversity, urban growth and scientific productivity. There was no discussion of physical structure in these examples because they concerned the counting of various quantities such as species, people, cities, words, and scientific publications. It is also interesting to examine the structure of complex physical phenomena, such as the familiar irregularity in the lightning flashes shown in Figure 1.13. The branches of the lightning bolt persist for fractions of a second and then blink out of existence. The impression left is verified in photographs, where the zigzag pattern of the electrical discharge is captured. The time scale for the formation of the individual zigs and zags is on the order of milliseconds and the space scales can be hundreds of meters. So let us turn our attention to webs having time scales of milliseconds, years or even centuries and spatial scales from millimeters to kilometers.

All things happen in space and time, and phenomena localized in space and time are called events. Publish a paper. Run a red light. It rains. The first two identify an occurrence at a specific point in time with an implicit location in space; the third implies a phenomenon extended in time over a confined location in space. But events are mental constructs that we use to delineate ongoing processes so that not everything happens at once. Publishing a paper is the end result of a fairly long process involving getting an idea about a possible research topic, doing the research, knowing when to gather results together into a paper, writing the paper, sending the manuscript to the appropriate journal, reading and responding to the referees' criticism of the paper, and eventually having the paper accepted for publication. All of these activities are part of the scientific publishing process, so it would appear that any modeling of the process should involve an understanding of each of the individual steps. Yet in the previous chapter we saw that Lotka was able to determine the distribution of papers published without such detailed understanding. This is both the strength and the weakness of statistical/probabilistic models. Such models enable the scientist to find patterns in data from complex webs, often without connecting that structure to the mechanisms known to be operating within the web.

In this chapter we explore ways of relating the complexity of a web to the form of the empirical distribution function. In fact we are interested in the inverse problem: given an empirical distribution, what is entailed about the underlying complex web producing the distribution? One tool that we find particularly useful is the geometrical notion of a fractal and consequently we devote some space to a discussion of fractal webs. It has been thirty years since Benoît Mandelbrot [44] introduced the notion of fractals into science to capture the spatial irregularity of the geometry of certain objects and the temporal burstiness of certain time series. This introduction touches on some of the applications that have been made since the publication of Mandelbrot's seminal books, but we leave for subsequent chapters how such fractal webs change in time. Modeling the dynamics of fractals is of fairly recent origin and much of that discussion breaks new ground, particularly with regard to the webs they describe.

Anyone who has lived in California for any length of time knows the disorientation resulting from the floor of a building suddenly lurching like the deck of a ship and coffee spilling from your cup. For some reason this is particularly disconcerting if the building is your home. One's first earthquake can be exciting and has convinced more than one person that the price of California's sunshine is too high. The magnitudes of these quakes and the time intervals between them have been recorded for over a hundred years so it is not surprising that quakes have empirical laws associated with both their size and their frequency. Earthquake magnitudes are measured by the Richter scale, the logarithm to the base 10 of the maximum amplitude of the detected motion, and the cumulative distribution of Californian earthquakes is given in Figure 2.1, which clearly follows the Gutenberg–Richter law. The linear horizontal scale is actually the exponent of the factor of ten and therefore this is a log–log plot of the data. It is evident that the distribution in the number of earthquakes of a given size versus the magnitude of the quake follows an inverse power law beyond the shaded region. However, the full data set can be fit using the hyperbolic distribution (1.41) derived using the maximum-entropy argument.

Another consideration that invariably arises after experiencing an earthquake is how long it will be before the "big one" hits. Will the devastating earthquake that will level a city occur next month, next year or ten years from now? Data have been collected on the time intervals between earthquakes of a given magnitude and are shown in Figure 2.2. One of the oldest geophysical laws is that of Omori, who in 1895 [55] determined that the time interval between earthquakes of a given magnitude follows an inverse power law with a slope of -1. With more recent data on California earthquakes it is possible to

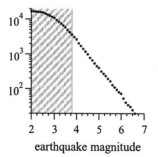

Figure 2.1.   The number of earthquakes in California from January 1910 to May 1992, as recorded in the Berkeley Earthquake Catalog. The data are taken from the National Geophysical Data Center www.ngdc.noaa.gov. Adapted with permission from [53].

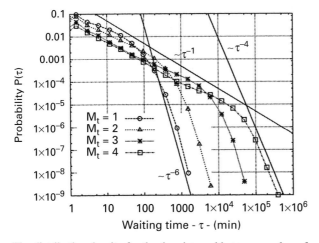

Figure 2.2. The distribution density for the time interval between quakes of a given magnitude $M \geq M_t = 1, 2, 3$ and 4. The initial inverse power law with slope -1 is Omori's law [55], but after a certain time delay the slope changes to -4. The data come from [73]. Reproduced with permission from [69].

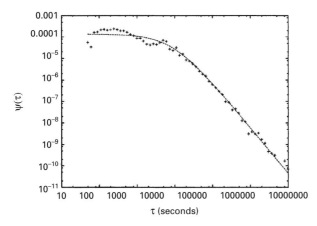

Figure 2.3. The distribution density of time intervals between solar-flare events is given by the crosses. The dashed curve is given by a fit to the data using (1.41) with normalization constant 30,567, $T = 8,422$ and power-law index $\beta = 2.14 \pm 0.05$ [25]. Reproduced with permission.

show that for very long times between quakes there is a transition from the slope of -1 to an inverse power law with slope -4. The location of the transition from the -1 slope to the -4 slope depends on the magnitude of the quake, with quakes of very large magnitude following Omori's law for very long times. Another way to view this moving of the transition region is that the larger the quake the longer it will be before another quake of that size occurs. This has led to estimates of the time until the occurrence of a given quake as being the inverse of the probability of a quake of that magnitude occurring.

Another physical phenomenon in which the distribution in the magnitude of the event as well as that in the time interval between events follows the hyperbolic distribution is that of solar flares. In Figure 2.3 the distribution of times between flares is recorded and is reproduced from Grigolini *et al.* [25]. The solar-flare waiting-time probability distribution function is determined to be the hyperbolic probability density function shown in Figure 2.3. The name waiting time is used because this is how long you wait, in a statistical sense, before another flare occurs. The data set consists of 7,212 hard-X-ray-peak flaring-event times obtained from the BATSE/CGRO (Burst and Transient Source Experiment aboard the Compton Gamma Ray Observatory satellite) solar-flare catalog list. The data cover a nine-year-long series of events from April 1991 to May 2000. The waiting-time probability distribution of flares is fitted with (1.41), where the stochastic variable τ in Figure 2.3 is the time between events. The fit yields the inverse power-law exponent $\beta = 2.14 \pm 0.05$.

The magnitudes of the solar-flare events themselves are measured in terms of the number of hard X-rays given off by flares. Figure 2.4 depicts the cumulative distribution of the peak gamma-ray intensity of solar flares and the relationship is strictly an inverse power law over nearly three factors of ten when the shaded region is ignored. However, when the shaped region is included these data can also be fit with the hyperbolic distribution (1.41), but this is not done here. Note the difference between the calendar time over which the data were recorded which was used to construct the intensity graph in Figure 2.4 and that used to generate the time-interval distribution in Figure 2.3. We expect the underlying mechanisms determining the magnitudes and times of solar-flare activity to be stable so it is not necessary to have the calendar time of the two data records coincide. We only require that the sequence of events be sufficiently long that we have good statistics.

These four examples of event-magnitude and event-time distributions do not show how the statistics are related to the spatial structure or to the dynamics of the phenomena being described. The complexity in earthquakes and solar flares is doubly evident in that both the size and the intervals are hyperbolic distributions. But earthquakes and flares are probably no more complicated than many other phenomena much closer to home. In fact it is not necessary to go outside the human body to observe structures that have the variety of scale so evident in the cosmological distributions. Neurons, for example, display this variety of scale in both space and time.

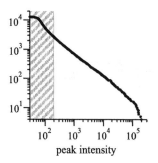

Figure 2.4.   The peak intensity of the gamma-ray spectrum produced by solar flares in counts per second, measured from Earth orbit between February 1980 and November 1989. Although the data are taken from the same archive as those used in Figure 2.3, the time periods used do not coincide [53]. Adapted with permission.

Neurons are complex in a variety of different ways. Consider their shape, which for no immediately apparent reason can be very complicated, ranging from a spindly structure of dendrites growing sparsely away from the soma to thickly entangled dendritic forms converging on compact blobs in space, and pretty much everything in between. Some neurons even have the branching structure of the lightning bolt displayed earlier. In Figure 2.5 the morphology of a representative sample of the kinds of neuron structures that are observed in the human brain and elsewhere in nature is depicted. The geometric concept of a fractal dimension is introduced to distinguish among the various forms shown.

There is little if any difference in structure, chemistry, or function between the neurons and their interconnections in man and those in a squid, snail, or leech. However, neurons can vary in size, position, shape, pigmentation, firing pattern and the chemical substances by which they transmit information to other cells. Here, in addition to the differences in the complex spatial structure, we are also interested in the firing patterns and the distribution of intervals between spikes for neurons of different morphologies. As Bassingthwaighte *et al.* [7] point out, certain cells are normally "silent" while others are spontaneously active. Some of the active ones generate regular action potentials, or nerve impulses, and others fire in recurrent brief bursts or pulse trains. These different

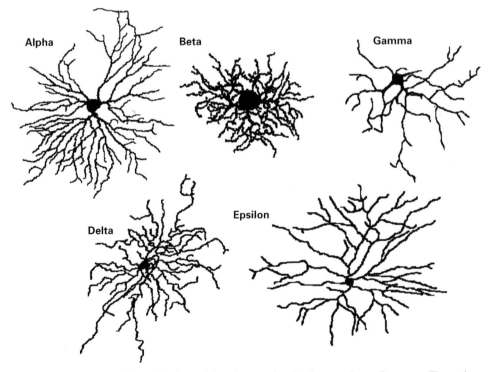

Figure 2.5. The fractal dimension is useful to characterize the shape or form of neurons. Those shown are for five different classes of ganglion cell in cat retina. When the neuron is less branched, the fractal dimension is closer to one, whereas when it is more branched, the fractal dimension is closer to two. Reproduced with permission from [33] Fig. 4.

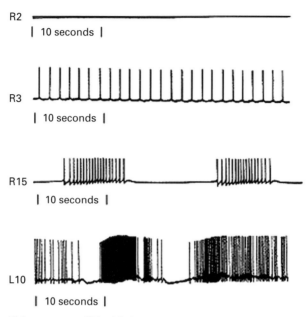

R2

| 10 seconds |

R3

| 10 seconds |

R15

| 10 seconds |

L10

| 10 seconds |

Figure 2.6.   Firing patterns of identified neurons in Aplysai's ganglion are portrayed. R2 is normally silent, R3 has a regular beating rhythm, R15 a regular bursting rhythm and L10 an irregular bursting rhythm. L10 is a command cell that controls other cells in the network. Reproduced with permission from [36] Fig. 4.

patterns result from differences in the types of ionic currents generated by the membrane of the cell body of the neurons. The variety of dynamics observed in neurons as depicted in Figure 2.6 has nearly as many scales as that in cell morphology.

It is probably worth speculating that the enhanced complexity in the firing pattern of L10 in Figure 2.6 may be a consequence of its greater role within the neuron network of which it is a part. The distribution in the time intervals between firings is also of interest, particularly for those that fire intermittently such as L10. It does not seem reasonable that the various kinds of complexity observed in these data, namely the irregular branching in space and the multiplicity of scales in time, are independent of one another. Interrelating these two kinds of complexity is one of the goals of network science. To reach this goal, let us begin exploring some of the modeling associated with complexity of branching. For this we go back a few hundred years to the ruminations of Leonardo da Vinci on natural trees and scaling in his notebooks [64].

2.1   Natural trees and scaling

The simplest type of branching tree is one in which a single conduit enters a vertex and two conduits emerge. The branching growth of such trees is called a dichotomous process and its final form is clearly seen in botanical trees, neuronal dendrites, lungs and arteries. In addition the patterns of physical networks such as lightning, river networks

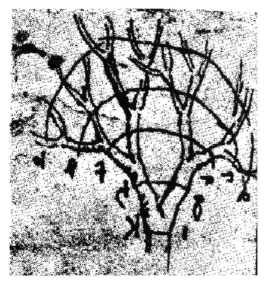

Figure 2.7.   A sketch of a tree from Leonardo da Vinci's *Notebooks*, PL.XXVII [64]. Note that da Vinci was relating the branches of equal generation number to make his association with flowing streams.

and fluvial landscapes have this tree-like structure. The quantification of branching through the construction of the mathematical laws that govern conduit size can be traced back to Leonardo da Vinci (1452–1519). In his *Notebooks* da Vinci wrote the following regarding Figure 2.7 [64]:

All the branches of a tree at every stage of its height when put together are equal in thickness to the trunk [below them]. All the branches of a water [course] at every stage of its course, if they are of equal rapidity, are equal to the body of the main stream.

Da Vinci also admonished his readers with the following statement:

Let no man who is not a Mathematician read the elements of my work.

This last statement increases in significance when we consider that da Vinci wrote it nearly two centuries before Galileo, the person who is generally given credit for establishing the importance of mathematics in modern science, crafted the algebra to describe his experiments.

The first sentence in the da Vinci quote is clarified in subsequent paragraphs of the *Notebooks*. With the aid of da Vinci's sketch reproduced in Figure 2.7, this sentence has been interpreted as follows: if a tree has a trunk of diameter d_0 that bifurcates into two limbs of diameters d_1 and d_2, the three diameters are related by

$$d_0^\alpha = d_1^\alpha + d_2^\alpha. \tag{2.1}$$

Simple geometrical scaling would yield the diameter exponent $\alpha = 2$, which corresponds to "rigid pipes" carrying fluid from one level of the tree to the next, while retaining a fixed cross-sectional area through successive generations of bifurcation. The diameter exponent for botanical trees was determined empirically by Murray in 1927

[51] to be insensitive to the kind of botanical tree and to have a value of 2.59 rather than 2. Equation (2.1) is referred to as Murray's law in the botany literature.

The significance of Murray's law was not lost on D'Arcy Thompson. In the second edition of his work *On Growth and Form* [83], Thompson argues that the geometric properties of biological webs can often be the limiting factor in the development and final function of an organism. This is stated in his general *principle of similitude*, which was a generalization of Galileo's observations regarding size discussed in the last chapter. Thompson goes on to argue that the design principle for biological systems is that of energy minimization. Minimization principles have been applied in related studies with mass and entropy sometimes replacing energy, as we did in the first chapter. The significance of the idea of energy minimization is explained in a subsequent chapter when we discuss dynamics.

The second sentence in the da Vinci quote is just as suggestive as the first. In modern language we interpret it as meaning that the flow of a river remains constant as tributaries emerge along the river's course. This equality of flow must hold in order for the water to continue moving in one direction instead of stopping and reversing course at the mouth of a tributary. Using the above pipe model and minimizing the energy with respect to the pipe radius yields $\alpha = 3$ in (2.1). Therefore the value of the diameter exponent obtained empirically by Murray falls between the theoretical limits of geometric self-similarity and hydrodynamic conservation of mass, $2 \leq \alpha \leq 3$. Here we consider how a geometric structure, once formed, can be analyzed in terms of scaling, and consequently how scaling can become a design principle for morphogenesis.

2.1.1   Two kinds of scaling

Suppose we have a tree, with the generations of the dichotomous branching indexed by n. Consider some property of the tree denoted by z_n in the nth generation, to which we impose a constant scaling between successive generations

$$z_{n+1} = \lambda z_n, \tag{2.2}$$

where λ is a constant. Note that the quantity being scaled could be the length, diameter, area or volume of the branch and that each of these has been used by different scientists at different times in modeling different trees. The solution to this equation can be expressed in terms of the size of the quantity at the first generation $n = 0$ by iterating (2.2) to obtain

$$z_n = \lambda^n z_0. \tag{2.3}$$

Consequently, we can write the solution to (2.2) as the exponential

$$z_n = e^{\gamma n} z_0, \tag{2.4}$$

where the growth rate is given in terms of the scaling parameter

$$\gamma = \ln \lambda. \tag{2.5}$$

The growth rate is positive for $\lambda > 1$ so that each successive generation is larger than the one preceding, implying exponential growth to infinity. The growth rate is negative for $0 < \lambda < 1$ so that each successive generation is smaller than the preceding one, implying exponential decay to zero.

In da Vinci's tree it is easy to assign a generation number to each of the limbs, but the counting procedure can become quite complicated in more complex networks like the generations of the bronchial tubes in the mammalian lung, the blood vessels in a capillary bed or the dendritic trees in a neuron. One form taken by the branching laws is determined by the ratio of the average diameters of the tubes from one generation to the next, which we use subsequently. But for now we introduce the general ratio

$$R_z = \frac{z_n}{z_{n+1}}. \tag{2.6}$$

In a geophysical context (2.6) is known as Horton's law for river trees and fluvial landscapes, where R_z can be the ratio of the number of branches in successive generations of branches, the average length of the river branch or the average area of the river segment at generation n. In any of these cases the parameter R_z determines a branching law and this equation implies a geometric self-similarity in the tree, as anticipated by Thompson.

Dodds and Rothman [17] view the structure of river webs as discrete sets of nested sub-webs built out of individual stream segments. In Figure 2.8 an example of stream ordering for the Mississippi basin is given. In the lower graphic is a satellite picture of the Mississippi basin and in the upper is a mathematical schematic representation of its branching. How this schematic representation of the real world is constructed is explained as follows [17]:

A source stream is defined as a section of stream that runs from a channel head to a junction with another stream ... These source streams are classified as the first order stream segments of the network. Next, remove these source streams and identify the new source streams of the remaining network. These are the second order stream segments. The process is repeated until one stream segment is left of order Ω. The order of the network is then defined to be Ω.

Note that Horton's law [31] is consistent with the simple scaling modeling adopted in Figure 2.8, but in addition requires a rather sophisticated method of counting to insure that the stream quantities are properly indexed by the stream-branch ordering. It should be emphasized that the ratios given are in terms of average quantities at sequential orders and consequently the observation that the ratios in (2.6) are constant requires the probability distributions associated with these geomorphological quantities to have interesting properties that we take up subsequently. Figure 2.9 depicts the empirical distribution of stream lengths for the Mississippi River where a straight line on bi-logarithmic graph paper would indicate that the distribution is an inverse power law beyond a certain point.

Here we demonstrate the first connection between the spatial complexity depicted in Figure 2.8 and an inverse power law characterizing some metric of the complex web. In Figure 2.9 the metric is the main stream length of the Mississippi River. However, we

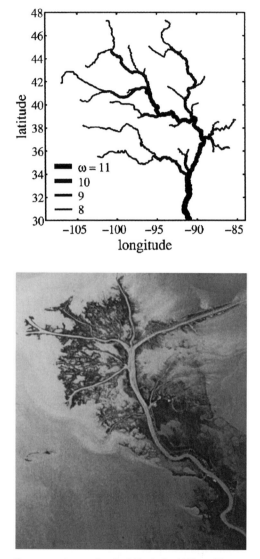

Figure 2.8.   Stream segments for the Mississippi River. The spherical coordinates of latitude and longitude are used and the scale corresponds to approximately 2,000 km along each axis. Top [18] and bottom [62] reproduced with permission.

do not want to restrict the discussion to the above geophysical context, so we consider an argument first exploited by Montroll and Shlesinger [50] in an economic context to determine how introducing increasing complexity into business webs can generate Pareto's law for the distribution of income.

Consider a complex phenomenon that for most of its range of values is described by $g(x)$, the basic distribution, which asymptotically falls to zero and has no long-range tail. We consider the normalized variable $\xi = x/\langle x \rangle$ with $\langle x \rangle$ the average value of the observed dynamic quantity X if the empirical tail of the distribution is neglected. Now we consider an iterated amplification process that induces a transition from familiar

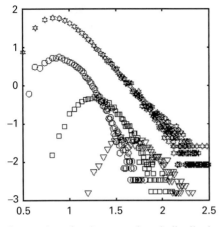

Figure 2.9. Summation of main stream-length distributions for the Mississippi River. Both axes are logarithmic, the unit of length is km and the vertical axis is the probability density with units of km^{-1}. Distributions of lengths at the nth generation l_n orders $n = 3$ (circles), $n = 4$ (squares) and $n = 5$ (triangles) are shown. The distributions sum together to give an inverse power-law tail (stars). The power-law distribution is vertically offset by a factor of ten to enhance its visibility, and is the summation of the distributions below as well as the distributions for order $n = 6$ main stream lengths [18]. Reproduced with permission.

statistical behavior, having a finite mean and variance, to inverse power-law behavior in which the second moment diverges and in some cases even the mean diverges.

Let $g(\xi)$ be amplified such that in addition to the mean value $\langle x \rangle$ it also contains a new mean value $N\langle x \rangle$ and that this new value occurs with probability p. In this way the probability of the dimensionless random variable being within an interval $(\xi, \xi + d\xi)$ changes:

$$g(\xi)d\xi \rightarrow g(\xi/N)d\xi/N.$$

In the second stage of amplification, which we assume occurs with probability p^2, the mean value becomes $N^2\langle x \rangle$. The new distribution function $G(\xi)$ that allows for the possibility of continuing levels of amplification is

$$G(\xi) = g(\xi) + \frac{p}{N}g(\xi/N) + \frac{p^2}{N^2}g(\xi/N^2) + \cdots. \tag{2.7}$$

The infinite series (2.7) may be reorganized to obtain

$$G(\xi) = g(\xi) + \frac{p}{N}G(\xi/N), \tag{2.8}$$

which is an implicit expression for the infinite series known as a renormalization-group relation. We note that (2.8) was first obtained using a different method by Novikov [54] in his study of intermittency in turbulent fluid flow [38] and later independently using the present method by Montroll and Shlesinger [50]. We also find this relation useful in the study of a number of biomedical phenomena such as the analysis of the architecture of the mammalian lung and the cardiac conduction network, both of which we discuss later [94].

It is interesting to note that this argument is similar to one presented in association with the data depicted in Figure 2.9. In that case the contributions from the increasing orders of branching within the web contribute to the total distribution of lengths with decreasing amplitudes. Consider the situation in which the dimensionless variable becomes very large, $\xi \to \infty$, and the basic function $g(\xi)$ asymptotically goes to zero faster than does $G(\xi)$, such that

$$\lim_{\xi \to \infty} \frac{g(\xi)}{G(\xi)} = 0. \tag{2.9}$$

In this asymptotic domain (2.8) can be approximated by

$$G(\xi) \approx \frac{p}{N} G(\xi/N). \tag{2.10}$$

Consequently, the asymptotic form of the distribution is determined by this simple scaling relation in the domain where (2.9) is valid. Equation (2.10) must provide the non-analytic part of the distribution since $g(\xi)$ is analytic by assumption, that is, the inverse power law is independent of the form of the basic distribution.

The solution to (2.10) is readily obtained by assuming a solution to the scaling equation of the form

$$G(\xi) = \frac{A(\xi)}{\xi^{\mu}}. \tag{2.11}$$

Note that this is how we solve differential equations. We assume the form of the solution with the appropriate number of constants to fit the initial and/or boundary conditions. Then we insert the assumed form of the solution (2.11) into the "dynamical" equation (2.10) to yield the relations for the amplitude

$$A(\xi) = A(\xi/N) \tag{2.12}$$

and the power-law index

$$\frac{1}{\xi^{\mu}} = \frac{p}{N} \frac{N^{\mu}}{\xi^{\mu}}. \tag{2.13}$$

The benign-looking (2.12) implies that the amplitude function is periodic in the logarithm of ξ with period $\ln N$ and consequently the general function can be expressed as the infinite but discrete Fourier series

$$A(\xi) = \sum_{n=-\infty}^{\infty} A_n \exp\left[\frac{i2\pi n \ln \xi}{\ln N}\right]. \tag{2.14}$$

The coefficients A_n are empirically determined by the problem of interest and very often only the $n = 0$ term contributes, that being a constant amplitude. The equation for the power-law index yields

$$\mu = 1 + \frac{\ln(1/p)}{\ln N} \tag{2.15}$$

and $\mu > 1$ since $p < 1$ and the scale factor $N > 1$. Consequently, the general solution to the scaling equation can be written as

$$G(\xi) = \sum_{n=-\infty}^{\infty} A_n \xi^{H_n},$$ (2.16)

where the scaling index is the complex quantity

$$H_n = -\mu + i \frac{2\pi n}{\ln N}.$$ (2.17)

In an economic context the general form of $A(\xi)$ given by (2.14) is simplified in that this function reduces to a constant, cf. Figure 1.6. Montroll and Shlesinger [50] determined that the Pareto index is given by $\mu \approx 1.63$ for the 1935–36 income distribution in the United States. If $pN > 1$, then $1 < \mu < 2$ and $G(\xi)$ represents the distribution function of a divergent branching (bifurcating) process in which the mean value is infinite. If $pN < 1$ but $pN^2 > 1$, then $2 < \mu < 3$ and the mean value of $G(\xi)$ is finite. However, the measure of the fluctuations about the mean value, namely the standard deviation, is infinite. These various ranges of the index are determined from the data.

Using the empirical value determined above, if we choose the probability of being in the amplifier class to be $p = 0.01$, then the amplification factor N would be approximately 16.8. As pointed out by Montroll and Shlesinger [50], this number is not surprising since one of the most common modes of significant income amplification in the United States is to organize a modest-sized business with on the order of 15–20 employees. This might also suggest the partitioning between relatively simple and truly complex networks, but that would lean too heavily on speculation.

It is clear that this argument is one possible explanation of the inverse power-law tail observed in Figure 2.9 and, using the data from Dodds and Rothman [18], the average value of the power-law index is estimated to be $\mu \approx 1.83$. This value of the scaling index is not too different from that obtained in the economic application. This kind of argument falls under the heading of renormalization and is taken up in some detail later because it is a powerful tool for identifying scaling fluctuations in erratic time series.

The geophysical applications of da Vinci's ideas are interesting, and in the next section we develop them further in a physiologic context. Da Vinci was quite aware that his notions were equally applicable in multiple domains, from the sap flowing up the trunk of a tree out to the limbs, to the water flowing in a branching stream. However, let us take a step back from the renormalization scaling argument used above and consider what classical scaling yields. Before we do this, however, it is interesting to see how his ideas reemerged in a variety of contexts from physiology to geophysics and are among the empirical laws recorded in Tables 2.1 and 2.2 where the disciplines of geophysics, physiology, economics, sociology, botany and biology have all had their share of inverse power laws. It should also be pointed out that in some phenomena recorded in these tables a better form for the distribution of the web variable is the hyperbolic, which is asymptotically an inverse power law.

Table 2.1. A list of empirical laws for complex webs is given according to their accepted names in the disciplines in which they were first developed. The power-law index α is different for each of the laws and is intended to be generic; the value listed is from the original reference and need not be the accepted value today.

Discipline	Law's name	Form of law
Anthropology		
1913 [4]	Auerbach	$\mathrm{Pr}(\text{city size rank } r) \propto 1/r$
1998 [65]	War	$\mathrm{Pr}(\text{intensity} > I) \propto 1/I^{\alpha}$
1978 [86]	$1/f$ Music	$\mathrm{Spectrum}(f) \propto 1/f$
Biology		
1992 [87]	DNA sequence	$\mathrm{Symbol\ spectrum}(\text{frequency } f) \propto 1/f^{\alpha}$
2000 [49]	Ecological web	$\mathrm{Pr}(k \text{ species connections}) \propto 1/k^{1.1}$
2001 [35]	Protein	$\mathrm{Pr}(k \text{ connections}) \propto 1/k^{2.4}$
2000 [34]	Metabolism	$\mathrm{Pr}(k \text{ connections}) \propto 1/k^{2.2}$
2001 [40]	Sexual relations	$\mathrm{Pr}(k \text{ relations}) \propto 1/k^{\alpha}$
Botany		
1883 [64]	da Vinci	Branching; $d_0^{\alpha} = d_1^{\alpha} + d_2^{\alpha}$
1922 [101]	Willis	No. of genera(No. of species N) $\propto 1/N^{\alpha}$
1927 [51]	Murray	$d_0^{2.5} = d_1^{2.5} + d_2^{2.5}$
Economics		
1897 [56]	Pareto	$\mathrm{Pr}(\text{income } x) \propto 1/x^{1.5}$
1998 [24]	Price variations	$\mathrm{Pr}(\text{stock price variations } x) \propto 1/x^3$
Geophysics		
1894 [55]	Omori	$\mathrm{Pr}(\text{aftershocks in time } t) \propto 1/t$
1933 [67]	Rosen–Rammler	$\mathrm{Pr}(\text{No. of ore fragments} < \text{size } r) \propto r^{\alpha}$
1938 [44]	Korčak	$\mathrm{Pr}(\text{island area } A > a) \propto 1/a^{\alpha}$
1945 [31]	Horton	No. of segments at n/No. of segments at $n+1 =$ constant
1954 [26]	Gutenberg–Richter	$\mathrm{Pr}(\text{earthquake magnitude} < x) \propto 1/x^{\alpha}$
1957 [27]	Hack	River length $\propto$ (basin area)$^{\alpha}$
1977 [44]	Richardson	Length of coastline $\propto 1/(\text{ruler size})^{\alpha}$
2004 [84]	Forest fires	Frequency density(burned area A) $\propto 1/A^{1.38}$
Information theory		
1999 [32]	World Wide Web	$\mathrm{Pr}(k \text{ connections}) \propto 1/k^{1.94}$
1999 [19]	Internet	$\mathrm{Pr}(k \text{ connections}) \propto 1/k^{\alpha}$

The listing of empirical laws in these tables is not intended to be exhaustive, any more than are the figures given throughout the book. The phenomena recorded are merely representative of the interconnectedness of the phenomena that we have historically partitioned into the various disciplines of science. What we hope to accomplish through such listings is to make the reader aware of the multitude of ways in which complex webs are manifest in phenomena that we read about and talk about every day, from the frequency of earthquakes around the world to the debate over the causes of global warming, from the reasons for economic upheaval to the causes of war and how we

Table 2.2. A list of empirical laws for complex webs is given according to their accepted names in the disciplines in which they were first developed. The power-law index α is different for each of the laws and is intended to be generic; the value listed is from the original reference and need not be the accepted value today.

Discipline	Law's name	Form of law
Physics		
1918 [70]	$1/f$ noise	Spectrum$(f) \propto 1/f$
2002 [25]	Solar flares	Pr(time between flares t) $\propto 1/t^{2.14}$
2004 [69]	Temperature anomalies	Pr(time between events t) $\propto 1/t^{2.14}$
Physiology		
1959 [61]	Rall	Neurons; $d_0^{1.5} = d_1^{1.5} + d_2^{1.5}$
1963 [76]	Mammalian vascular network	Veins and arteries; $d_0^{2.7} = d_1^{2.7} + d_2^{2.7}$
1963 [90]	Bronchial tree	$d_0^3 = d_1^3 + d_2^3$
1973 [48]	McMahon	Metabolic rate(body mass M) $\propto M^{0.75}$
1975 [103]	Radioactive clearance	Pr(isotope expelled in time t) $\propto 1/t^{\alpha}$
1987 [93]	West–Goldberger	Airway diameter(generation n) $\propto 1/n^{1.25}$
1991 [30]	Mammalian brain	Surface area $\propto$ volume$^{0.90}$
1992 [77]	Interbreath variability	No. of breaths(interbreath time t) $\propto 1/t^{2.16}$
1993 [58]	Heartbeat variability	Power spectrum(frequency f) $\propto f$
2007 [23]	EEG	Pr(time between EEG events) $\propto 1/t^{1.61}$
2007 [13]	Motivation and addiction	Pr(k behavior connections) $\propto 1/k^{\alpha}$
Psychology		
1957 [75]	Psychophysics	Perceived response(stimulus intensity x) $\propto x^{\alpha}$
1963 [71]	Trial and error	Reaction time(trial N) $\propto 1/N^{0.91}$
1961 [29]	Decision making	utility(delay time t) $\propto 1/t^{\alpha}$
1991 [3]	Forgetting	Percentage correct recall(time t) $\propto 1/t^{\alpha}$
2001 [20]	Cognition	Response spectrum(frequency f) $\propto 1/f^{\alpha}$
2009 [37]	Neurophysiology	Pr(phase-locked interval $< \tau$)$\propto 1/\tau^{\alpha}$
Sociology		
1926 [41]	Lotka	Pr(No. of papers published rank r) $\propto 1/r^2$
1949 [104]	Zipf	Pr(word has rank r) $\propto 1/r$
1963 [16]	Price	Pr(citation rank r) $\propto 1/r^3$
1994 [8]	Urban growth	Population density(radius R) $\propto 1/R^{\alpha}$
1998 [88]	Actors	Pr(k connections) $\propto 1/k^{2.3}$

think about these various things. What we hope is apparent to the reader is that when phenomena become complex their distributions become suspiciously similar.

2.1.2 Physiologic trees

Physiology is one area where the classical scaling approach has been applied, specifically to the branching of the bronchial tree. The physiologist Rohrer [66] assumed that the bronchial tree is homogeneous and that the volume of air in the nth generation of a bronchial tube, denoted by V_n, is divided equally between the two daughter tubes in the next generation, $2V_{n+1}$. Unlike in the more general case considered by da Vinci, the value in each daughter branch was assumed by him to be the same. Rohrer argued that, since the bronchial-tube diameters decrease by the same proportion from one generation to the next, the diameter should decrease exponentially with generation number as determined in the previous section. If we assume that the tube length scales in the same way as the tube diameter, that is, $l_{n+1} = \lambda l_n$, this argument yields for the volume

$$V_n = \pi d_n^2 l_n = 2V_{n+1} \tag{2.18}$$

and

$$V_{n+1} = \lambda^3 V_n. \tag{2.19}$$

Consequently, if volume is conserved from one generation to the next the scaling parameter must satisfy the relation

$$2\lambda^3 = 1 \tag{2.20}$$

and the optimal scaling parameter is given by

$$\lambda = 2^{-1/3}. \tag{2.21}$$

Over half a century after Rohrer [66] formulated his principle, the same exponential scaling relationship was reported by Weibel and Gomez in their classic paper on bronchial airways [91]. We refer to the exponential reduction in the diameter of the airway with generation number as classical scaling and find in the following section that it applies to the bronchial tree only in a limited domain.

The same scaling parameter was obtained by Wilson [102], who explained the proposed exponential decrease in the average diameter of a bronchial tube with generation number by showing that this is the functional form for which a gas of given composition can be provided to the alveoli with minimum metabolism or entropy production in the respiratory musculature. He assumed that minimum entropy production is the design principle for biological systems to carry out a given function. It is also noteworthy that if we set the daughter diameters in da Vinci's equation (2.1) equal to one another we obtain

$$d_0 = 2^{1/\alpha} d_1. \tag{2.22}$$

Using the value of α given by the energy minimization argument, we determine that the size of the diameter is reduced by $2^{-1/3}$ in each successive bifurcation. Consequently we obtain exactly the same scaling result for the diameter of the bronchial airway from da Vinci as from scientists nearly half a millennium later.

In classical scaling the index is $\alpha = 3$ so that the branching structure fills the available space. A less space-filling value of the scaling index is obtained for the arterial system, where it has empirically been determined that $\alpha = 2.7$. For general non-integer branching index, the scaling relation (2.22) defines a fractal tree. Such trees have no characteristic scale length and were first organized and discussed as a class by Mandelbrot [44], the father of fractals. The classical approach relied on the assumption that biological processes, like their physical counterparts, are continuous, homogeneous and

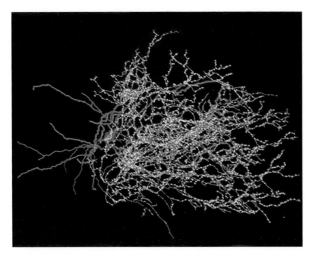

Figure 2.10. The complex pattern of the neuronal web is seen to interleave the brain mass and interconnect spatially distant regions of the brain.

regular. However, most biological networks, and many physical ones, are discontinuous, inhomogeneous and irregular and are necessarily that way in order to perform a particular function, such as gas exchange in lungs and arteries.

The branching trees of neurons such as those depicted in Figure 2.10 interleave the brain and form the communication web within the body. In the neurophysiology literature the radius of the branches from one bifurcation to the next was given in 1959 by Rall's law [61], where $\alpha = 1.5$. In this case the ratio of the number of branches in successive generations is given by

$$R = \frac{N_n}{N_{n+1}} = 2^{1/\alpha} \tag{2.23}$$

independently of branching number. More recent measurements of the scaling index, at each generation of dendritic branching, revealed a change with generation number, that is, the single ratio parameter R discussed in the previous section is replaced with a parameter that is dependent on generation number. This non-constant ratio coefficient implies that Thompson's principle of similitude is violated.

The dendrite branches of a neuron can take weeks to grow under the best of conditions, and the spatial scales are millimeters to meters in length. These space and time scales can be compared with the hundreds of meters or kilometers of the lightning bolt or river bed. However, even with the differences in the space and time scales the structures are remarkably similar to one another.

It is reasonable to point out that classical scaling disregards the irregular surfaces and structures seen in the heart, intestine and brain and the corresponding time series generated by their dynamics. The classical approach relies on the assumption that the processes of interest, whether biological, social or physical, are continuous, homogeneous and regular. It is quite remarkable then that this assumption that is so often made is very quickly shown by experiment and observation to be invalid. Most complex webs are discontinuous, inhomogeneous and irregular. The variable, complicated structure and behavior of living webs do not seem to converge on some regular pattern, but instead maintain or increase their variability as data collection is increased. Consequently, new models are required to characterize these kinds of complex webs. Herein we explore how the concepts of fractals, non-analytic mathematical functions, the fractional calculus, non-ergodic stochastic processes and renormalization-group transformations provide novel approaches to the study of physiologic form and function, as well as a broad-based understanding of complex webs. Do not be discouraged if you do not recognize these technical terms; most experts do not. They will become familiar to you in due course.

This is one of those forks in the road that ought to be flagged but is usually part of a smooth transition from what we know and understand into what we do not know and do not understand; a stepping from the sunlight into the shade and eventually into total darkness. It is a decision point for most scientists who historically formed a career involving the dynamics of differential equations, continuous differentiable functions, predictability and a sense of certainty. The scientists who made this choice developed the disciplines of biophysics, sociophysics, psychophysics and so on. These disciplines were birthed to make the complexity in biology, sociology, psychology and so on compatible with the scientific paradigm of physics. The other choice led to the road less-traveled, where scientists have grappled with complexity directly; first in the physical sciences, producing such successes as the renormalization and scaling models of phase transitions. More recently there has been success in understanding complexity outside the physical domain. One area having some success is the physiology briefly discussed in this section, which emphasizes the deviation from the traditional modeling of living webs. G. Werner commented that this fork is a rupture from classical to fractal scaling and deserves signaling with alarm bells.

Complexity is perhaps the most obvious characteristic of physiologic networks and consequently living webs provide a rich experimental data base with which to test mathematical models of complexity. Modern biology is challenged by the richness of physiologic structure and function and has not as yet been able to capture this variability in a single law or set of laws. For the sake of this preliminary discussion, consider the bronchial tree as a paradigm for anatomic complexity. Subsequently we investigate the dynamics underlying physiologic time series, but let's focus on the static situation for the time being.

West and Goldberger [93] point out that two paradoxical features of the bronchial tree are immediately evident to the untrained eye, see Figure 2.11. The first is the extreme variability of tube lengths and diameters; the second is the high level of organization limiting this variability. One labeling convention in the literature is to number successive bifurcations of the bronchial tree as "generations." The first generation of tubes in the counting scheme adopted in this context includes just two members, the left and right mainstem bronchi, which branch off from the trachea. The second generation consists of four tubes, and so forth. One observes that from one generation to the next the tube sizes vary, tending to get shorter and narrower with each generation. However, this variability is not restricted to comparisons between generations; tubes vary markedly also within

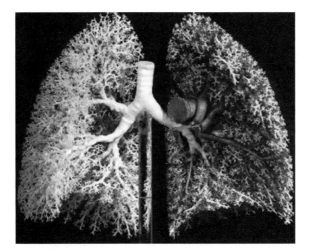

Figure 2.11.   A silicone cast of a human lung. Note the balance between apparently local randomness and the global organization within which it is contained.

any given generation. For example, after eight generations there are $2^8 = 256$ tubes with a broad distribution of tube lengths and diameters and this procedure continues for another twelve generations. Consequently, the statistical distribution of sizes with generation number becomes an important consideration in understanding how the lung functions.

One's second impression stands in sharp contrast to the first and that is how well the inter-generational and intra-generational variabilities of bronchial trees are organized. The bronchial tree, despite its asymmetries, so evident in Figure 2.11, clearly follows some ordering principles. There is a pattern underlying the variability of sizes: lung structure appears roughly similar from one generation of tubes to the next. This paradoxical combination of variability and order emerges from the fractal model of bronchial architecture [92]. The details of this model we leave for subsequent discussion, but we note here that this model suggests a mechanism for the organized variability inherent in physiologic structure and function. The essential concept underlying this kind of "constrained randomness" is scaling. It is true that the general notion of scaling was well established in biology through the work of Thompson and others; however, the kind of scaling we have in mind here does not follow the classical argument given earlier. But before we turn to these more modern ideas let us look at some other types of scaling.

The discussion of classical scaling up to this point is consistent with the data and the analysis of Weibel and Gomez for ten generations of the bronchial tree. However, a remarkable systematic deviation from the exponential scaling behavior is found for the remaining ten generations as depicted in Figure 2.12. Weibel and Gomez attributed this deviation to a change in the flow mechanism in the bronchial tree from that of minimum resistance to that of molecular diffusion. West *et al.* [92] observed that this change in the average diameter can equally well be explained without recourse to such a change in flow properties. We postpone this discussion until we have discussed the idea of fractals in some detail. However, we can show that the data reveal a different kind of scaling.

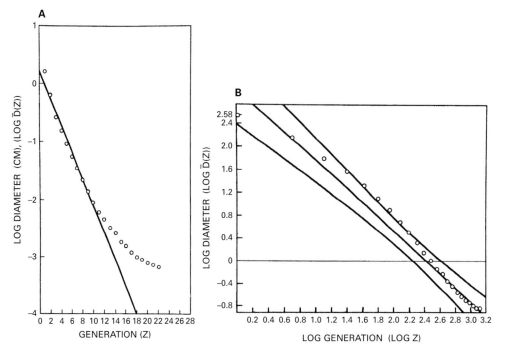

Figure 2.12. The human-lung-cast data of Weibel and Gomez for twenty-three generations are indicated by the circles and the prediction using classical scaling for the average diameter is given by the straight-line segment on the left. The fit is quite good until $z = 10$, after which there is a systematic deviation of the anatomic data from the theoretical curve [91]. The same data as on the left are plotted on log–log graph paper on the right and fit to an inverse power law. Adapted from [93].

Recall that the classical scaling argument neglects the variability in the linear scales at each generation and uses only average values for the lengths and diameters. This distribution of linear scales at each generation accounts for the deviation in the average diameter from a simple exponential form. The problem is that the seemingly obvious classical scaling sets a characteristic scale size. This will clearly fail for complex webs where no characteristic scale exists. We find that the fluctuations in the linear sizes are inconsistent with simple scaling, but are compatible with more general scaling theory. This is essentially the renormalization-group argument that was given earlier.

At each generation z of the bronchial tree the diameter of the bronchial tube takes on a distribution of values. The average diameter $d(z)$ of the tube as a function of the generation number is given by the solution to the renormalization-group relation (2.16) with a complex power-law index (2.17). The best fit to the bronchial-tube data is given in Figure 2.12 for both classical scaling (on the left) and fractal (renormalization) scaling (on the right). We see that the inverse power law with $n = 0$ in the index (2.17) captures the general behavior of the scaling over all generations. The apparent modulation of the general scaling can be fit by using the $n = 1$ in (2.17) so that the scaling index becomes

Table 2.3. Bronchial length ratios. Method A; values were computed from the ratio of short/long conjugate tube lengths for each bifurcation for which measurements were available in bronchial generations one through seven. The overall mean and variance were calculated for these seven generations using the number of ratios indicated. Method B; the sums of all short and all long branches in bronchial generations one to seven were obtained. The ratio of these sums was then calculated for each generation. The overall mean and variance were calculated for these seven generations [22]. Reproduced with permission.

Lung cast	Method A	Method B
Human	0.65 ± 0.02 ($n = 226$)	0.62 ± 0.03
Dog	0.67 ± 0.02 ($n = 214$)	0.67 ± 0.02
Rat	0.63 ± 0.03 ($n = 100$)	0.55 ± 0.05
Hamster	0.62 ± 0.02 ($n = 107$)	0.59 ± 0.07

complex and captures the modulation with period $\ln N$ [72]. In the next section we argue that this result is representative of many physiologic phenomena and should not be restricted to bronchial airways.

It is worthwhile to point out that in addition to the classical and fractal (renormalization) scaling discussed above other kinds of scaling appear in physiologic webs. For example, the Fibonacci scaling introduced in the problem set has been observed in a number of biological and anatomic relations. The historical observation that the ratio of the total height to the vertical height of the navel in humans approximates the golden mean has been verified in systematic studies [15]. We can also apply this reasoning to the asymmetric branching within the mammalian lung. We assume that the daughter branches in the bronchial tree have the lengths α and β and hypothesize that these lengths satisfy Fibonacci scaling. To test this hypothesis Goldberger *et al.* [22] used a data set of detailed morphometric measurements of silicone-rubber bronchial casts from four mammalian species. The results of analysis of these data for human, dog, rat and hamster lungs are shown in Table 2.3. In this table we see a consistency with the hypothesis that the Fibonacci ratio is a universal scaling ratio in the bronchial tree.

Results such as these suggest a regularity underlying the variability found in natural phenomena. We often lose sight of this variability due to our focus on the regularity and controllability that modern society imposes on our thinking and our lives. Consequently, to understand natural phenomena we must examine the variability together with the regularity. The phrase "thinking outside the box" attempts to prompt thinking outside the averaged boundaries in order to capture the variability.

2.2   Physiology in fractal dimensions

Physical science contains a number of assumptions that are so basic to the phenomena being examined that their significance is often lost and it is not obvious what the assumptions have to do with what is being measured. The nature of measurement and how metrics determine the knowability of a phenomenon is the domain of epistemology. But most scientists are not particularly interested in philosophy; a scientist's or engineer's concern is typically restricted to what one can record using a piece of measuring equipment, such as a ruler or thermometer, or even a finger on a pulse. The philosophical discussion is traditionally ignored, being pushed into the realm of metaphysics, and the scientist/engineer moves on. It is only when the traditional definition of a quantity leads to inconsistencies in experiment or theory that the working scientist decides to reexamine the assumptions upon which measurements or calculations are based. In the last half of the twentieth century Benoît Mandelbrot literally forced scientists to reexamine the way they measure lengths in space and intervals in time. His introduction of the geometric concept of the fractal provided a new tool with which to examine the physical, social and life sciences to better understand the world. In his monograph [45] Mandelbrot merged mathematical, experimental and scientific arguments that, taken together, undermined the traditional world picture of the physical, social and life sciences. For example, it had been accepted for over a century that celestial mechanics and physical phenomena are described by smooth, continuous and unique analytic functions. This belief was part of the conceptual and mathematical infrastructure of the physical sciences. The changes of physical processes in time were modeled by webs of dynamical equations and the solutions to such equations were thought to be continuous and differentiable at all but a finite number of points. Therefore the phenomena being described by these equations were themselves also thought to have the properties of continuity and differentiability. Thus, the solutions to the equations of motion such as the Euler–Lagrange equations, or Hamilton's equations, were believed to be analytic functions and to represent physical phenomena in general. Perhaps more importantly, the same beliefs were adopted by the social and life sciences as well.

Mandelbrot presented many examples of physical, social and biological phenomena that cannot be properly described using the traditional tenets of dynamics from physics. The functions required to explain these complex phenomena have properties that for a hundred years were thought to be mathematically pathological. He argued that, rather than being pathological, these functions capture essential properties of reality and therefore are better descriptors of the world than are the traditional analytic functions of theoretical physics. The fractal concept shall be an integral part of our discussions, sometimes in the background and sometimes in the foreground, but always there. At this point, reversing perspective, let us state that the mathematical pathology attributed to what are now called fractals can be taken as one of a number of working definitions of complexity in that a complex web often has fractal topology and the corresponding measured time series have fractal statistics.

2.2.1   Fractals

We make use of a simple example of a fractal process in order to develop some insight into the meaning of scaling relations and to develop nomenclature. There are three distinct kinds of fractals. Some are purely geometric, where the geometric form of an object at one scale is repeated at subsequent scales. Then there are the dynamics fractals where the geometric self-similarity is replaced by the repetition of structure across time rather than space. Finally, there are fractal statistics in which the properties of the distribution function are repeated across multiple scales of either space or time, or both. There are many books on each kind of fractal, so here we give only a brief overview of each type. This should provide enough information to stimulate the imagination and describe one or more of the phenomena where data have a statistical distribution of the hyperbolic form. One of the things we learn is that Gaussian statistics have scaling behavior, perhaps the simplest statistical scaling. But more significantly the hyperbolic statistics we discussed in the first chapter have a richer scaling behavior. The hyperbolic statistics in space indicate a kind of clustering associated with the fluctuations, so there are clusters within clusters within clusters, and it is this clustering behavior that is associated with the fractal property. Most of the complex phenomena of interest to us have this scaling, which is fundamentally different from the scaling in normal statistics.

We previously introduced the notion of self-similarity and for geometric fractals self-similarity is based on the idea of a reference structure repeating itself over many scales, telescoping both up and down in size. It is interesting to use this scaling argument to test our intuitive definition of dimension.

A one-dimensional unit line segment can be covered by N line segments of length r, the linear scale interval or ruler size. A unit square can be covered by N areas r^2 and finally a unit cube can be covered (filled) by N cubes of volume r^3. Note that in each of these examples smaller objects of the same geometric shape as the larger object are used to cover the larger object. This geometric equivalence is the basis of the notion of self-similarity. In particular, the linear scale size of the covering objects is related to the number of self-similar objects required to cover an object of dimension D by

$$r = 1/N^{1/D}. \tag{2.24}$$

This relation can be inverted by taking the logarithm of both sides of the equation and defining the dimension of the object by the equation

$$D = \frac{\log N}{\log(1/r)}. \tag{2.25}$$

This expression is mathematically rigorous only in the limit of vanishingly small linear scale size $r \to 0$. This is where the discussion becomes very interesting. Although (2.25) defines the dimension of a self-similar object, there is nothing in the definition that guarantees that D is an integer, such as 1, 2 or 3. In fact, in general D is not an integer.

It is one thing to define a non-intuitive quantity such as a non-integer dimension. It is quite another to construct an object having such a dimension. So let us examine how to construct a geometric object that has a non-integer dimension. Consider the unit line segment depicted in Figure 2.13 over which mass points are uniformly distributed. Now we remove the middle third of the line segment and redistribute the mass of the removed section along the remaining two segments so that the total mass of the resulting set remains constant. At the next stage, cut the middle third out of each of these two line segments, and again redistribute the mass of the removed sections along the

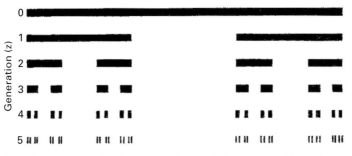

Figure 2.13.   A Cantor set constructed from infinitely many trisecting operations of the unit interval is depicted for the first six generations. The fractal dimension of the resulting "fractal dust" is 0.6309.

remaining line segments so that none is lost. In this example the parameter a is the ratio of the total mass to the mass of each segment after one trisecting operation. Since each segment receives half the mass of its parent $a = 2$. The second parameter b is the ratio of the length of the original line to the length of each remaining segment. Since this is a trisection $b = 3$. The parameter a gives us an idea of how quickly the mass is being concentrated and the parameter b gives us a comparable idea of how quickly the available space is being thinned out.

Using the definition (2.25) we see that, for the trisection Cantor set at each stage n, the number of line segments increases by two so that $N = 2^n$; also the scaling ratio decreases by a factor of 1/3 at each stage so that the ruler size becomes a function of n, $r = 1/3^n$. Consequently, the fractal dimension is given by

$$D = \frac{\log[2^n]}{\log[3^n]} = \frac{\log 2}{\log 3} = 0.6309, \tag{2.26}$$

independently of the branching or generation number n. In general $N = a^n$ and $r = 1/b^n$, so the fractal dimension D of the resulting Cantor set is the ratio of the logarithms,

$$D = \frac{\log a}{\log b} \tag{2.27}$$

in terms of scaling parameters. Note that this is the same ratio as that previously encountered in our discussion of scaling.

As West and Goldberger [93] point out, there are several ways in which one can intuitively make sense of such a fractional dimension. Note first of all that in this example D is greater than zero but less than one, the dimension of a line in Euclidean geometry. This makes sense when one thinks of the Cantor set as a physical structure with mass: it is something less than a continuous line, yet more than a vanishing set of points. Just how much less and how much more is given by the ratio of the two parameters. If a were equal to b, the structure would not change no matter how much the original line were magnified; the mass would clump together as quickly as the length scaled down and a one-dimensional line would appear on every scale. However, if a were greater than b a branching or flowering object would appear, namely an object that develops finer and finer structures under magnification such as the fractal trees

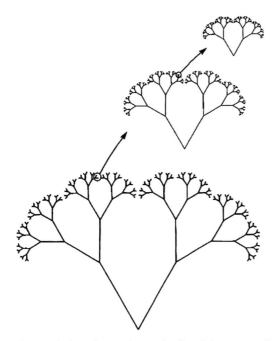

Figure 2.14. Geometric fractals constitute a family of shapes containing infinite levels of detail. In the fractals reproduced here, the tip of each branch continues branching over many generations, on smaller and smaller scales, and each magnified, smaller-scale structure is similar to the larger form (self-similarity). As the fractal dimension increases between one and two, the tree sprouts new branches more and more vigorously.

shown in Figure 2.14, which bursts out of a one-dimensional space but does not fill a two-dimensional Euclidean plane. Again the precise character of the bursting depends on the value of D. If the fractal dimension is barely above one the tree is wispy and broom-like; as the dimension increases from one to two, the canopy of branches becomes more and more lush.

Fractal objects are ubiquitous in the physical world, being seen, for example, in the self-similar accretions of cumulus clouds, in turbulent fluid flows, in the folds of the surface of the human brain and in the patterning of magnetic spins when iron reaches a critical temperature. The organic, tree-like fractal shown in Figure 2.14 bears a striking resemblance to many physiologic structures.

Compared with a smooth, classical geometric form, a fractal curve (surface) appears wrinkled. Furthermore, if the wrinkles of a fractal are examined through a microscope more wrinkles become apparent. If these wrinkles are now examined at higher magnification, still smaller wrinkles (wrinkles on wrinkles on wrinkles) appear, with seemingly endless levels of structure emerging. The fractal dimension provides a measure of the degree of irregularity. A fractal structure as a mathematical entity has no characteristic scale size and so the emergence of irregularity proceeds downward to ever smaller scales and upward to every larger scale. A natural fractal, on the other hand, always ends at some smallest scale as well as at some largest scale and whether or not being fractal is a useful concept depends on the size of the interval over which the process appears to be scale-free. We document this association subsequently, but now let us turn from geometric fractals to fractal distributions.

One way to picture a distribution having a fractal scaling property is to imagine approaching a mass distribution from a great distance. At first, the mass will seem to be in a single great cluster. As one gets closer, the cluster is seen to be composed of smaller clusters. However, upon approaching each smaller cluster, it is seen that each subcluster is composed of a set of still smaller clusters and so on. It turns out that this apparently contrived example in fact describes a number of physical, social and biological phenomena that we discuss in due course.

Let us now look at the distribution of mass points depicted in Figure 2.15. The total mass $M(R)$ is proportional to R^d, where d is the dimension of the space occupied by the masses. In the absence of facts to the contrary it is reasonable to assume that the point masses are uniformly distributed throughout the volume and that d, the dimension

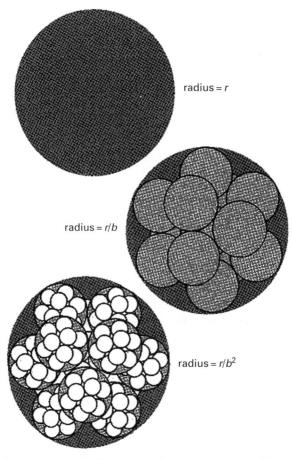

radius = r

radius = r/b

radius = r/b^2

Figure 2.15.   Here we schematically represent how a given mass can be non-uniformly distributed within a given volume in such a way that the volume occupied by the mass has a fractal dimension given by (2.31). The parameter b gives the scaling from the original sphere of radius R and the parameter a gives the scaling from the original total mass M. Adapted from [93].

associated with the mass distribution, is equal to the Euclidean dimension E of the physical space, for example, in three-dimensional space $d = E = 3$.

Suppose that on closer inspection of Figure 2.15 the mass points are seen to be not uniformly distributed in space, but instead clumped in distinct spheres of size R/b each having a mass that is $1/a$ smaller than the total mass. Here a and b are independent parameters each greater than unity. Thus, what had initially appeared as a huge globe filled uniformly with sand (mass points) turns out to resemble one filled with basketballs, each of the basketballs being filled uniformly with sand. We now examine one of these smaller spheres (basketballs) and find that instead of the mass points being uniformly distributed in this reduced region the sphere consists of still smaller spheres, each of radius R/b^2 and each having a mass $1/a^2$ smaller than the total mass. Now again the image changes so that the basketballs appear to be filled with baseballs, and each baseball is uniformly filled with sand. It is now assumed that this procedure of constructing spheres within spheres can be telescoped indefinitely to obtain

$$M(R) = \lim_{n\to\infty} a^n M\left(\frac{R}{b^n}\right). \tag{2.28}$$

Relation (2.28) is one of those equations that looks formidable, but is actually easy to solve when looked at physically. Physically the mass of this object must be finite, and one way to accomplish this is by requiring conditions on the parameters that yield a finite value for the total mass in the limit of n becoming infinitely large. Recall that the mass increases as the dth power of the radius so that

$$M(R) \propto \lim_{n\to\infty} a^n \left(\frac{R}{b^n}\right)^d. \tag{2.29}$$

Thus, the total mass is finite if the parameters satisfy the relation

$$\left(\frac{a}{b^d}\right)^n = 1 \tag{2.30}$$

independently of n and therefore the dimension d must satisfy the relation

$$d = \frac{\ln a}{\ln b}. \tag{2.31}$$

The index of the power-law distribution of mass points within the volume can therefore be distinct from the Euclidean dimension of the space in which the mass is embedded, that is, in general $d \neq E = 3$. Consequently, $d = D$, the fractal dimension.

The fractal idea also spills over from the geometry in Figure 2.15 into the realm of statistics. We do not go into detail here to explain fractal statistics, but observe that fractional exponents arise in the analysis of time series stemming from complex dynamic webs, and such time series and the webs that generate them are among the main topics of this book. The fractional index, either in time or in displacements, is the basis of interpretation of these time-series data and of the associated probability distribution functions in terms of the fractal dimension. Although the arguments given above are geometric, they apply with equal validity to the trace of time series and therefore to the time series themselves, often describing processes that have a temporal fractal dimension. These are called fractal time series and when they are also irregular, which is to say random, we call them random fractal time series.

Let us consider how the fractal idea can be married to the notion of probability. We do this by constructing a fractal random process for which the moments, such as the arithmetic mean, may or might not exist. It does seem that it ought to be possible to repeat a given experiment a large number of times, take the measurements of a given quantity, and obtain the data sequence $Q_1, Q_2, \ldots, Q_N$, sum the measured values, $\sum_{j=1}^{N} Q_j$, divide by the number of data points N, and thereby determine their average value $\overline{Q}$,

$$\overline{Q} = \frac{1}{N} \sum_{j=1}^{N} Q_j. \tag{2.32}$$

This procedure gives us the arithmetic mean of the measured process. In statistical theory, this one realization of the measured values is one sample taken from the entire population of all the possible experimental values. The mean determined from this sample of data is called the sample mean and the mean of the entire population is called the population mean. The sample mean can always be determined using the available data to carry out the averaging. The population mean, on the other hand, is a theoretical concept, and the sample mean provides an estimate for it. If the population mean is finite, then as we collect more and more data the sample mean converges to a fixed value that is identified as the best estimate of the population mean. This need not be the case, however. If, for example, as we collect and analyze more data the value of the sample mean keeps increasing, we must conclude that the population mean of the process does not exist, which is to say that the mean is not finite. This is exactly what happens for some fractal processes. Bassingthwaighte *et al.* [7] point out that, for a fractal process, as more data are analyzed,

... rather than converge to a sensible value, the mean continues to increase toward ever larger values or decrease toward ever smaller values.

The nineteenth-century social philosopher Quetelet proposed that the mean alone can represent the ideals of a society in politics, aesthetics and morals. His view involved, in the words of Porter [60], "a will to mediocrity." Like Cannon in medicine [12], Quetelet's notion involved a tendency for the enlightened to resist the influences of external circumstances and to seek a return to a normal and balanced state. The idea that a property of society could be defined, but not have a finite average, would have been anathema to him as well as to most other nineteenth-century scientists. We mention the few exceptions who saw beyond the normal distribution in due course.

For a concrete example of such a fractal statistical process consider the following game of chance. A player tosses a coin and the bank (a second player with a great deal of money) agrees to pay the first player $2 if a head occurs on the first toss, $4 if the first head occurs on the second toss, $8 if the player obtains the first head on the third toss, and so on, doubling the prize every time a winning first head is delayed an additional toss. What we want to know is the player's expected winnings in this game.

The chance of a head on the first toss is 0.5 since either side of the coin is equally likely to be face up. The chance of a tail followed by a head is $0.5 \times 0.5 = 0.25$; because the two events are independent the probability of both occurring is the product of the separate probabilities. The chance of a tail occurring on the first $n - 1$ tosses and then a head on the nth toss is $0.5^{n-1} \times 0.5 = 0.5^n$. Here again the independence of the events requires the formation of the product of the n independent probabilities. The prize if there is no head until the nth toss is therefore given by

$$\frac{1}{2}(\$2) + \frac{1}{2^2}\left(\$2^2\right) + \frac{1}{2^3}\left(\$2^3\right) + \cdots = \$1 + \$1 + \$1 + \cdots = \$n$$

which increases linearly with increasing n. Thus, the expected winnings of the player will be $\$n$ for n tosses of the coin. But n is arbitrary and can always become larger, in particular it can be made arbitrarily large or infinite. This is the type of process that is depicted in Figure 2.16, where the more the number of tosses the greater is the average value. In fact, there is no well-defined expectation value since there is no limiting value for the mean. An excellent account of such games can be found in Weaver [89].

Many more natural data sets manifest this lack of convergence to the mean than most scientists were previously willing to believe. Nicolaus Bernoulli and his cousin Daniel pointed out that this game of chance led to a paradox, which arises because the banker argues that the expected winnings of the player are infinite so the ante should be very large. On the other hand, the player argues that if we consider a large number of plays then half of them will result in a player winning only $\$1$, so the ante should be small. This failure to agree on an ante is the paradox, the St. Petersburg paradox, named after the journal in which Daniel Bernoulli published his discussion of the game. We emphasize that there is no characteristic scale associated with the St. Petersburg game, but there is a kind of scaling; let's call it the St. Petersburg scaling. Here two parameters are adjusted to define the game of chance: the frequency of occurrence decreases by a factor of two with each additional head and the size of the winnings increases by a factor of two with each additional head. The increase and decrease therefore compensate for one another. This is the general structure for scaling; the scale of a given event increases in

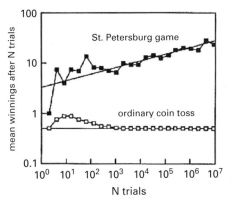

Figure 2.16. Playing the St. Petersburg game millions of times on a computer indicates that its local average diverges to infinity [7]. Reproduced with permission.

size to the base b, but the relative frequency of occurrence of that event decreases to base a. The index for the scaling is typically expressed in terms of the parameters a and b; in fact it is in terms of the parameter $\mu = \ln a / \ln b$ as found earlier.

One measure of play is, as we said, the expected outcome of a wager. If the jth wager is w_j the expected winnings are determined by summing over all the wagers

$$\langle W \rangle = \sum_{j=1}^{\infty} w_j p_j,$$

which for the St. Petersburg game diverges. To circumvent this infinity Daniel Bernoulli argued that not everyone should wager in the same way since the subjective value or utility associated with a unit of currency depends on one's total wealth. Consequently, he proposed estimating the "winnings" by using the average utility of a wager

$$\langle U \rangle = \sum_{j=1}^{\infty} U\left(w_j\right) p_j,$$

not the wager itself. It is the average utility that one ought to maximize, not the expected winnings. The mathematical wrangling over the resolution of the St. Petersburg paradox lasted another two centuries, but that need not concern us here, and we move on.

2.2.2 Temporal scaling

Scaling relates the variations of time series across multiple time scales and has been found to hold empirically for a number of complex phenomena, including many of physiologic origin. The experimentalist sees the patterns within fluctuating time series as correlations produced by interactions that mitigate the randomness. Scientists with a bias towards theory recognize time series as solutions to equations of motion, sometimes with the web of interest coupled to the environment and other times with the web having nonlinear dynamics with chaotic solutions. All these perspectives, as well as those of the computer modeler, need to be considered in order to understand complex webs, particularly since there is not a unique way to generate the fractal properties discussed in the previous section. As a matter of fact, most biological phenomena are described by the "pathological" behavior just discussed, and it is common to use life models as paradigms of complex webs.

One way physicists understand the source of scaling relations in experimental data is through the use of renormalization groups that tie together the values of certain functions at different time scales. An unknown function $Z(t)$ that satisfies a relation of the form [74]

$$Z(bt) = aZ(t) \qquad (2.33)$$

is said to scale. However, this equation expresses a dynamical relation, when t is the time, in the same sense as a differential equation relates a dynamical variable with its derivatives and consequently the scaling relation has a solution. It is worth emphasizing here that, although (2.33) is analogous to an equation of motion, it may describe a function that is continuous everywhere but which has no derivatives. Such functions are discussed subsequently.

Following the approach taken previously, we assume a trial solution of the form

$$Z(t) = A(t)t^{\mu}, \tag{2.34}$$

and insert it into (2.33). As before we determine the separate equalities for the scaling parameters and obtain the power-law index as the ratio of the scaling parameters

$$\mu = \frac{\ln a}{\ln b}, \tag{2.35}$$

the same ratio as we found in the discussion of the distribution of mass within the large sphere. The unknown function $A(t)$ may be written in general as the Fourier series (2.14) and is a function periodic in the logarithm of time with a period $\ln b$.

In the literature $Z(t)$ is called a homogeneous function [5]. The scaling index given by (2.35) is related to the fractal dimension obtained from the geometric argument given for the distribution of mass within a sphere (2.31) and elsewhere. The homogeneous solution to the scaling relation can be written as

$$Z(t) = \sum_{k=-\infty}^{\infty} A_k t^{H_k}, \tag{2.36}$$

where the complex scaling index is given by

$$H_k = -\mu + i2\pi \frac{k}{\ln b}. \tag{2.37}$$

The complex exponent in (2.36) indicates a possible periodic component of the homogeneous function superposed on the more traditional power-law scaling. There exist phenomena that have a slow modulation ($k = 1$); recall that we saw the suggestion of this in the scaling of the bronchial tubes and we discuss the fractal nature of such time series more fully in subsequent chapters. For the moment it is sufficient to see that we can have a process that is fractal in time just as we have a process that is fractal in space.

Functions of the form (2.36) can be continuous everywhere but are not necessarily differentiable anywhere. Certain of these continuous non-differentiable functions have been shown to be fractal. In his investigation of turbulence in 1926 Richardson observed that the velocity field of the atmospheric wind is so erratic that it probably cannot be described by an analytic function. In his paper [63], Richardson asked "does the wind possess a velocity?," addressing the non-differentiability of the wind field. He suggested a Weierstrass function as a candidate to represent the wind field. Scientists have since come to realize that Richardson's intuition was superior to nearly a century of analysis regarding the nature of turbulence.

A generalization of Weierstrass' original function has the form

$$W(t) = \sum_{n=-\infty}^{\infty} \frac{1}{a^n}[1 - \cos(b^n t)], \tag{2.38}$$

where a and b are real parameters and this form is chosen to simplify some of the algebra. Using the properties of trigonometric functions it is easy to see that the generalized Weierstrass function (GWF) $W(t)$ satisfies the scaling relation (2.33) and consequently shares the scaling properties of the renormalization solution (2.36). The non-differentiability of the GWF arises because each new term in the series inserts an order of magnitude more wiggles (in base b) on a previous wiggle, but each new wiggle is smaller in amplitude by an order of magnitude (in base a). Recall Bernoulli's game of chance in which the scale of the wager increases by a factor b for an outcome whose likelihood decreases by a factor of $1/a$. This Bernoulli scaling appears in the Weierstrass function as well.

Weierstrass was the teacher of Cantor. Later, as a colleague, Cantor was interested in questions of singularities and the various kinds of infinities; the apprentice challenged the master to construct a function that was continuous everywhere but was nowhere differentiable. The resulting Weierstrass function became the first non-analytic function having subsequent utility in the physical sciences. The scaling of the GWF does not in itself guarantee that it is a fractal function even though it is non-analytic, but Mauldin and Williams [47] examined its formal properties and concluded that for $b > 1$ and $b > a > 0$ the dimension is essentially given by

$$D = 2 - \mu, \tag{2.39}$$

a result established numerically somewhat earlier by Berry and Lewis [10].

Now we restrict the discussion to a special class of complex phenomena having the scaling properties found in a large number of data sets. This focus has the advantage of providing a tractable analytic treatment for a wide class of webs where all or almost all the stated properties are present. Consider the homogeneous function $Z(t)$ introduced above. We use the renormalization-group argument to define the scaling observed in the averages of an experimental time series with long-time *memory*.[1] Suppose the second moment of a stochastic process $Q(t)$ having long-time memory is given by

$$\left\langle Q(bt)^2 \right\rangle = b^{2\mu} \langle Q(t)^2 \rangle, \tag{2.40}$$

where the angle brackets denote an average over an ensemble of realizations of the fluctuations in the time series. Although it is probably not clear why such a function should exist, or why it is appropriate to discuss such functions here, let us assure you that this scaling is closely connected to the scaling we have observed in the data of the last two chapters.

For the same process whose second moment is given by (2.40) a different but related scaling is given by the autocorrelation function

$$C(bt) = b^{2\mu-2}C(t). \tag{2.41}$$

---

[1] In this chapter we use the intuitive term "memory" to introduce the mathematical concept of time correlation. A more formal definition of memory is provided subsequently by introducing memory functions as kernels in integro-differential equations for the fractional calculus.

The autocorrelation function is defined by

$$C(\tau) = \frac{\langle Q(t+\tau)Q(t) \rangle}{\langle Q^2(t) \rangle} \tag{2.42}$$

for a zero-centered process, that is, a process with $\langle Q(t) \rangle = 0$. It is worthwhile to point out that the two scaling relations imply that the underlying process is stationary in time, meaning that it depends only on the difference between the two times, not on the times separately, which can be expressed formally as $\langle Q(t_1)Q(t_2) \rangle = \langle Q(t_1 - t_2)Q(0) \rangle$.

Finally, the spectral density for this time series is given by the Fourier transform of the autocorrelation function,

$$S(\omega) = \mathcal{FT}[C(t); \omega] \equiv \int_{-\infty}^{\infty} e^{i\omega t} C(t) dt. \tag{2.43}$$

Introducing a scaling parameter into (2.43), inserting the scaling relation (2.41) and changing variables, $z = bt$, yields

$$S(b\omega) = \int_{-\infty}^{\infty} e^{i\omega z} C\left(\frac{z}{b}\right) \frac{dz}{b} = \frac{1}{b^{2\mu-1}} \int_{-\infty}^{\infty} e^{i\omega z} C(z) dz$$

and subsequently the scaling behavior

$$S(b\omega) = b^{1-2\mu} S(\omega). \tag{2.44}$$

The solutions to each of these three scaling equations, the second moment (2.40), the autocorrelation function (2.41) and the spectrum (2.44), are of precisely the algebraic form (2.33) of the solution to the renormalization-group relation, with the modulation amplitude fixed at a constant value.

The scaling properties given here are usually assumed to be the result of long-time memory in the underlying statistical process. Beran [9] discusses these power-law properties of the spectrum and autocorrelation function, as well as a number of other properties involving long-time memory for discrete time series. Here the result corresponds to a second moment of the form

$$\langle Q(t)^2 \rangle \propto t^{2\mu}, \tag{2.45}$$

which for $\mu = 1/2$ corresponds to ordinary or classical diffusion and for $\mu \neq 1/2$ this is the second moment for an *anomalous* diffusive process. We discuss both classical and anomalous diffusion subsequently in terms of both dynamics and statistical distributions. But for the time being we are content with the realization that such processes exist and are observed in nature.

2.2.3 Autocorrelation functions and fractal dimensions

Before ending this section let us make a few observations about the generality of the autocorrelation functions, the interpretation of the scaling indices and the properties of fractal dimensions. Consider an autocorrelation function in terms of a dimensionless variable of the following form:

$$C(t) = \frac{1}{(1 + t^\alpha)^{\beta/\alpha}} \tag{2.46}$$

(see Gneiting and Schlather [21] for a complete discussion of this autocorrelation function and the implications of its form). Any combination of parameters $0 < \alpha \leq 2$ and $\beta > 0$ is allowed, in which case (2.46) is referred to as the Cauchy class of autocorrelation functions. Now consider two asymptotic limits.

In the short-time limit the autocorrelation function can be expanded in a Taylor series to obtain the power-law form

$$C(t) \approx 1 - \frac{\beta}{\alpha} |t|^\alpha \text{ as } t \to 0, \tag{2.47}$$

for the range of parameter values given above. The autocorrelation functions in this case are realizations of a random process in an E-dimensional Euclidean space that has a fractal dimension given by

$$D = E + 1 - \alpha/2 \tag{2.48}$$

with probability unity [21]. In the one-dimensional case ($E = 1$) the power spectrum corresponding to (2.47) is the inverse power law

$$S(\omega) \approx \frac{1}{|\omega|^{\alpha+1}} \text{ as } \omega \to \infty. \tag{2.49}$$

Note that the $t \to 0$ limit of the autocorrelation function corresponds to the asymptotic limit as $\omega \to \infty$ of the spectrum since the two are Fourier-transform pairs. Consequently, the inverse power-law spectrum obtained in this way can be expressed as a straight-line segment on bi-logarithmic graph paper. The line segment corresponds to the spectrum and has a slope related to the fractal dimension in (2.48) by

$$\alpha + 1 = 5 - 2D. \tag{2.50}$$

At the long-time extreme, the autocorrelation function (2.46) collapses to

$$C(t) \approx \frac{1}{|t|^\beta} \text{ as } t \to \infty, \tag{2.51}$$

where the inverse power-law indicates a long-time memory when $\beta > 0$. In this case we introduce the Hurst exponent

$$2H = 2 - \beta, \text{ where } 2 \geq \beta > 0 \tag{2.52}$$

and the index H was introduced by Mandelbrot [44] to honor the civil engineer Hurst who first investigated processes with such long-term memory. Here again the Fourier transform of the autocorrelation function yields the power spectrum

$$S(\omega) \propto |\omega|^{\beta-1} = \frac{1}{|\omega|^{2H-1}}, \tag{2.53}$$

which is still an inverse power law for the index in the range given above or equivalently for the Hurst exponent in the range $0.5 \leq H \leq 1$.

We cannot over-emphasize that the fractal dimension D and the Hurst exponent H can vary independently of one another as indicated above in the two asymptotic regimes. From this example we see that the fractal dimension is a local property of time series ($t \rightarrow 0$), whereas the Hurst exponent is a global property of time series ($t \rightarrow \infty$) and, although not proven here, these are general properties of D and H [21]. It is in this asymptotic regime that the fractal dimension and the Hurst index are related by $D = 2 - H$. Consequently, the scaling behavior of the second moment does not always give the Hurst exponent as is claimed all too often in the literature and methods of analysis have been developed to distinguish between the short-time and long-time scaling properties of time series. We take this question up again at the appropriate places in our discussion.

2.3   Measuring what we cannot know exactly

There is no general theory describing the properties of complex webs, or, more precisely, there is a large number of theories depending on whether the web is biological, informational, social or physical. Part of the reason for the different theories has to do with the assumptions about web properties which are systematically made so that the discipline-specific problem being studied becomes mathematically tractable. At the same time it is often argued that the observed properties of these webs should not strongly depend on these assumptions, at least in some asymptotic limit. Two examples of where this asymptotic independence is almost universally believed to be true are in the choices of initial and/or boundary conditions.

One similar assumption that appears to have only incidental significance, but which will prove to be pivotal in our discussion, has to do with the preparation of a web and the mathematical specification of that preparation. Consider a set of dynamical equations describing some physical phenomenon, whose complete solution is given when one specifies the initial state of the web. The initial state consists of the values for all the dynamic variables in the web taken at the initial time t_0, a procedure that is completely justified when solving deterministic equations. Note that this specification of the initial state of the web together with the analytic solution to the equations of motion determines the future evolution of the web. It is generally assumed that the initial state specifies when the web is prepared as well as when the measurement of the web is initiated. We find that neither of these assumptions is necessarily valid. While criticism of the assumptions of classical determinism is widely known, and leads to the adoption of stochastic approaches, the study of time lags between the preparation and observation of physical quantities leads to the fairly new idea of *aging*.

In simple webs, such as physical networks described by Hamiltonians, it is well known that the dynamical variables can be determined to any desired degree of accuracy, including the initial state. A real web, on the other hand, is different. By real web we mean one that does not necessarily satisfy the assumptions necessary for a Hamiltonian description and can therefore have dynamics whose description is not obvious. The state of a real web cannot be determined to arbitrary accuracy; finite-temperature effects ultimately limit the reliability of a measurement and therefore limit any specification of the displacement and momentum of the constituent particles. Ultimately the displacements and momenta of the particles are random, and are described by a probability density function (pdf), not a trajectory. Even without introducing restrictions imposed by thermodynamics, the results of experiments are not reproducible: the observed output changes from experiment to experiment, regardless of how careful the preparation of the initial state is or how skilled the experimenter is. The initial state is not exactly reproducible and consequently experimental observation is slightly altered. The collection of results from a series of ostensibly identical experiments gives rise to the notion of a bundle of experimental values, characterized by an ensemble distribution function. The classical notion of the existence of an ensemble distribution of a collection of experimental results historically led to the law of frequency of errors and the functional form of the normal distribution discussed in Chapter 1.

Historically, the law of errors was the first attempt to characterize complex physical webs in a formal way and through it one of the most subtle concepts in physics was introduced: the existence and role of randomness in measurement. Randomness is associated with our inability to predict the outcome of an experiment, say the flipping of a coin or the rolling of a die. It also applies to more complicated phenomena, for example, the outcome of an athletic contest or, more profoundly, the outcome of a medical procedure such as the removal of a tumor. From one perspective the unknowability of such events has to do with the large number of elements in the web; there are so many, in fact, that the behavior of the web ceases to be predictable. On the other hand, the scientific community now knows that having only a few dynamical elements in a web does not insure predictability or knowability. It has been demonstrated that the irregular time series observed in such disciplines as biology, chemical kinetics, economics, meteorology, physics, physiology and on and on, are at least in part due to chaos. Technically the term chaos refers to the sensitive dependence of the solution to a set of nonlinear, deterministic, dynamical equations on initial conditions. The practical meaning of chaos is that the solutions to deterministic nonlinear equations look erratic and may pass all the traditional tests for randomness even though the solutions are deterministic. In subsequent chapters we examine more closely these ideas having to do with nonlinear dynamics and statistics. For the time being let us examine the properties of the non-normal statistics of interest.

2.3.1   Inverse power laws of Zipf and Pareto

Zipf's law, an empirical law formulated using mathematical statistics [104], refers to the fact that many types of data studied in the physical and social sciences can be approximated with an inverse power-law distribution, one of a family of related discrete power-law probability distributions. The law is named after the linguist George Kingsley Zipf, who first proposed it in 1949, though J. B. Estoup appears to have noticed the regularity before Zipf [46]. This law is most easily observed by scatterplotting the data, with the axes being log(rank order) and log(frequency). For example, in Figure 1.12 the word "the" as described in the above scatterplot would have the coordinates $x = \log(1)$, $y = \log(69{,}971)$. The data conform to Zipf's law to the extent that the plotted points appear to fall along a single straight-line segment.

Formally, let L be the number of elements in the data set, let r be the rank of a data element when ordered from the most to the least frequent, and let η be the value of the exponent characterizing the distribution. Zipf's law then predicts that, out of a population of L elements, the frequency of elements of rank r, $f(r; \eta, L)$, is given by

$$f(r; \eta, L) = \frac{1/r^{\eta}}{\sum_{n=1}^{L}(1/r^{\eta})}. \tag{2.54}$$

Thus, in the example of the frequency of words in the English language given in the first chapter, L is the number of words in the English language and, if we use the classic version of Zipf's law, the exponent η is one. The distribution function $f(r; \eta, L)$ is the fraction of the time the rth most common word occurs within a given language. Moreover, it is easily seen from its definition that the distribution is normalized, that is, the predicted frequencies sum to one:

$$\sum_{r=1}^{L} f(r; \eta, L) = 1. \tag{2.55}$$

The simplest case of Zipf's law is a "$1/f$ function." Given a set of Zipfian distributed frequencies, sorted from most common to least common, the second most common frequency will occur $1/2$ as often as the first. The third most common frequency will occur $1/3$ as often as the first. The nth most common frequency will occur $1/n$ as often as the first. However, this cannot hold exactly, because items must occur an integer number of times: there cannot be 2.5 occurrences of a word. Nevertheless, over fairly wide ranges, and to a fairly good approximation, many natural and social phenomena obey Zipf's law.

Mathematically, it is not possible for the classic version of Zipf's law to hold exactly if there are infinitely many words in a language, since the sum of all relative frequencies in the denominator (2.54) is equal to a harmonic series that diverges with diverging N:

$$\sum_{r=1}^{\infty} \frac{1}{r} = \infty. \tag{2.56}$$

In the English language, the frequencies of the approximately 1,000 most frequently used words are empirically found to be approximately proportional to $1/r^{\eta}$, where η is just slightly above one. As long as the exponent η exceeds 1, it is possible for such a law to hold with infinitely many words, since if $\eta > 1$ the sum no longer diverges,

$$\zeta(\eta) = \sum_{r=1}^{\infty} \frac{1}{r^{\eta}} < \infty, \tag{2.57}$$

where ζ is the Riemann zeta function.

Just why data from complex webs, such as those of Auerbach, Lotka, Willis and others shown in Chapter 1, conform to the distribution of Zipf is a matter of some controversy. Contributing to this controversy is the fact that the Zipf distribution arises in randomly generated texts having no linguistic structure, thereby suggesting to some that, in linguistic contexts, the law may be a statistical artifact [39, 43].

Zipf's law states that, given some corpus of natural-language utterances, the frequency of any word is inversely proportional to its rank in the frequency table. Thus, the most frequent word occurs approximately twice as often as the second most frequent word, which occurs twice as often as the fourth most frequent word, etc. For example, in Claude Shannon's application of Zipf's law to language in Figure 1.12 "the" is the most frequently occurring word, and by itself accounts for nearly 7% of all word occurrences (69,971 out of slightly over a million). True to Zipf's law, the second-place word "of" accounts for slightly over 3.5% of words (36,411 occurrences), followed by "and" (28,852). Only 135 vocabulary items are needed to account for half the words used.

In summary, Zipf's law is illustrated by the log–log diagram of Figure 2.17 for a variety of integer power-law indices.

Now let us establish a connection between Zipf's law and Pareto's law [57], also shown in Chapter 1. The distinction between the two laws is that the former is expressed in terms of rank and the latter is expressed in terms of the value of the variate. The variate is the variable that characterizes the dynamics in the underlying dynamic web. What makes the Pareto distribution so significant is the sociological implications that Pareto and subsequent generations of scientists were able to draw from its form [56]. For example, he identified a phenomenon that later came to be called the Pareto principle, that being that 20% of the people owned 80% of the wealth in western countries. It actually turns out that fewer than 20% of the population own more than 80% of the wealth, and this imbalance between the two groups is determined by the inverse power-law index. The actual numerical value of the partitioning is not important for the present discussion: what is important is that the imbalance exists. Pareto referred to this as "the law of unequal distribution of results" and thought that it was a predictable imbalance

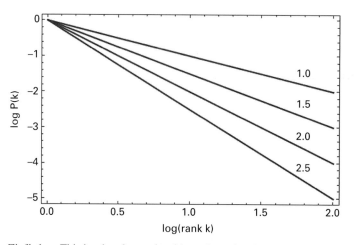

Figure 2.17.   Zipf's law. This is a log–log scale with total number $L = 10$. Note that the slope η is identified with the inverse power-law index.

to be found in a variety of phenomena. This imbalance is ultimately interpretable as the implicit unfairness found in complex webs.

Let us set the words of a text in order of frequency of appearance, and rank them assigning the rank $r = 1$ to the most frequently used word, the rank $r = 2$ to the word with the highest frequency after the first, and so on. The function $W(r)$ denotes the number of times the word of rank r appears. Zipf found

$$W(r) = \frac{K}{r^\eta},\tag{2.58}$$

where $\eta \approx 1$ and K is simply determined from (2.54).

Let us imagine that the number of times a word appears, W, can be interpreted as the wealth of that word. This makes it possible to define the probability $\Psi(W)$, namely the probability that a wealth larger than W exists. According to Pareto the distribution of wealth (actually Pareto's data were on the distribution of income, but we shall not dwell on the distinction between income and wealth here) is

$$\Psi(W) = \frac{A}{W^k}.\tag{2.59}$$

The distribution density $\psi(W)$ is given by the derivative of the probability with respect to the web variable W,

$$\psi(W) = -\frac{d}{dW}\Psi(W),\tag{2.60}$$

which yields

$$\psi(W) = \frac{B}{W^a},\tag{2.61}$$

with the normalization constant given by

$$B = kA\tag{2.62}$$

and the new power-law index related to the old by

$$a = k + 1.\tag{2.63}$$

Now let us take (2.58) into account in the distribution of wealth. Imagine that we randomly select, with probability density $p(r)$, a word of rank r from the collection of words with distribution number given by (2.58). A relation between the wealth variable and a continuous version of the rank variable is established using the equality between the probability of realizing wealth in the interval $(W, W + dW)$ and having the rank in the interval $(r, r + dr)$,

$$\psi(W)dW = p(r)dr.\tag{2.64}$$

We are exploring the asymptotic condition $r \gg 1$, and this makes it possible for us to move from the discrete to the continuous representation. The equality (2.64) generates a relation between the two distribution densities in terms of the Jacobian between the two variates,

$$\psi(W) = p(r) \left| \frac{dr}{dW} \right|. \tag{2.65}$$

The wealth and rank variables are related by (2.58), which can be inverted to yield

$$r = \left(\frac{K}{W} \right)^{1/\eta}, \tag{2.66}$$

and thus (2.65) becomes

$$\psi(W) = p(r) \frac{K^{1/\eta}}{\eta} \frac{1}{W^{1/\eta+1}}. \tag{2.67}$$

On examining (2.61), this yields the power-law index

$$\alpha = 1 + \frac{1}{\eta} \tag{2.68}$$

if $p(r)$ is independent of r and therefore independent of W. Note that earlier we adopted the symbol μ for the index of the inverse power-law distribution density ψ. We have already remarked that $\mu = 2$ is an important value, because it corresponds to the boundary between the region $\mu = a > 2$, where the mean value of W can be defined, and the region $\mu = a < 2$, where this mean value diverges to infinity.

To define $p(r)$ we must take into account that in a real language the number of possible words is finite, and we define that length with the symbol L. As a consequence the maximum wealth W_{max} is given by

$$W_{max} = \frac{K}{L^\eta}. \tag{2.69}$$

We set the uniform distribution for randomly choosing a word to be

$$p(r) = \frac{1}{L}, \tag{2.70}$$

thereby yielding the normalization constant

$$B = \frac{1}{L} \frac{K^{1/\eta}}{\eta}. \tag{2.71}$$

Now we can express Pareto's law (2.59) in terms of Zipf's law (2.58). In fact, using (2.63) and (2.68), we have

$$k = \mu - 1 = \frac{1}{\eta}, \tag{2.72}$$

and, using (2.62), we have

$$A = \frac{K^{1/\eta}}{L}. \tag{2.73}$$

An example of the empirical relation between the probability and the probability density is shown in Figure 2.18, depicting the number of visitors to Internet sites through the America On Line (AOL) server on a single day. The proportion of sites visited is $\psi(W)$, W is the number of visitors and the power-law index $\mu = 2.07$ is given by the best-fit slope to the data.

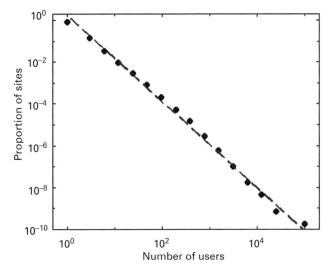

Figure 2.18. The number of visitors to Internet sites through the AOL search engine on a December day in 1997. The plot corresponds to the density function $\psi(W)$, with W being the number of visitors in this case [1]. Adapted with permission.

2.3.2 Chaos, noise and probability

But from where does uncertainty emerge? A nonlinear web with only a few degrees of freedom can have chaotic solutions and therefore may generate random patterns. So we encounter the same restrictions on our ability to know and understand a web when there are only a few dynamical elements as when there are a great many, but for very different reasons. Let us refer to random processes, which are produced by the unpredictable influence of the environment on the web of interest as noise. In this case the environment is assumed to have an infinite number of elements, all of which we do not know, but they are coupled to the web of interest and perturb it in a random, that is, unknown, way [68, 96]. By way of contrast, chaos is a consequence of the nonlinear, deterministic interactions in isolated dynamical webs, resulting in the erratic behavior of limited predictability. Chaos is an implicit property of complex dynamic webs, whereas noise is a property of the environment in contact with such webs of interest. Chaos can therefore be controlled and predicted over short time intervals, whereas noise can neither be predicted nor be controlled except perhaps through the way the environment is coupled to the web.

One of the formal mathematical constructs involving dynamical webs is ergodic theory, a branch of mathematics that seeks to prove the equivalence between ensemble averages and time averages. To determine a time average we imagine that the recurrence time of a trajectory to its initial state is not exact, but has a certain error. The trajectory is not really coming back to the same initial state, but is as close to that state as one may want it to be. In this way, if the phase space is partitioned into boxes, every box may contain an infinite number of recursions, and this property is independent of the box size. Ergodic theory affirms that all the cells in the available phase space are "visited" by the network dynamics. The relative amount of time the web trajectory spends in cell i is t_i and a metric of cell p_i is proportional to this time,

$$p_i = \frac{t_i}{t_T}, \tag{2.74}$$

where t_T is the total duration of the trajectory. If the probability of the trajectory being in that cell does not change in time, as the length of the trajectory becomes infinitely long, the probability is metrically invariant. This invariance property is realized for ergodic webs that may, but need not, reach equilibrium in a finite time.

Physical theories describe the dynamics of microscopic and macroscopic webs that conserve probability in the sense that once a probabilistic point of view is adopted the total probability remains unchanged in time. Consequently, if the local probability in phase space increases, it must decrease somewhere else in phase space. Recall that the phase space has all the dynamical variables describing the complex web as independent axes. In classical mechanics, if one selects a restricted volume in the phase space and uses each point in this volume as an initial condition for the equations of motion, then the evolution of the volume may change its geometry, but not the magnitude of the volume itself. Equilibrium, however, means that the volume itself does not change. You may recall that this was the basis of Boltzmann's definition of entropy (1.27). So how is this invariance property compatible with ergodicity? How is all of the available phase space explored as the web evolves?

To achieve equilibrium the volume of initial states develops twisted whorls and long slender tendrils that can become dense in the available phase space. Like the swirls of cream in your morning coffee, these tendrils of probability interlace the phase space, making the coffee uniformly tan. This property is called mixing and refers to thermodynamically closed networks. Not all dynamical webs exhibit mixing, but mixing is required in order for the dynamics of a closed web to reach equilibrium, starting from a non-equilibrium initial condition. The notion of a closed system or network was convenient in physics in order to retain a Hamiltonian description of an isolated mechanical system such as a harmonic oscillator or more elaborate nonlinear dynamical webs. Such a dynamic web could be opened up and allowed to interact with the environment. If the interaction satisfies certain energy-balance conditions the composite system is considered to be thermodynamically closed. However, this is not the case in general. A web that is open to the environment may receive energy and not necessarily give energy back, thereby becoming unstable. These concepts of open and closed, linear and nonlinear, dynamical webs are subtle and are discussed in subsequent chapters.

Let us denote the random dynamic variable by $Q(t)$ and its statistics by the probability density function $p(q)$, so that q is the corresponding phase-space variable. Consequently, the average of the random variable is determined by integrating over the range of the variate R,

$$\langle Q \rangle = \int_R q p(q) dq, \tag{2.75}$$

and the second moment is realized by the integral

$$\left\langle Q^2 \right\rangle = \int_R q^2 p(q) dq, \tag{2.76}$$

resulting in the time-independent variance

$$\sigma^2 = \left\langle q^2 \right\rangle - \langle q \rangle^2. \tag{2.77}$$

Notice that we distinguish the average determined by the pdf from the average determined from the discrete values of the dynamic variable given by (2.32), where the sum is over the discrete time index. If the measurements are taken at equally spaced time points that are sufficiently close together we could use the continuous time average given by

$$\overline{Q} = \frac{1}{t} \int_0^t Q(t') dt'. \tag{2.78}$$

The ergodic hypothesis requires that $\langle Q \rangle = \overline{Q}$ after sufficiently long times.

If the random variable is determined by an ensemble of trajectories, then the statistics depend on time, and the probability density function also depends on time, $p(q,t)$. Consequently the variance is also time-dependent:

$$\sigma(t)^2 = \left\langle q^2; t \right\rangle - \langle q; t \rangle^2. \tag{2.79}$$

The divergence of the variance with increasing amounts of data (for time series this means longer times) is a consequence of the scaling property of fractals. In this context scaling means that the small irregularities at small scales are reproduced as larger irregularities at larger scales. These increasingly larger fluctuations become apparent as additional data are collected. Hence, as additional data from a fractal object or process are analyzed, these ever larger irregularities increase the measured value of the variance. If the limit continues to increase, the variance continues to increase; that is, the variance becomes infinite. We lose the variance as a reliable measure of the data, but scaling implies that there is a fractal dimension that can replace the variance as a new measure; the variance is replaced with a scaling index.

The increased variability indicated by the divergence of the variance has been observed in the price of cotton [45], the measured density of microspheres deposited in tissue to determine the volume of blood flow for each gram of heart tissue [6] and the growth in the variance of the mean firing rate of primary auditory nerve fibers with increasing length of time series [81, 82]. Many other examples of the application of this concept in physiologic contexts may be found in reviews [7, 95]. We shall discuss a select subset of these examples in detail later.

It is sometimes confusing, so it should be stated explicitly that fractal random processes are indicated by two distinct kinds of inverse power laws, namely inverse power-law spectra and inverse power-law pdfs. The inverse power-law spectrum, or equivalently the corresponding autocorrelation function, indicates the existence of long-term memory, far beyond that of the familiar exponential. In probability the inverse power law indicates the contribution of terms that are quite remote from the mean, even when the mean exists. The preceding discussion of the Zipf and Pareto distributions concerned time-independent inverse power-law complex webs. It is simple to show that these distributions scale as

$$\psi(bW) = b^{-\mu}\psi(W), \tag{2.80}$$

with the parameter μ referred to as the scaling index, the scaling parameter or any of a number of other terms. We interpret this scaling as meaning that increasing the resolution of the dynamic variable by a factor $b > 1$ is equivalent to retaining the present resolution and decreasing the amplitude of the distribution by $b^{-\mu} < 1$ for $\mu > 1$.

2.4   Scaling physiologic time series

The previous subsections have stressed the importance of scaling, so now let us look at some data. The question is how to determine from data whether the web from which the data have been taken has the requisite scaling behavior and is therefore a candidate for this kind of modeling. One method of analysis that is relatively painless to apply is determined by the ratio of the variance to the mean for a given level of resolution and the ratio is called the relative dispersion. The relation of the relative dispersion to the level of aggregation of the data from a given time series is readily obtained. In fact it is possible to use the expression for the relative dispersion to write a relation between the variance and the mean that was obtained empirically by Taylor [78] in his study of the growth in the number of biological species, that being

$$\mathrm{Var}\, Q^{(n)} = a\,\overline{Q^{(n)}}^{b}, \tag{2.81}$$

where n denotes the level of resolution.

Taylor sectioned off a large field into plots and in each plot sampled the soil for the variety of beetles present. In this way he was able to determine the distribution in the number of new species of beetle that was spatially distributed across the field. From the empirical distribution he calculated the mean and variance in the number of new species. Subsequently he partitioned the field into smaller plots and carried out his procedure again, calculating a new mean and variance. After n such partitionings Taylor plotted the variances against the means for the different levels of resolution and found a relation of the form (2.81). In the ecological literature a graph of the logarithm of the variance versus the logarithm of the average value is called a *power curve*, which is linear in logarithms of the two variables and b is the slope of the curve:

$$\log \mathrm{Var}\, Q^{(n)} = \log a + b \log \overline{Q^{(n)}}. \tag{2.82}$$

Taylor [78] developed his ideas for spatial distributions in which the resolution is increased between successive measurements. However, for fractal data sets it should not matter whether we increase the resolution or decrease it; the scaling behavior should be the same. Consequently we expect that aggregating the time-series data should allow us to utilize the insight obtained by Taylor in his study of speciation.

Taylor was able to interpret the curves he obtained from processing speciation data in a number of ways using the two parameters. If the slope of the curve and the intercept are both equal to one, $a = b = 1$, then the variance and mean are equal to one another. This equality holds only for a Poisson distribution, which, when it occurred, allowed him to interpret the number of new species as being randomly distributed over the field, with the number of species in any one plot being completely independent of the number of species in any other plot. If, however, the slope of the curve was less than unity, $b < 1$, the number of new species appearing in the plots was interpreted as being quite regular, like the trees in an orchard. Finally, if the slope of the variance versus average was greater than one, $b > 1$, the number of new species was interpreted as being clustered in space, like disjoint herds of sheep grazing in a meadow.

Of particular interest to us here is the mechanism postulated nearly two decades later to account for the experimentally observed allometric relation [79]:

We would argue that all spatial dispositions can legitimately be regarded as resulting from the balance between two fundamental antithetical sets of behavior always present between individuals. These are, repulsion behavior, which results from the selection pressure for individuals to maximize their resources and hence to separate, and attraction behavior, which results from the selection pressure to make the maximum use of available resources and hence to congregate wherever these resources are currently most abundant.

It is the conflict between the attraction and repulsion, emigration and immigration, which produces the interdependence of the spatial variance and the average population density. The kind of clustering observed in the spatial distribution of species number, when the slope of the power curve is greater than one, is consistent with an asymptotic inverse power-law distribution of the underlying data set. Furthermore, the clustering or clumping of events is due to the fractal nature of the underlying dynamics. Recall that Willis, some forty years before Taylor, established an empirical inverse power-law form of the number of species belonging to a given genus. Willis used an argument associating the number of species with the size of the area they inhabit. It was not until the decade of the 1990s that it became clear to more than a handful of experts that the relationship between an underlying fractal process and its space-filling character obeys a scaling law [44, 45]. It is this scaling law that is reflected in the allometric relation between the variance and average.

Note that Taylor and Woiwod [80] were able to extend Taylor's original argument from the stability of the population density in space, independent of time, to the stability of the population density in time, independent of space. Consequently, the stability of the time series, as measured by the variance, can be expressed as a power of the time-series average. With this generalization in hand we can apply Taylor's ideas to time series.

2.4.1   Fractal heartbeats

Mechanisms producing the observed variability in the time interval from one heartbeat to the next apparently arise from a number of sources. The sinus node (the heart's natural pacemaker) receives signals from the autonomic (involuntary) portion of the nervous system, which has two major branches: the parasympathetic, whose stimulation decreases the firing rate of the sinus node, and the sympathetic, whose stimulation increases the firing rate of the sinus node's pacemaker cells. The influence of these two branches of the nervous system produces a continual tug-of-war on the sinus node, one decreasing and the other increasing the heart rate. It has been suggested that it is this tug-of-war that produces the fluctuations in the heart rate of healthy subjects, and we note the similarity to the competitive mechanism postulated by Taylor and Taylor [79]. Consequently, these fluctuations in the interbeat interval are termed heart-rate variability (HRV) and strongly influence the heart's ability to respond to everyday disturbances that can affect its rhythm. The clinician focuses on retaining the balance in regulatory impulses from the vagus nerve and sympathetic nervous system and in this effort requires a robust measure of that balance. A quantitative measure of HRV such as the fractal dimension serves this purpose.

The beating of the human heart is typically recorded on a strip chart called an electrocardiogram, where the maximum of the pulse is denoted by R and the interbeat time interval is denoted by the time between these maxima. The RR intervals therefore define the time series of interest. A typical HRV time series for a young healthy adult male is depicted in Figure 2.19. The variability in the time between heartbeats is clear from this figure; the variation in the time intervals between heartbeats is relatively small, the mean being 1.0 s and the standard deviation being 0.06 s.

In Figure 2.20 the logarithm of the standard deviation is plotted versus the logarithm of the average value for the HRV time series depicted in Figure 2.19. At the left-most position the data point indicates the standard deviation and average, using all the data

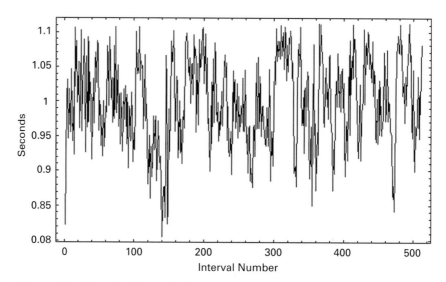

Figure 2.19.   A time series of heartbeat intervals of a healthy young adult male is shown. It is clear that the variation in the time interval between beats is relatively modest but not negligible [100]. Reproduced with permission.

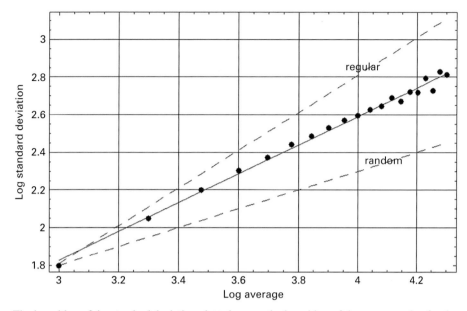

Figure 2.20. The logarithm of the standard deviation plotted versus the logarithm of the average value for the heartbeat interval time series for a young adult male, using sequential values of the aggregation number. The solid line is the best fit to the aggregated data points and yields a fractal dimension of $D = 1.24$ midway between those of the curve for a regular process and that for an uncorrelated random process [100]. Reproduced with permission.

points. Moving from left to right, the next data point is constructed from the time series with two nearest-neighbor data points added together; and the procedure is repeated moving right until the right-most data point has a block of twenty data points added together. The solid line is the best linear regression on the data and represents the scaling with a positive slope of 0.76. We can see that the slope of the HRV data is midway between those of the dashed curves depicting an uncorrelated random process (slope = 1/2) and one that is deterministically regular (slope = 1).

Figure 2.20 implies that heartbeat intervals do not form an uncorrelated random sequence. This implication is tested by randomly changing the order of the time intervals and, since no intervals are added or deleted, the statistics of the intervals is not changed by such a change in ordering. The correlation, on the other hand, depends on the ordering of the data points, so the scaling is lost by this shuffling process. This is shown in Figure 2.21, where the information (entropy) is plotted versus the logarithm of time and the functional form

$$S(t) = - \int dx \, p(x, t) \log_2 p(x, t) = \delta \log_2 t + \text{constant} \qquad (2.83)$$

is obtained when the probability density scales as

$$p(x, t) = \frac{1}{t^\delta} F\left(\frac{x}{t^\delta}\right). \qquad (2.84)$$

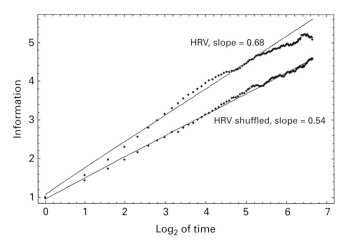

Figure 2.21.   The entropy or information for the HRV time series calculated using (2.83) is graphed versus the logarithm of time. The difference between the slopes of the shuffled and unshuffled data sets is evident [100]. Reproduced with permission.

The HRV data are used to construct a histogram for the probability density that can be found in most texts on data analysis [42]. For the time being we note that the scaling parameter δ would be 0.5 if the time series were an uncorrelated random process. For the original data set the probability density is found empirically to scale in time with an index $\delta = 0.68$ and when the data are shuffled the statistics do not change but the scaling index drops to $\delta = 0.54$. A rigorous determination of the latter scaling index would require doing the shuffling a number of times and taking the average of the resulting indices. This was not done here because the shuffling was done for illustrative purposes and with the value obtained it is clear that the average would converge to the uncorrelated value of 0.5.

   The long-range correlation implied by the scaling is indicative of an organizing principle for complex webs that generates fluctuations across a wide range of scales. Moreover, the lack of a characteristic scale helps prevent excessive mode-locking that would restrict the functional responsiveness of the cardiovascular network. We propose later a dynamical model using a fractional Langevin equation that is different from the random-walk model proposed by Peng *et al.* [58], but which does encompass their interpretation of the functionality of the long-range correlations.

2.4.2   Fractal breaths

Now let us turn our attention to the apparently regular breathing as you sit quietly reading this book. We adopt the same perspective as that used to dispel the notion of "regular sinus rhythm." The adaptation of the lung by evolution may be closely tied to the way in which the lung carries out its function. It is not accidental that the cascading branches of the bronchial tree become smaller and smaller; nor is it good fortune alone that ties the dynamics of our every breath to this biological structure. We argue that, like the heart, the lung is made up of fractals, some dynamic, like breathing itself, and others now static, like its anatomic structure. However, both kinds of fractals lack a characteristic scale and a simple argument establishes that such lack of scale has evolutionary advantages [97].

Breathing is a function of the lungs, whereby the body takes in oxygen and expels carbon dioxide. The smooth muscles in the bronchial tree are innervated by sympathetic and parasympathetic fibers, much like the heart, and produce contractions in response to stimuli such as increases in concentration of carbon dioxide, decreases in concentration of oxygen and deflation of the lungs. Fresh air is transported through some twenty generations of bifurcating airways of the lung, during inspiration, down to the alveoli in the last four generations of the bronchial tree. At this tiny scale there is a rich capillary network that interfaces with the bronchial tree for the purpose of exchanging gases with the blood.

As with heartbeats, the variability of breathing using breath-to-breath time intervals is called breathing-rate variability (BRV), to maintain a consistent nomenclature. An example of a typical BRV time series on which scaling calculations are subsequently based is shown in Figure 2.22. Note the change in time scale due to the heart rate being nearly a factor of five higher than the breathing rate. It is not apparent on comparing the time series in Figure 2.19 with that in Figure 2.22 that HRV and BRV time series scale in the same way, but they do [99].

The logarithm of the aggregated variance is plotted versus the logarithm of the aggregated mean in Figure 2.23 for the BRV data. At the extreme left of the graph all the data points are used to calculate the variance and mean ($m = 1$). At the extreme right of the graph the aggregated quantities use $m = 10$ data points. The aggregate is stopped at $m = 10$ because of the small number of data points in the breathing sequence. The solid curve is the best least-square fit to the aggregated BRV data and has a slope of 0.86, which is the scaling index. It is clear that the BRV time series falls well within the range of the regular and uncorrelated random time series.

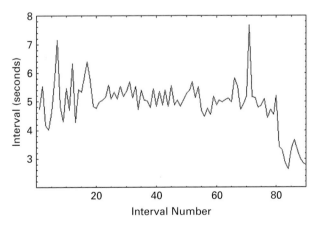

Figure 2.22. A typical time series from one of the eighteen subjects in the study conducted by West *et al.* [99], recorded while the subject was at rest, for the interbreath intervals (BRV) [100]. Reproduced with permission.

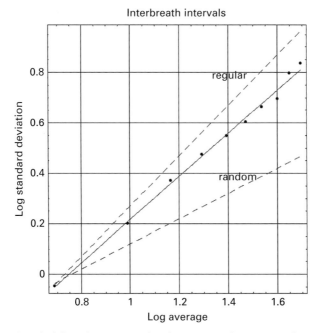

Figure 2.23.   A typical fit to the aggregated variance versus the aggregated mean for BRV time series obtained by West *et al.* [99]. The points are calculated from the data and the solid curve is the best least-square fit to the data with slope = 0.86. It is evident that the allometric relation (2.82) does indeed fit the BRV data extremely well and lies within the regular and random extremes [100]. Reproduced with permission.

BRV has led to a revolutionary way of utilizing mechanical ventilators, which historically facilitate breathing after an operation and have a built-in periodicity. Mutch *et al.* [52] redesigned the ventilator using an inverse power-law spectrum to vary the ventilator's motion in time. They demonstrated that the variable ventilator-supported breathing produces an increase in arterial oxygenation over that produced by conventional control-mode ventilators. This comparison indicates that the fractal variability in breathing is not the result of happenstance, but is an important property of respiration. A reduction in variability of breathing reduces the overall efficiency of the respiratory network.

Altemeier *et al.* [2] measured the fractal characteristics of ventilation and determined not only that local ventilation and perfusion are highly correlated, but also that they scale. Finally, Peng *et al.* [59] analyzed the BRV time series for forty healthy adults and found that, under supine, resting and spontaneous breathing conditions, the time series scale. This result implies that human BRV time series have "long-range (fractal) correlations across multiple time scales."

2.4.3   Fractal steps

Humans must learn how to walk and, if other creatures had to consciously do so as well, the result might be that given by the anonymously written poem:

> A centipede was happy quite,
> Until a frog in fun
> Said, "Pray, which leg comes after which?"
> This raised her mind to such a pitch,
> She lay distracted in the ditch
> Considering how to run.

Walking is one of those things that is learned in early childhood, but is done later with little or no thought, day in and day out. We walk confidently with a smooth pattern of strides and on flat level ground there is no visible variation in gait. This apparent lack of variability is remarkable considering that walking is created by the loss of balance, as pointed out by Leonardo da Vinci in his treatise on painting. Da Vinci considered walking to be a sequence of fallings, so it should come as no surprise that there is variability in this sequence of falling intervals, even if such variability is usually masked.

The regular gait cycle, so apparent in everyday experience, is no more regular than "normal sinus rhythm" or breathing. The subtle variability in the stride characteristics of normal locomotion was first discovered by the nineteenth-century experimenter Vierordt [85], but his findings were not followed up for over a century. The random variability he observed was so small that the biomechanical community has historically considered these fluctuations to be biological noise. In practice this means that the fluctuations in gait were thought to contain no information about the underlying motor-control process. The follow-up experiment to quantify the degree of irregularity in walking was finally done in the middle of the last decade by Hausdorff *et al.* [28]. Additional experiments and analyses were subsequently done by West and Griffin [98], which both verified and extended the earlier results.

Walking is a complex process, since the locomotor system synthesizes inputs from the motor cortex, the basal ganglia and the cerebellum, as well as feedback from vestibular, visual and proprioceptive sources. The remarkable feature of the output of this complex web is that the stride pattern is stable in healthy individuals, but the duration of the gait cycle is not fixed. As with normal sinus rhythm, for which the interval between successive beats changes, the time interval for a gait cycle fluctuates in an erratic way from step to step. The gait studies carried out to date concur that the fluctuations in the stride-interval time series exhibit long-time inverse power-law correlations indicating that the phenomenon of walking is a self-similar fractal activity; see West [100] for a review.

Walking consists in a sequence of steps and the corresponding time series is made up of the time intervals for these steps. The stride-interval time series for a typical subject is shown in Figure 2.24, where it is seen that the variation in time interval is on the order of 3%–4%, indicating that the stride pattern is very stable. The stride-interval time series is referred to as stride-rate variability (SRV) for consistency with the other two time series we have discussed. It was the stability of SRV that historically led investigators to decide that they could not go far wrong by assuming that the stride interval is constant and that the fluctuations are merely biological noise.

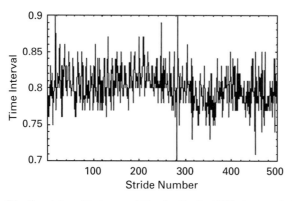

Figure 2.24. The time interval between strides for the first 500 steps made by a typical walker in an experiment [98], from [100]. Reproduced with permission.

Using an SRV time series of 15 minutes, from which the data depicted in Figure 2.24 were taken, we apply the allometric aggregation procedure to determine the relation between the aggregated standard deviation and aggregated average of the time series as shown in Figure 2.25. In the latter figure, the line segment for the SRV data is, as we did with the other data sets, contrasted with an uncorrelated random process (slope 0.5) and a regular deterministic process (slope 1.0). The slope of the data curve is 0.70, midway between the two extremes of regularity and uncorrelated randomness. So, as in the cases of HRV and BRV time series, we again find the erratic physiologic time series to represent random fractal processes.

The question regarding the source of the second-moment scaling is again raised. Is the scaling due to long-range correlations or is it due to non-standard statistics? We again use histograms to construct the probability density and the associated diffusion entropy for the SRV data. The entropy is plotted as a function of the logarithm of time in Figure 2.26, where we see the probability scaling index $\delta = 0.69$ for the original SRV data and the index drops to $\delta = 0.46$ for the shuffled data. This scaling suggests that the walker does not smoothly adjust stride from step to step, but instead there is a number of steps over which adjustments are made followed by a number of steps over which the changes in stride are completely random. The number of steps in the adjustment process and the number of steps between adjustment periods are not independent. The results of a substantial number of stride-interval experiments support the universality of this interpretation, see, for example, West [100], for a review.

Finally, it should be pointed out that people use essentially the same control network, namely maintaining balance, when they are standing still as they do when they are walking. This suggests that the body's slight movements around the center of mass of the body in any simple model of locomotion would have the same statistical behavior as the variability observed during walking. These movements are called postural sway and have given rise to papers with such interesting titles as "Random walking during

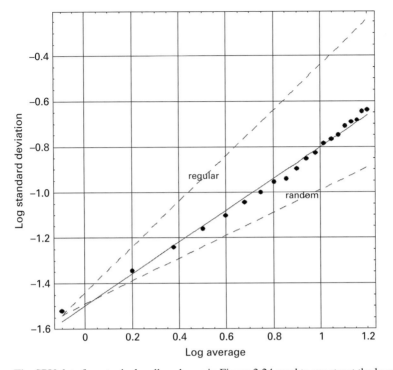

Figure 2.25. The SRV data for a typical walker shown in Figure 2.24 used to construct the logarithm of the aggregated variance and the logarithm of the aggregated mean (indicated by the dots), starting with all the data points at the lower left, to the aggregation of twenty data points at the upper right. The SRV data curve lies between the extremes of uncorrelated random noise (lower dashed curve), and regular deterministic process (upper dashed curve) with a fractal dimension of $D = 1.30$ [100]. Reproduced with permission.

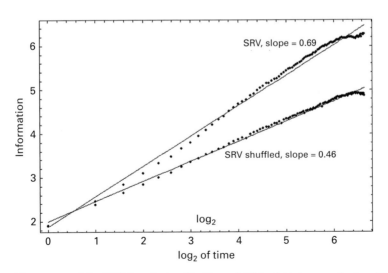

Figure 2.26. The entropy of the SRV data depicted in Figure 2.24 is plotted versus the logarithm of time using (2.83) for the original data (slope 0.69) and the shuffled data (slope 0.46) [100]. Reproduced with permission.

quiet standing" by Collins and De Lucca [14]. Subsequently it was determined that postural sway is really chaotic [11], so one might expect that there exists a relatively simple dynamical model for balance regulation that can be used in medical diagnosis. Here again the fractal dynamics can be determined from the scaling properties of postural-sway time series and it has been established that a decrease of postural stability is accompanied by an increase of fractal dimension.

2.5   Some afterthoughts

In this chapter we have discussed a number of interesting ideas, but the emphasis has been on power-law scaling and the relation of that scaling to fractal processes both theoretical and observational. It is useful to state explicitly the different ways in which the power law has been used.

First of all, the power law refers to the power spectrum for random time series that have fractal dimensions, namely those whose second moment may diverge. The autocorrelation function for such processes increases algebraically with time so that its Fourier transform, the power spectral density, decreases as an inverse power law with frequency. The scaling index in this case characterizes the degree of influence present events have on future events. The greater the index the more short term is the influence.

Second of all, the term inverse power law has also been applied to probability densities that asymptotically behave as inverse power laws in the web variable. These were also called hyperbolic distributions and the variates are hyperbolic random variables. The statistical behavior of such variables is very different from normal random processes. In the "normal" case the data being acquired appear with increasing frequency in the vicinity of the average value with very few data points falling very far away. In many of the hyperbolic cases we mentioned the initial data are acquired locally with a width determined by the parameter a. Every so often, however, a value of size b is realized and then a cluster of data points of size a in that region is accumulated. Even more rarely a data point of size b^2 occurs and a cluster of values of size a in that region is accumulated. In this way an ever-expanding set of data is accumulated with holes between regions of activity and whose overall size is determined by the inverse power-law index $\mu = \log a / \log b$. The values of the web variable out in the tail of the distribution dominate the behavior of such processes.

In the spatial domain the inverse power law is manifest through data points that resemble clumps of sheep grazing in a meadow as opposed to the regular spacing of trees in an orchard. In the time domain the heavy-tailed phenomena give rise to a pattern of intermittency. However, even in a region of activity, magnification reveals irregularly spaced gaps between times of activity having the same intermittent behavior down to smaller and smaller time scales.

We closed this chapter with a review of how the notions of complexity reveal the underlying structure in physiologic time series such as the time intervals between heart beats, as well as those between breaths, and the variability of strides during normal walking. We established that such physiologic time series are fractal; they satisfy scaling, and the scaling represents a kind of long-time memory in complex physiologic webs.

2.6   Problems

2.1 The Fibonacci sequence

Fibonacci scaling has been observed in the dimensions of the Parthenon as well as in vases and sculptures of the same period. The golden mean was made famous by the mathematician Leonardo Fibonacci (Filius Bonaci), also known as Leonardo of Pisa, who was born in 1175 AD. The problem he solved was as follows. Given a sequence of integers such that an integer is the sum of the two preceding integers, what is the limiting ratio of successive integers? The algebraic expression of the problem given the sequence of integers labeled by a subscript is

$$u_{n+1} = u_n + u_{n-1}.$$

Show that u_{n+1}/u_n equals ϕ as n becomes large.

2.2 An infinitely long line

Consider a curve that connects two points in space. It is not unreasonable to assume that the length of the curve connecting the two points is finite if the Euclidean distance between the two points is finite. However, let us modify the argument for the Cantor set in Figure 2.13 and not concentrate on the mass, but instead increase the length of the line segment by not removing the trisected region, and replacing this region with a cap that is twice the length of the interval removed. In this way the unit line segment after one generation is 4/3 units long. If each of the four line segments in the curve is now replaced with the trisecting operation the second-generation curve has a length of 16/9. What are a and b in this scaling operation and consequently what is the fractal dimension of the line segment? If the operation is repeated an infinite number of times, determine the dimension of the resulting curve. Show that the length of the curve given by $L(r) = Nr = N^{1-D}$ diverges as $r \to 0$.

2.3 The generalized Weierstrass function

(a) Show that the GWF given by (2.38) satisfies the scaling relation (2.33). Solve the scaling equation to obtain the coefficients in the solution (2.36).

(b) The GWF is so elegant pedagogically because it allows the student to explore explicitly all the scaling concepts being described. Write a computer program to evaluate the series (2.38) for $a = 4$ and $b = 8$. (1) Graph the function separately for the intervals $-1 < n < 1$, $-2 < n < 2$ and $-3 < n < 3$ and discuss the resulting curves. (2) Compare the graph with $0 < W < 0.8$ and $0 < t < 1$ with the graph with $0 < W < 0.2$ and $0 < t < 0.12$ and discuss. (3) Superpose the algebraic increase in W obtained from the renormalization-group relation with the numerical solution, keeping in mind that $\mu = \log a / \log b = 2/3$. How do you interpret this comparison?

References

[1] L. A. Adamic, "Zipf, power-laws, and Pareto – a ranking tutorial," http://www.hpl.hp.com/research/idl/papers/ranking/ranking.html.

[2] W. A. Altemeier, S. McKinney and R. W. Glenny, "Fractal nature of regional ventilation distribution," *J. Appl. Physiol.* **88**, 1551–1557 (2000).

[3] J. R. Anderson and L. J. Schooler, "Reflections of the environment in memory," *Psych. Sci.* **2**, 396–408 (1991).

[4] F. Auerbach, "Das Gesetz der Bevölkerungskonzentration," *Petermanns Mitteilungen* **59**, 74 (1913).

[5] G. I. Barenblatt, *Scaling Phenomena in Fluid Mechanics*, Cambridge: Cambridge University Press (1994).

[6] J. B. Bassingthwaighte and J. H. G. M. van Beek, "Lightning and the heart: fractal behavior in cardiac function," *Proc. IEEE* **76**, 693 (1988).

[7] J. B. Bassingthwaighte, L. S. Liebovitch and B. J. West, *Fractal Physiology*, Oxford: Oxford University Press (1994).

[8] M. Batty and P. Longley, *Fractal Cities*, San Diego, CA: Academic Press (1994).

[9] J. Beran, *Statistics for Long-Memory Processes*, New York: Chapman & Hall (1994).

[10] M. V. Berry and Z. V. Lewis, "On the Weierstrass–Mandelbrot fractal function," *Proc. Roy. Soc. Lond. Ser. A* **370**, 459–484 (1980).

[11] J. W. Blaszczyk and W. Klonowski, "Postural stability and fractal dynamics," *Acta Neurobiol. Exp.* **61**, 105–112 (2001).

[12] W. B. Cannon, *The Wisdom of the Body*, New York: W. W. Norton (1932).

[13] R. A. Chambers, W. K. Bickel and M. N. Potenza, "A scale-free systems theory of motivation and addiction," *Neurosci. Biobehav. Rev.* **31**, 1017–1045 (2007).

[14] J. J. Collins and C. J. De Lucca, "Random walking during quiet standing," *Phys. Rev. Lett.* **73**, 764–767 (1994).

[15] T. A. Cook, *The Curves of Life*, New York: Dover (1979); originally published by Constable and Co. (1914).

[16] D. J. de Solla Price, *Little Science, Big Science*, New York: Columbia University Press (1963).

[17] P. S. Dodds and D. H. Rothman, "Unified view of scaling laws for river networks," *Phys. Rev. E* **49**, 4865 (1999).

[18] P. S. Dodds and D. H. Rothman, "Geometry of river networks II: Distribution of component size and number," *Phys. Rev. E* **63**, 016115 (2001).

[19] M. Faloutsos, P. Faloutsos and C. Faloutsos, "On the power-law relationships of the Internet dynamics," *Comput. Commun. Rev.* **29**, 251 (1999).

[20] D. L. Gilden, "Cognitive emissions of $1/f$ noise," *Psych. Rev.* **108**, 33–56 (2001).

[21] T. Gneiting and M. Schlather, "Stochastic models that separate fractal dimensions and the Hurst effect," *SIAM Rev.* **46**, 269–282 (2004).

[22] A. L. Goldberger, B. J. West, V. Bhargava and T. Dresselhaus, "Bronchial asymmetry and Fibonacci scaling," *Experientia* **41**, 1–5 (1985).

[23] P. Gong, A. R. Nikolaev and C. van Leeuwen, "Scale-invariant fluctuations of the dynamical synchronization in human brain electrical activity," *Phys. Rev. E* **76**, 011904 (2007).

[24] P. Gopikrishnan, M. Meyer, L. A. N. Amaral and H. E. Stanley, "Inverse cubic law for the distribution of stock price variations," *Eur. Phys. J. B* **3**, 139–140 (1998).

[25] P. Grigolini, D. Leddon and N. Scafetta, "The diffusion entropy and waiting-time statistics and hard X-ray solar flares," *Phys. Rev. E* **65**, 046203 (2002).

[26] B. Gutenberg and C. F Richter, *Seismicity of the Earth and Associated Phenomena*, 2nd edn., Princeton, NJ: Princeton University Press (1954), pp. 17–19.

[27] J. T. Hack, "Studies of longitudinal stream profiles in Virginia and Maryland," *U.S. Geol. Surv. Prof. Paper* **294-B**, 45 (1957).

[28] J. M. Hausdorff, C.-K. Peng, Z. Ladin *et al.*, "Is walking a random walk? Evidence for long-range correlations in stride interval of human gait," *J. Appl. Physiol.* **78** (1), 349–358 (1995).

[29] R. J. Herrnstein, "Relative and absolute strength of response as a function of frequency of reinforcement," *J. Exp. Anal. Behav.* **29**, 267–272 (1961).

[30] M. A. Hofman, "The fractal geometry of the convoluted brain," *J. Hirnforsch.* **32**, 103–111 (1991).

[31] R. E. Horton, "Erosional development of streams and their drainage basins: hydrophysical approach to quantitative morphology," *Bull. Geol. Soc. Am.* **56**, 275–370 (1945).

[32] B. A. Huberman and L. A. Adamic, "Growth dynamics of the World-Wide Web," *Nature* **401**, 131 (1999).

[33] H. F. Jelinek and E. Fernandez, "Neurons and fractals: how reliable and useful are calculations of fractal dimensions?," *J. Neurosci. Methods* **81**, 9–18 (1998).

[34] H. Jeong, B. Tombor, R. Albert, Z. N. Oltvai and A.-L. Barabási, "The large-scale organization of metabolic networks," *Nature* **407**, 651 (2000).

[35] H. Jeong, S. P. Mason, A.-L. Barabási and Z. N. Oltvai, "Lethality and centrality in protein networks," *Nature* **411**, 41 (2001).

[36] E. R. Kandel, "Small systems of neurons," in *Mind and Behavior*, eds. R. L. Atkinson and R. C Atkinson, San Francisco, CA: W. H. Freeman and Co. (1979).

[37] M. G. Kitzbichler, M. L. Smith, S. R. Christensen and E. Bullmore, "Broadband criticality of human brain network synchronization," *PLoS Comp. Biol.* **5**, 1–13, www.ploscombiol.org (2009).

[38] A. N. Kolmogorov, "Local structure of turbulence in an incompressible liquid for very large Reynolds numbers," *Comptes Rendus (Dokl.) Acad. Sci. URSS (N.S.)* **26**, 115–118 (1941); "A refinement of previous hypotheses concerning the local structure of turbulence in a viscous incompressible fluid at high Reynolds number," *J. Fluid Mech.* **13**, 82–85 (1962).

[39] W. Li, "Random texts exhibit Zipf's-law-like word frequency distribution," *IEEE Trans. Information Theory* **38** (6), 1842–1845 (1992).

[40] F. Lilerjos, C. R. Edling, L. A. N. Amaral, H. E. Stanley and Y. Aberg, "The web of human sexual contacts," *Nature* **411**, 907 (2001).

[41] A. J. Lotka, "The frequency distribution of scientific productivity," *J. Wash. Acad. Sci.* **16**, 317 (1926).

[42] S. Mallat, *A Wavelet Tour of Signal Processing*, 2nd edn., San Diego, CA: Academic Press (1999).

[43] B. B. Mandelbrot, "On the theory of word frequencies and on related Markovian models of discourse," in *Structures of Language and Its Mathematical Aspects*, ed. R. Jacobson, New York: American Mathematical Society (1961).

[44] B. B. Mandelbrot, *Fractals, Form and Chance*, San Francisco, CA: W. F. Freeman (1977).

[45] B. B. Mandelbrot, *The Fractal Geometry of Nature*, San Francisco, CA: W. J. Freeman (1983).

[46] C. D. Manning and H. Schütze, *Foundations of Statistical Natural Language Processing*, Cambridge, MA: MIT Press (1999), p. 24.

[47] R. D. Mauldin and S. C. Williams, "On the Hausdorff dimension of some graphs," *Trans. Am. Math. Soc.* **298**, 793 (1986).

[48] T. McMahon, "Size and shape in biology," *Science* **179**, 1201–1204 (1973).

[49] J. M. Montoya and R. V. Solé, "Small world pattern in food webs," arXiv: cond-mat/0011195.

[50] E. W. Montroll and M. F. Shlesinger, "On $1/f$ noise and other distributions with long tails," *PNAS* **79**, 337 (1982).

[51] C. D. Murray, "A relationship between circumference and weight and its bearing on branching angles," *J. General Physiol.* **10**, 725–729 (1927).

[52] W. A. C. Mutch, S. H. Harm, G. R. Lefevre *et al.*, "Biologically variable ventilation increases arterial oxygenation over that seen with positive end-expiratory pressure alone in a porcine model of acute respiratory distress syndrome," *Crit. Care Med.* **28**, 2457–2464 (2000).

[53] M. E. J. Newman, "Power laws, Pareto distributions and Zipf's law," *Contemp. Phys.* **46**, 323–351 (2005).

[54] E. A. Novikov, "Mathematical model for the intermittence of turbulence flow," *Sov. Phys. Dokl.* **11**, 497–499 (1966).

[55] F. Omori, "On the aftershocks of earthquakes," *J. College Sci., Imperial Univ. Tokyo* **7**, 111–200 (1894).

[56] V. Pareto, *Cours d'Economie Politique*, Lausanne and Paris (1897).

[57] V. Pareto, *Manuale di Economia Politica*, Milan (1906); French edn., revised, *Manuel d'Economie Politique*, Paris (1909); English translation, *Manual of Political Economy* (1971).

[58] C.-K. Peng, J. Mietus, J. M. Hausdorff *et al.*, "Long-range anticorrelations and non-Gaussian behavior of the heartbeat," *Phys. Rev. Lett.* **70**, 1343–1346 (1993).

[59] C. K. Peng, J. Metus, Y. Li *et al.* "Quantifying fractal dynamics of human respiration: age and gender effects," *Ann. Biomed. Eng.* **30**, 683–692 (2002).

[60] T. M. Porter, *The Rise of Statistical Thinking 1820–1900*, Princeton, NJ: Princeton University Press (1986).

[61] W. Rall, "Theory of physiological properties of dendrites," *Ann. New York Acad. Sci.* **96**, 1071 (1959).

[62] S. Redner, "Networking comes of age," *Nature* **418**, 127–128 (2002).

[63] L. F. Richardson, "Atmospheric diffusion shown on a distance–neighbor graph," *Proc. Roy. Soc. Lond. Ser. A* **110**, 709–725 (1926).

[64] J. P. Richter, *The Notebooks of Leonardo da Vinci*, Vol. 1, New York: Dover (1970); unabridged edition of the work first published in London in 1883.

[65] D. C. Roberts and D. L. Turcotte, "Fractality and self-organized criticality of wars," *Fractals* **6**, 351–357 (1998).

[66] F. Rohrer, *Pflügers Arch.* **162**, 225–299; English translation "Flow resistance in human air passages and the effect of irregular branching of the bronchial system on the respiratory process in various regions of the lungs," *Translations in Respiratory Physiology*, ed. J. B. West, Stroudsberg, PA: Dowden, Hutchinson and Ross Publishers (1975).

[67] P. Rosen and E. Rammler, "Laws governing the fineness of powdered coal," *J. Inst. Fuel* **7**, 29–36 (1933).

[68] D. Ruelle, *Chance and Chaos*, Princeton, NJ: Princeton University Press (1991).

[69] N. Scafetta and B. J. West, "Multiscaling comparative analysis of time series and a discussion on 'earthquake conversations' in California," *Phys. Rev. Lett.* **92**, 138501 (2004).

[70] W. Schottky, "Über spontane Stromschwankungen in verschiedenen Elektrizitätsleitern," *Ann. Phys.* **362**, 541–567 (1918).

[71] R. Seibel, "Discrimination reaction time for a 1023-alternative task," *J. Exp. Psych.* **66**, 215–226 (1963).

[72] M. Shlesinger and B. J. West, "Complex fractal dimension of the bronchial tree," *Phys. Rev. Lett.* **67**, 2106 (1991).

[73] Southern California Earthquake Center, http://www.sceddc.scec.org/ftp/SCNS/.

[74] H. E. Stanley, *Introduction to Phase Transitions and Critical Phenomena*, Oxford: Oxford University Press (1979).

[75] S. S. Stevens, "On the psychophysical law," *Psychol. Rev.* **64**, 153–181 (1957).

[76] N. Suwa, T. Nirva, H. Fukusawa and Y. Saski, "Estimation of intravascular blood pressure by mathematical analysis of arterial casts," *Tokoku J. Exp. Med.* **79**, 168–198 (1963).

[77] H. H. Szeto, P. Y. Cheng, J. A. Decena *et al.*, "Fractal properties of fetal breathing dynamics," *Am. J. Physiol.* **262** (*Regulatory Integrative Comp. Physiol.* **32**) R141–R147 (1992).

[78] L. R. Taylor, "Aggregation, variance and the mean," *Nature* **189**, 732–735 (1961).

[79] L. R. Taylor and R. A. J. Taylor, "Aggregation, migration and population mechanics," *Nature* **265**, 415–421 (1977).

[80] L. R. Taylor and I. P. Woiwod, "Temporal stability as a density-dependent species characteristic," *J. Animal Ecol.* **49**, 209–224 (1980).

[81] M. C. Teich, "Fractal character of auditory neural spike trains," *IEEE Trans. Biomed. Eng.* **36**, 150–160 (1989).

[82] M. C. Teich, R. G. Turcott and S. B. Lowen, "The fractal doubly stochastic Poisson point process as a model for the cochlear neural spike train," in *The Mechanics and Biophysics of Hearing*, eds. P. Dallos, C. D. Geisler, J. W. Matthews, M. A. Ruggero and C. R. Steele, New York: Springer (1990), pp. 345–361.

[83] D. W. Thompson, *On Growth and Form*, Cambridge: Cambridge University Press (1917); 2nd edn. (1963).

[84] D. L. Turcotte and B. D. Malamud, "Landslides, forest fires, and earthquakes: examples of self-organized critical behavior," *Physica A* **340**, 580–589 (2004).

[85] H. Vierordt, *Über das Gehen des Menschen in gesunden und kranken Zuständen nach selbstregistrierenden Methoden*, Tübingen (1881).

[86] R. V. Voss and J. Clark, "1/f Noise in music: music from 1/f noise," *J. Acoust. Soc. Am.* **63**, 258–263 (1978).

[87] R. Voss, "Evolution of long-range fractal correlations and 1/f noise in DNA base sequences," *Phys. Rev. Lett.* **68**, 3805 (1992).

[88] D. J. Watts and S. H. Strogatz, "Collective dynamics of 'small-world' networks," *Nature* **393**, 440 (1998).

[89] W. Weaver, *Lady Luck: The Theory of Probability*, New York: Dover (1963).

[90] E. R. Weibel, *Morphometry of the Human Lung*, New York: Academic Press (1963).

[91] E. R. Weibel and D. M. Gomez, "Architecture of the human lung," *Science* **137**, 577 (1962).

[92] B. J. West, V. Bhargava and A. L. Goldberger, "Beyond the principle of similitude: renormalization in the bronchial tree," *J. Appl. Physiol.* **60**, 1089–1098 (1986).

[93] B. J. West and A. Goldberger, "Physiology in fractal dimensions," *American Scientist* **75**, 354 (1987).

[94] B. J. West, *Fractal Physiology and Chaos in Medicine*, Singapore: World Scientific (1990).

[95] B. J. West and W. Deering, "Fractal physiology for physicists: Lévy statistics," *Phys. Rep.* **246**, 1–100 (1994).

[96] B. J. West, *The Lure of Modern Science*, Singapore: World Scientific (1995).

[97] B. J. West, "Physiology in fractal dimensions: error tolerance," *Ann. Biomed. Eng.* **18**, 135–149 (1990).

[98] B. J. West and L. Griffin, "Allometric control of human gait," *Fractals* **6**, 101–108 (1998); "Allometric control, inverse power laws and human gait," *Chaos, Solitons & Fractals* **10**, 1519–1527 (1999).

[99] B. J. West, L. A. Griffin, H. J. Frederick and R. E. Moon, "The independently fractal nature of respiration and heart rate during exercise under normobaric and hyperbaric conditions," *Respiratory Physiol. & Neurobiol.* **145**, 219–233 (2005).

[100] B. J. West, *Where Medicine Went Wrong*, Singapore: World Scientific (2006).

[101] J. C. Willis, *Age and Area: A Study in Geographical Distribution and Origin of Species*, Cambridge: Cambridge University Press (1922).

[102] T. A. Wilson, "Design of bronchial tree," *Nature* **213**, 668–669 (1967).

[103] M. E. Wise, "Skew distributions in biomedicine including some with negative power of time," in *Statistical Distributions in Scientific Work*, Vol. 2, eds. G. P. Patil, S. Kotz, and J. K. Ord, Dordrecht: D. Reidel (1975), pp. 241–262.

[104] G. K. Zipf, *Human Behavior and the Principle of Least Effort: An Introduction to Human Ecology*, Cambridge, MA: Addison-Wesley (1949).

3 Mostly linear dynamics

One strategy for predicting an improbable but potentially catastrophic future event is to construct a faithful model of the dynamics of the complex web of interest and study its extremal properties. But we must also keep in mind that the improbable is not the same as the impossible. The improbable is something that we know can happen, but our experience tells us that it probably will not happen, because it has not happened in the past. Most of what we consider to be common sense or probable is based on what has happened either to us in the past or to the people we know. In this and subsequent chapters we explore the probability of such events directly, but to set the stage for that discussion we examine some of the ways webs become dynamically complex, leading to an increase in the likelihood of the occurrence of improbable events. The extremes of a process determine the improbable and consequently it is at these extremes that failure occurs. Knowing a web's dynamical behavior can help us learn the possible ways in which it can fail and how long the recovery time from such failure may be. It will also help us answer such questions as the following. How much time does the web spend in regions where the likelihood of failure is high? Are the extreme values of dynamical variables really not important or do they actually dominate the asymptotic behavior of a complex web?

A web described by linear deterministic dynamics is believed to be completely predictable, even when a resonance between the internal dynamics and an external driving force occurs. In a resonance a fictitious web can have an arbitrarily large response to the excitation and we discuss how such singular behavior is mitigated in the real world. The idea is that, given complete information about a linear web, we can predict, with absolute certainty, how that information determines the future. However, we never have complete information, especially when the number of variables becomes very large, so we also examine how uncertainty emerges, spreads and eventually dominates what we can know about the future.

On the other hand, even with complete information a nonlinear web is only partially predictable, since nonlinear dynamical equations typically have chaotic solutions. The formal definition of chaos is a sensitive dependence of the solution to the equation of motion on initial conditions. What this means is that an arbitrarily small change in how a web is prepared results in a very different final state. The predictability of chaotic outcomes and the control of phenomena manifesting chaos are examined in the following chapter with a view towards at least anticipating the unpredictable, such as the occurrence of the next heart attack; see for example [45]. Chaos is mentioned in this chapter on linearity so the reader does not fall into the trap of thinking that random fluctuations are always the result of what we do not know or cannot measure. Random fluctuations can also result from nonlinear dynamics.

We start the discussion of dynamics with the simplest web and incrementally increase its complexity until we exhaust the modes of analysis available for understanding its linear dynamics. It is not our intention to provide a complete discussion of all these techniques, but rather to indicate how particular kinds of complexity defy traditional methods of analysis and therefore require new techniques for their understanding. We did some of that in the first two chapters by introducing non-analytic functions and pointing out that because they were non-differentiable they could not be solutions to differential equations of motion. In this chapter, among other things, we suggest one way to study the dynamics of fractals using the fractional calculus, but postpone discussion of that calculus until Chapter 5.

In this chapter we examine ways of incrementally increasing the complexity of a web, beginning with the linear dynamics of mechanical networks that can be solved exactly. Such finite-dimensional webs can be put in contact with infinite-dimensional environments, whose influence on the web is to introduce a random force. In this way the web is described by a set of stochastic differential equations. This approach was used by Langevin at the beginning of the last century to describe the physical phenomenon of diffusion in terms of the forces moving the diffusing particles around. Of particular interest is the long-time influence of the correlations in the random force on the web dynamics. We do not restrict our focus to physical webs, but instead use diffusion as a generic concept to examine the behavior of aggregates of people, bugs and information packets as well as particles. In this way we examine how a web's dynamics responds to both external and internal potentials.

Two approaches have been developed to describe the dynamics of complex webs in physics, namely the one mentioned above, which involves the equations of motion for the dynamical web variables, and a second technique that examines the dynamics of the corresponding probability density. The equations for the dynamical variables trace out the behavior of a single trajectory of the complex web, whereas the probability density determines how an ensemble of such trajectories evolves over time in phase space. The equations for the ensemble distribution function in physics are called Fokker–Planck equations (FPEs) and are essentially equivalent to the Langevin equation in describing the web dynamics. Here again we discuss how the FPEs can be used outside the physics domain to study such things as the social phenomenon of decision-making. We discuss in some detail a model of choosing between two alternatives using a process of aggregating uncertain information until one is sufficiently certain to make a decision.

Another way to increase the complexity of the web is through the statistical properties of the random force. With this in mind we start from the simple Poisson process, which we subsequently learn describes the statistics of random networks, and generalize the statistics to include the influence of memory. As an example of the importance of memory we examine the phenomenon of failure through the use of time-dependent failure rates, the application of inverse power-law survival probabilities and autocorrelation functions.

3.1   Hamilton's equations

The most ubiquitous model in theoretical physics is the harmonic oscillator. The oscillator is the basis of understanding a wide variety of phenomena ranging from water waves to lattice phonons and all the -*ons* in quantum mechanics. The stature of the lowly harmonic oscillator derives from the notion that complex phenomena, such as celestial mechanics, have equations of motion that can be generated by means of the conserved total energy and a sequence of weak perturbations that modulate the dominant behavior. When the conditions appropriate for using perturbation theory are realized, the harmonic oscillator is often a good lowest-order approximation to the dynamics.

Of course, not all linear dynamical webs are Hamiltonian in nature. Some have dissipation leading to an asymptotic state in which all the energy has been bled out. Physically the energy extracted from the state of interest has to appear elsewhere, so dissipation is actually a linear coupling of one web to another, where the second web is absorbing the energy the first web is losing. In simplest terms the second web must be much larger than the first in order that it does not change its character by absorbing the energy. Such a second web is referred to in statistical physics as a heat bath and we shall spend some time discussing its properties.

Consider the total energy of a dynamical web described by the displacements $\mathbf{q} = \{q_1, q_2, \ldots, q_N\}$ and momenta $\mathbf{p} = \{p_1, p_2, \ldots, p_N\}$ such that the total energy E defines the Hamiltonian $E = H(\mathbf{q}, \mathbf{p})$. In a conservative network the energy is constant so that, indicating the variation in a function F by δF, we have for a conservative network

$$\delta H(\mathbf{q}, \mathbf{p}) = 0. \tag{3.1}$$

The variation of the Hamiltonian is zero; however, the Hamiltonian does change due to variations in the momenta and displacements. What occurs is that the variations in momenta produce changes in the kinetic energy and variations in displacement produce changes in the potential energy. The overall constant-energy constraint imposes the condition that the separate changes in energy cancel one another out, so that energy is transferred from kinetic to potential and back again without change in the total energy. The general variation given by (3.1) can be replaced by the time derivative, which also vanishes, giving rise to

$$\frac{dH}{dt} = \sum_{k=1}^{N} \left[\frac{\partial H}{\partial q_k} \frac{dq_k}{dt} + \frac{\partial H}{\partial p_k} \frac{dp_k}{dt} \right] = 0, \tag{3.2}$$

where the Hamiltonian is not an explicit function of time. In the situation where each of the indexed variables is independent we have each term in the sum being separately equal to zero, resulting in the relations

$$\frac{dq_k}{dt} = \frac{\partial H}{\partial p_k}, \tag{3.3}$$

$$\frac{dp_k}{dt} = -\frac{\partial H}{\partial q_k}, \tag{3.4}$$

which are called Hamilton's equations, and the sign convention given here makes the resulting equations of motion compatible with Newton's dynamical equations for an N-variable dynamical web. We should emphasize that Hamilton's equations of motion are usually constructed from much more general conditions than what we have presented here and are a formal device for producing Newton's equations of motion. For a general Hamiltonian of the form of the kinetic plus potential energy,

$$H(\mathbf{p}, \mathbf{q}) = \sum_{k=1}^{N} \frac{p_k^2}{2m_k} + V(\mathbf{q}), \tag{3.5}$$

Hamilton's equations reduce to the following more familiar form for particles of mass m_k:

$$\frac{dq_k}{dt} = \frac{p_k}{m_k},$$
$$\frac{dp_k}{dt} = -\frac{\partial V(\mathbf{q})}{\partial q_k}. \tag{3.6}$$

The first equation is the definition of the canonical momentum and the second equation is Newton's force law. Let us now restrict our analysis to the simplest of non-trivial dynamical systems, the one-dimensional linear harmonic oscillator.

Consider the Hamiltonian for a linear web given by H_0 in terms of the canonical variables,

$$H = H_0(q, p). \tag{3.7}$$

A simple one-dimensional oscillator of mass m and elastic spring constant κ has the Hamiltonian

$$H_0 = \frac{1}{2}\left(\frac{p^2}{m} + \kappa q^2\right). \tag{3.8}$$

Hamilton's equations are, of course, given by

$$\frac{dp}{dt} = -\frac{\partial H_0}{\partial q}; \qquad \frac{dq}{dt} = \frac{\partial H_0}{\partial p}, \tag{3.9}$$

which, when combined, provide the equation of motion for the linear harmonic oscillator,

$$\frac{d^2q}{dt^2} + \omega_0^2 q = 0, \tag{3.10}$$

where the natural frequency of the oscillator is

$$\omega_0 = \sqrt{\frac{\kappa}{m}}. \tag{3.11}$$

It is well known that, according to Hooke's law, the force required to stretch a spring by a displacement q is κq and consequently Newton's third law yields the restoring force $-\kappa q$.

The formal solution to the linear equation of motion (3.10) is given by

$$q(t) = A_0 \cos[\omega_0 t + \phi_0], \tag{3.12}$$

with the amplitude A_0 and phase ϕ_0 determined by the oscillator's initial conditions

$$A = \sqrt{q^2(0) + \frac{p^2(0)}{\omega_0^2}}, \tag{3.13}$$

$$\tan \phi_0 = -\frac{p(0)}{\omega_0 q(0)}. \tag{3.14}$$

Figure 3.1(a) depicts the harmonic-oscillator potential with two distinct energy levels shown as flat horizontal line segments. Figure 3.1(b) is the (p, q) phase space for the oscillator and the two curves are energy-level sets, namely curves of constant energy for the two energy levels in Figure 3.1(a). The energy-level sets are ellipses for $\omega_0 > 1$ and circles for $\omega_0 = 1$.

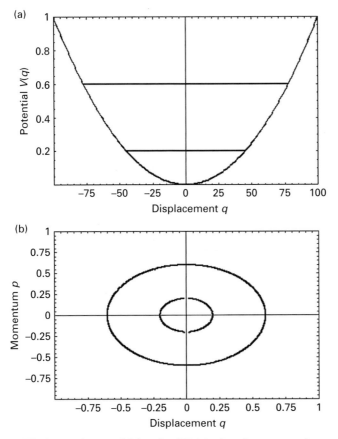

Figure 3.1. (a) The harmonic potential function $V(q)$ is plotted versus q and two energy levels are denoted. (b) The orbits for the energy level depicted in (a) are shown in the (q, p) phase space.

The simple linear harmonic oscillator is the building block from which most of the theoretical constructs of complex physical webs are constructed. To understand the overwhelming importance the linearity concept has had in science, consider the traditional style of thought used in physics and established by Sir Isaac Newton. Let us examine how he answered one of the outstanding physics questions of his day: what is the speed of sound in air? Our purpose is to show how Newton the physicist thought about difficult problems for which he had no exact solution (indeed, he did not even have an equation of motion), and how the success of his way of thinking influenced subsequent generations of scientists to think in the same way.

As you probably know, sound travels as a wave through media and is described by a wave equation. However, the wave equation is a partial differential equation and such mathematical expressions had not been invented when Newton considered the speed of sound *circa* 1686. He did know from experiment that sound was a wave phenomenon and he argued that a standing column of air could be modeled as a linear harmonic chain of oscillators. The image consists of air molecules of equal masses arranged along a line and locally interacting with other air molecules by means of linear elastic forces to keep the molecules localized. The movement of an air molecule is therefore modeled as a linear harmonic oscillator, and the air column as a linear superposition of such molecules. The linear nature of the model allowed Newton to reason to the solution of the equation of motion without explicitly constructing the equation. He deduced that the speed of sound in air is

$$u = \sqrt{\frac{P}{\rho}}, \tag{3.15}$$

where P is the pressure of the wave and ρ is the mass density of the air. Using the isothermal value of elasticity in air, which is the pressure itself, he obtained a speed of sound in air of 945 feet/s, a value 17% below the observed value of 1142 feet/s.

Lagrange severely criticized Newton's argument (1759) and presented a much more rigorous statement of the problem. His approach was to write out explicit equations for the motion of the individual molecules of air and then take the continuum limit to obtain the wave equation, which he then solved. He must have been very surprised when after all his analysis he obtained the same speed of sound as Newton had. This is how the matter stood until 1816 when Laplace, doing no calculation of his own, replaced the elastic constant of the springs used by Newton and Lagrange by one that assumes the sound wave moves so rapidly that no heat is released in the process. With this replacement Laplace obtained essentially exact agreement with experiment. This should alert the potential scientist to the following dictum: think before calculating. Thus, the modeling of the complicated continuum process of sound propagation in air by a discrete linear harmonic chain was in fact successful! This success established a precedent that has become the backbone of modeling in science.

Consider a simple material consisting of molecules interacting by means of the potential function $V(\mathbf{q})$. For the material to remain solid the molecules must all remain at or near their equilibrium positions; they are nearly stationary with respect to one another.

For simplicity, assume that this material is a physical spring made up of N particles and the potential function is represented by the Taylor expansion

$$V(\mathbf{q}) = V_0 + \sum_{j=1}^{N} \left.\frac{\partial V}{\partial q_j}\right|_0 \left(q_j - q_j^0\right) + \sum_{j,k=1}^{N} \frac{1}{2}\left.\frac{\partial^2 V}{\partial q_j\, \partial q_k}\right|_0 \left(q_j - q_j^0\right)\left(q_k - q_k^0\right) + \cdots.$$

(3.16)

If the particles are near equilibrium, the net force acting on each one vanishes, resulting in

$$\left.\frac{\partial V}{\partial q_j}\right|_0 = 0, \quad j = 1, 2, \ldots, N.$$

(3.17)

The first term in (3.16) is a constant and defines the reference potential, which for convenience is chosen to be zero. The leading-order term in the Taylor expansion of the potential function is therefore the third one for small displacements and has the quadratic form

$$V(\mathbf{q}) = \sum_{j,k=1}^{N} \frac{1}{2}\left.\frac{\partial^2 V}{\partial q_j\, \partial q_k}\right|_0 \left(q_j - q_j^0\right)\left(q_k - q_k^0\right)$$

(3.18)

in terms of the oscillator displacements from their equilibrium positions and the symmetric coupling coefficients.

The Hamiltonian for the N particles of our spring is therefore, with the equilibrium displacements of the individual particles set to zero, or equivalently, with the variables shifted to the new set $\mathbf{q}' = \mathbf{q} - \mathbf{q}^0$,

$$H(\mathbf{p}, \mathbf{q}') = \frac{1}{2}\sum_{j=1}^{N}\left[\frac{p_j^2}{m} + \sum_{k=1}^{N} V_{jk} q_j' q_k'\right],$$

(3.19)

and the quantities V_{jk} are the derivatives of the potential with respect to q_j and q_k at equilibrium. From Hamilton's equations we arrive at the equations of motion written in vector form (suppressing the primes),

$$m\frac{d^2\mathbf{q}}{dt^2} + \mathbf{V}\mathbf{q} = 0,$$

(3.20)

where the matrix $\mathbf{V}$ has elements given by V_{jk}. The main ideas can be developed by restricting considerations to a one-dimensional lattice and considering only nearest-neighbor interactions, like Newton did in his model of the air column. Imposing the simplification of nearest-neighbor interactions, we have the string of oscillator equations

$$m\frac{d^2 q_j}{dt^2} = \kappa[q_{j+1} - 2q_j + q_{j-1}], \quad j = 1, 2, \ldots, N,$$

(3.21)

where κ is the elastic force constant of an individual spring (oscillator). The solution to this equation is determined by the boundary conditions selected and the initial conditions. A particularly simple and convenient form of the solution for the nth oscillator's complex displacement is

$$q_n(t) = A e^{i(n\theta - \omega t)}, \tag{3.22}$$

which, when substituted into (3.21), yields the dispersion relation between the frequency of motion and the spring constant:

$$\omega^2 = \frac{2\kappa}{m}[1 - \cos\theta]. \tag{3.23}$$

Of course, the value of the angle θ is determined by the boundary conditions. If we assume that the string of particles is wrapped into a necklace so that the last particle couples to the first, then $q_{N+1} = q_1$ and this forces the phase in the dispersion relation to satisfy the requirement

$$e^{iN\theta} = 1,$$

implying that the phase takes on a discrete number of values

$$\theta_l = l\frac{2\pi}{N}, \quad l = 1, 2, \ldots, N. \tag{3.24}$$

The dispersion relation when indexed to the phase values becomes

$$\omega_l^2 = \frac{2\kappa}{m}\left[1 - \cos\left(l\frac{2\pi}{N}\right)\right]; \tag{3.25}$$

other conditions on the phase-dispersion relation can be derived from other boundary conditions. The general solution (3.22) can be written in terms of these phases as

$$q_n(t) = \sum_{l=1}^{N} A_l e^{i(n\theta_l - \omega_l t)}, \tag{3.26}$$

where the expansion coefficients are determined by the initial conditions obtained by the inverse discrete Fourier transform

$$A_l = \frac{1}{N}\sum_{n=1}^{N} e^{-in\theta_l} q_n(0). \tag{3.27}$$

Note that the solution of each particle displacement is dependent on the initial displacement of all the other particles in the chain due to the coupling. This is the simplest manifestation of the propagation of influence through a web and establishes a way of viewing how the elements of a web interact and transmit information from one member to another. Of course, to make such a claim is a bit facile on our part because we know in advance what the important variables are for a linear physical network, but we do not necessarily know that about social or physiologic webs.

In the absence of certainty on how to represent linear web dynamics it is worth emphasizing the important matter of representation. Consider an even simpler linear case than the one described above, namely a set of N *uncoupled* oscillators with distinct masses and elastic constants. The Hamiltonian in the momentum–displacement canonical representation is

$$H(\mathbf{p}, \mathbf{q}) = \frac{1}{2} \sum_{j=1}^{N} \left[\frac{p_j^2}{m_j} + \kappa_j q_j^2 \right]. \tag{3.28}$$

We now consider the transformation variables for each oscillator in the form

$$a_j = \frac{1}{\sqrt{2}} \left[q_j + i c_j p_j \right],$$

$$a_j^* = \frac{1}{\sqrt{2}} \left[q_j - i c_j p_j \right], \tag{3.29}$$

where c_j is a constant to be determined. Substituting these new variables into (3.28) yields the new Hamiltonian

$$H'(\mathbf{a}, \mathbf{a}^*) = \sum_{j=1}^{N} c_j \left[\left(m_j \omega_j^2 - \frac{1}{m_j c_j^2} \right) \left(a_j^2 + a_j^{*2} \right) + 2 \left(m_j \omega_j^2 + \frac{1}{m_j c_j^2} \right) a_j a_j^* \right] \tag{3.30}$$

and selecting the constants $c_j = 1/(m_j \omega_j)$ reduces the new Hamiltonian to the form

$$H'(\mathbf{a}, \mathbf{a}^*) = \sum_{j=1}^{N} \omega_j a_j a_j^*. \tag{3.31}$$

The transformation (3.29) is canonical because it preserves Hamilton's equations in the form

$$\frac{da_j}{dt} = -i \frac{\partial H'}{\partial a_j^*},$$

$$\frac{da_j^*}{dt} = i \frac{\partial H'}{\partial a_j} \tag{3.32}$$

and the variable a_j is independent of its complex conjugate a_j^*. The equation of motion for each oscillator in the case of N independent oscillators is given by

$$\frac{da_j}{dt} = -i \omega_j a_j, \tag{3.33}$$

whose solution is

$$a_j(t) = a_j(0) e^{-i\omega_j t} \tag{3.34}$$

together with its complex conjugate.

Here we have reduced the linear harmonic oscillator to its essential characteristics. It is a process with a constant amplitude and initial phase specified by the initial displacement

$$A_j \equiv |a_j(0)|, \tag{3.35}$$

$$\phi_j \equiv -\frac{\operatorname{Im} a_j(0)}{\operatorname{Re} a_j(0)} \tag{3.36}$$

and a time-dependent phase

$$a_j(t) = A_j e^{-i(\omega_j t + \phi_j)}, \tag{3.37}$$

which cycles through the same values periodically. The time it takes each oscillator to go through a cycle is different and depends on the oscillator frequency $\tau_j = 2\pi/\omega_j$, the oscillator period.

3.2   Linear stochastic equations

One strategy for developing an understanding of complex webs is to start from the simplest dynamical network that we have previously analyzed and embellish it with increasingly more complicated mechanisms. The linear harmonic oscillator is the usual starting point and it can be extended by introducing an external driver, a driver represented by a function that can be periodic, random or completely general. You are asked to examine the response of a linear harmonic oscillator to a periodic driver in Problem 3.1. But now let us turn our attention to the response of a linear web to the influence of random forces. Here we continue to follow the physics approach and partition the universe into the web of interest and the environment (everything else).

We cannot know the physical properties of the environment in a finite experiment because the environment can be indefinitely complex. As a result of this unknowability the environment is not under our control in any given experiment, only knowledge of the much smaller-dimensional web is accessible to us. Consequently, the environment's influence on the experimental web is unpredictable and unknowable except in an average way when the experiment is repeated again and again. It is this repeatability of the experiment that provides the data showing the different ways the environment influences the network. Functionally it is the ensemble distribution function that captures the patterns common to all the experiments in the ensemble, as well as the degree of difference from one experiment to the next.

One way we can model this uncertainty effect is by assuming that the linear oscillators of the last section are pushed and pulled in unknown ways by the environment before we begin to track their dynamics. Consequently, we assume that the initial phase of each oscillator ϕ_j is uniformly distributed on the interval $0 \leq \phi_j \leq 2\pi$ and therefore a function of the total oscillator displacements,

$$F(t) = \sum_{j=1}^{N} A_j \cos[\omega_j t + \phi_j], \tag{3.38}$$

is random. This was one of the first successful models to describe the dynamical displacement of a point on the surface of the deep ocean [31] in the middle of the last century. Here A_j is the amplitude of a water wave, ω_j is its frequency and the displacement of the ocean's surface at a point in space (3.38) is the linear superposition of these waves, each with a phase that is shifted randomly with respect to one another, ϕ_j. This is sometimes called the random-phase approximation. This function was apparently first used by Lord Rayleigh at the end of the nineteenth century to represent the total sound received from N incoherent point sources [40]. Note that this function $F(t)$ consists of a large number of identically distributed random variables, so that if the variance of the surface displacement is finite the central limit theorem (CLT) predicts that $F(t)$ has normal statistics. The physics literature is replete with this kind of phenomenology to describe the statistical behavior of observations when the exact dynamics of the web was not in hand. We return to this point repeatedly in our discussions.

Describing the statistical properties of physical webs phenomenologically was not very satisfying to theorists in statistical physics, who looked around for a more systematic way to understand physical uncertainty. The most satisfying explanation of statistical ambiguity, of course, should spring from a mechanical model and that meant the equations of Hamilton or Newton. Or at least that was the thinking in the statistical-physics community. So let us sketch the kind of argument that was constructed.

We construct a network of linear equations following the discussion of Bianucci *et al.* [8] and write

$$\frac{dw(t)}{dt} = \xi(t),$$

$$\frac{d\xi(t)}{dt} = -\Delta^2 w(t) + v(t),$$

$$\frac{dv(t)}{dt} = -\Delta_1^2 \xi(t) + \pi(t), \qquad (3.39)$$

$$\frac{d\pi(t)}{dt} = -\Delta_2^2 v(t) + \cdots$$

$$\vdots$$

It should be stressed that this linear model is completely general and constitutes an infinite hierarchy of equations, where for the present discussion $w(t)$ is the web variable and all the other variables are associated with the soon-to-be-ignored degrees of freedom in the dynamics and constitute the environment or heat bath. Why we use the nomenclature of a heat bath for the environment will become clear shortly. Subsequently the web variable is identified with the velocity of a Brownian particle, as it so often is in physical models.

The dynamics of an isolated linear environment are described by the subset of equations

$$\frac{d\xi(t)}{dt} = v(t),$$

$$\frac{dv(t)}{dt} = -\Delta_1^2 \xi(t) + \pi(t), \qquad (3.40)$$

$$\frac{d\pi(t)}{dt} = -\Delta_2^2 v(t) + \cdots$$

$$\vdots$$

where the coupling coefficients Δ_1^2, Δ_2^2, ..., can all be different and in (3.40) we have detached the environment from the web variable of interest. On comparing the sets of equations (3.39) and (3.40) it is clear that the heat bath perturbs the web variable $w(t)$

by means of the doorway variable $\xi(t)$, which in turn is perturbed by the "back-reaction term" $-\Delta^2 w(t)$. The equation set (3.40) consequently models the unperturbed bath, whereas the set (3.39) includes the energy exchange between the two networks.

The solution to the perturbed equations of motion for the doorway variable can be expressed immediately in terms of Green's function $K(t)$ for the unperturbed linear network as

$$\xi(t) = \xi_0(t) - \Delta^2 \int_0^t dt' \, K(t - t') w(t'), \tag{3.41}$$

where $\xi_0(t)$ is the solution to the equations of motion for the unperturbed bath and the kernel depends only on this unperturbed solution. Inserting this equation into that for the web variable yields

$$\frac{dw(t)}{dt} = \xi_0(t) - \Delta^2 \int_0^t dt' \, K(t - t') w(t'), \tag{3.42}$$

resulting in an integro-differential equation for the web variable that in physics is called the generalized Langevin equation under certain conditions. The unperturbed solution $\xi_0(t)$ is interpreted as a random force because of the uncertainty of the initial state of the environment. The integral term is interpreted as a dissipation with memory when the kernel has the appropriate properties. In the present notation the fluctuation–dissipation relation has the general form

$$\langle \xi_0(t)\xi(0)\rangle_0 = \Delta^2 \langle w^2 \rangle_{eq} K(t), \tag{3.43}$$

where the subscript zero denotes an average taken over the distribution of the initial state of the bath starting at $\xi(0)$ and the subscript eq denotes an average taken over an equilibrium distribution of fluctuations of the web variable. The kernel is the response function of the doorway variable to the perturbation $-\Delta^2 w(t)$ and the relation (3.43) indicates that the kernel is proportional to the autocorrelation function of the doorway variable

$$K(t) = \frac{\langle \xi^2 \rangle_0}{\Delta^2 \langle w^2 \rangle_{eq}} \Phi_\xi(t). \tag{3.44}$$

In this simple model the dissipation is given by the integral in (3.42) and the dissipative nature of the kernel (3.44) is a consequence of the infinite number of degrees of freedom of the heat bath. Poincaré proved that the solution to such a set of linear equations would periodically return to its initial state if the equations were generated by a Hamiltonian. Consequently the doorway variable would be periodic, but the large number of degrees of freedom makes the Poincaré recurrence time extremely long. This practical "irreversibility" leads to a characteristic correlation time τ_c for the autocorrelation function of the doorway variable and therefore of the kernel $K(t)$,

$$\tau_c \equiv \int_0^\infty dt \frac{\langle \xi_0(t)\xi\rangle_0}{\langle \xi^2 \rangle_0} = \int_0^\infty dt \, \Phi_\xi(t). \tag{3.45}$$

Assume a time-scale separation such that the web variable changes in time much more slowly than does the bath and therefore does not change appreciably over the correlation time τ_c. Using this slowness property we can approximate the web variable under the integral in (3.42) as being dependent solely on the present time and extend the upper limit on the integral to infinity without significant change to the integrated value of the response function,

$$\frac{dw(t)}{dt} = \xi_0(t) - \Delta^2 \int_0^\infty dt' \, K(t')w(t). \tag{3.46}$$

The integral can therefore be approximated as the constant

$$\gamma = \Delta^2 \int_0^\infty dt' \, K(t') \equiv \Delta^2 \chi, \tag{3.47}$$

resulting in the linear Langevin equation

$$\frac{dw(t)}{dt} = -\gamma w(t) + \xi_0(t), \tag{3.48}$$

where γ is the dissipation parameter and χ is the stationary susceptibility of the doorway bath variable ξ. The susceptibility in (3.47) is the same as that obtained for a constant perturbation applied suddenly at time $t = 0$.

In this linear model irreversibility is a consequence of the infinite number of degrees of freedom in the bath. However, the statistical fluctuations are explicitly introduced into the solution by the assumption that the initial state of the bath (environment) is not known and consequently the initial state of the bath is specified by a probability density. The Langevin equation (3.48) is valid only if we assume that the initial conditions of an ensemble of realizations of the bath are normally distributed. The width of this Gauss distribution is imposed by our technical judgement and determines the temperature of the bath. It is clear that in this approach thermodynamics is introduced by fiat through the initial conditions, and that the infinite number of variables of the (linear) bath simply preserves microscopic thermodynamics up to finite times. Note that the linearity and the initial normality yield Gauss statistics for the variable $\xi_0(t)$ at all times: this in turn implies that the stochastic force acting on the web variable is Gaussian as well. The fact of the matter is that most physical phenomena modeled by (3.48) do not satisfy the conditions under which the strictly linear model was constructed and yet the linear Langevin equation often does quite well in describing their thermodynamic behavior.

Unfortunately there does not exist a systematic procedure to include all the ways the environment can interact with an arbitrary dynamical web and so the desirable situation outlined above has not as yet been attained. One way this program has been pursued has been to make the evolution of the web uncertain by inserting a random force into the equations of motion. This procedure saves the applicability of physical theories, but it also introduces the seed of subjectivity into the evolution of the web. It requires that we develop mathematical tricks to treat webs with an infinite number of degrees of freedom like the environment, where the global properties of the overwhelming majority of degrees of freedom are chosen in a subjective manner, mostly for computational convenience. One such trick is to assume that the environment is already in the state of thermodynamic equilibrium before it interacts with the web, and that it remains in equilibrium throughout the interaction because of its enormous size. Therefore it is assumed that one knows nothing of the environment at the start of the experiment and that one can learn nothing about the environment from the results of the experiment, thereby making a virtue out of a vice.

3.2.1   Classical diffusion

Another approach to understanding uncertainty is through examining phenomena having statistical fluctuations as one of their essential features, such as that of diffusion. Classical diffusion was correctly understood for the first time in 1905 by Einstein, and is a physical realization of a stochastic process. Diffusive motion is a consequence of a heavy particle embedded in a fluid of lighter particles being buffeted on all sides by the large number of lighter particles. The response of the heavy particle is to move in an erratic way through space because of the random imbalance in the forces being applied to its surface. This motion was seen by the botanist Robert Brown, who, in the late 1820s, observed through his microscope pollen motes suspended in water undergoing the most peculiar motion depicted in Figure 3.2. The mote appeared to have an internal force that caused it to lurch first this way and then that way, with no apparent goal in mind. This erratic motion is today called Brownian motion, after Brown, even though he did not know the cause of the motion.

The motion depicted in Figure 3.2 is that of a heavy particle in a fluid of lighter particles where the location of the heavy particle is recorded every few seconds. The picture shows the actual experimental data recorded by Perrin [39]. The path shown in Figure 3.2 is obtained by recording the position of the heavy (mesoscopic) particle at equal time intervals and then connecting these points by straight lines to aid the eye in seeing the magnitude and direction of the particle's motion during the time interval. The straight lines have no other significance; they do not record or imply where the mote is

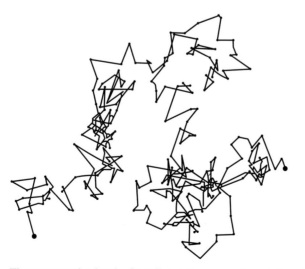

Figure 3.2.   The coarse-grained path of a pollen mote as seen through the microscope of Perrin [39].

or where it was between the recorded points. The lines might be thought of as "possible" or "average" paths of the heavy particle rather than the actual path.

The random motion depicted in Figure 3.2 can be modeled by the dynamics of the heavy particle using Newton's force law

$$M\frac{d\mathbf{V}}{dt} = \mathbf{F}(t), \tag{3.49}$$

where M is the mass of the heavy particle, $\mathbf{V}$ is its velocity and $\mathbf{F}$ is the vector sum of all the forces acting on it. For simplicity we set $M = 1$, reduce the problem to one dimension and restrict all the forces to be those generated by collision with the lighter fluid particles, so that (3.49) reduces to

$$\frac{dV}{dt} = F(t). \tag{3.50}$$

This is of the form (3.39) for a free particle being buffeted about by the fluid.

For additional simplicity we require the average of the random force to vanish, but physical arguments establish that for Brownian motion the average force is given by the hydrodynamic back-reaction of the fluid to the motion of the heavy particle. This fluid back-reaction is the Stokes drag produced by the heavy particle pulling ambient fluid along with it as it moves through the fluid, yielding an average force proportional to the velocity of the heavy particle:

$$\langle F(t)\rangle = -\lambda V(t). \tag{3.51}$$

On replacing the force in (3.50) by its average value (3.51) plus random fluctuations, the equation for the Brownian motion of a heavy particle becomes

$$\frac{dV}{dt} = -\lambda V(t) + \eta(t), \tag{3.52}$$

where $\eta(t)$ is a zero-centered random force and is of the form (3.48). This mode of argument was introduced into the physics literature by Langevin [29] and resulted in physical stochastic differential equations. These equations were the first indication that the continuous nature of physical phenomena might not be as pervasive as had been believed in traditional theoretical physics. But then physicists as a group never let rigorous mathematics stand in the way of a well-argued physical theory. A new calculus involving stochastic processes had to be developed in order to rigorously interpret (3.52).

It is probably time to step back and consider what we have. The source of the above Langevin equation is undoubtedly physics and the attempt to model complex physical interactions of a simple web with the environment. But the form of the equation is actually a great deal more. Suppose that V is not the velocity of a Brownian particle, but is the variable describing the dynamics of the web aggregating information from a number of sources. If the sources are incoherent, such as the collisions, then the incremental information is random. The term linear in the dynamical variable V in an information context may be considered to be a filter on the noise. Consider the Fourier transform of a function

$$\tilde{F}(\omega) = \mathcal{FT}\,[F(t); \omega] \equiv \int_{-\infty}^{\infty} dt\, F(t)e^{i\omega t} \tag{3.53}$$

so that the Fourier transform of the solution to (3.52) can be expressed as

$$\tilde{V}(\omega) = \frac{\tilde{\eta}(\omega)}{\lambda - i\omega} \tag{3.54}$$

when the initial velocity is set to zero. The spectrum of the response is consequently expressed in terms of the spectrum of the incoherent information

$$\left\langle |\tilde{V}(\omega)|^2 \right\rangle = \frac{\left\langle |\tilde{\eta}(\omega)|^2 \right\rangle}{\lambda^2 + \omega^2}, \tag{3.55}$$

where the brackets denote an averaging process that for the moment we leave unspecified. It is clear that for high-frequency waves, $\lambda \ll \omega$, the spectrum of the excitation is suppressed by ω^{-2} and the web response to the noise is suppressed. For the low-frequency waves, $\lambda \gg \omega$, the frequency of the response is proportional to that of the excitation and consequently this is a low-pass filter, that is, the low-frequency information is passed without distortion. There are other kinds of filters as well, but here we thought it important to bring to your attention the fact that Langevin stochastic differential equations arise in a variety of contexts, not just in Brownian motion.

Returning now to the Langevin equation, let us study some of the properties of the solutions to such equations. When the inertial force, the left-hand side of (3.52), is negligible, the velocity equation can be replaced with one for the distance traveled (displacement) $Q(t)$:

$$\frac{dQ(t)}{dt} = \xi(t), \tag{3.56}$$

and the dissipation parameter is used to normalize the random force $\xi(t) = \eta(t)/\lambda$. The solution to the stochastic equation of motion (3.56) is formally given by

$$Q(t) = Q(0) + \int_0^t \xi(t')dt', \tag{3.57}$$

from which it is clear that the simple web response undergoes motion with erratic fluctuations induced by the random force. Using the traditional language of statistical physics, we do not restrict our analysis to the behavior of single trajectories, namely the solutions to the equation of motion, but rather examine the mean values of quantities such as $\langle Q(t)^n \rangle$, where n is an integer and the average is the nth moment of the displacement. In the above presentation the random noise was constructed to be zero-centered, so that

$$\langle Q(t) \rangle_\xi = Q(0), \tag{3.58}$$

where the subscript denotes that the indicated average is taken over an ensemble of realizations of the random force.

The second moment of the solution to the Langevin equation is given in terms of an average over an ensemble of realizations of the initial condition of the Brownian particle and consequently over an ensemble of fluctuations of the random force

$$\left\langle Q(t)^2 \right\rangle_\xi = \int_0^t dt_1 \int_0^t dt_2 \langle \xi(t_1)\xi(t_2) \rangle_\xi + Q(0)^2 + 2 \int_0^t dt_1 \langle \xi(t_1)Q(0) \rangle_\xi. \quad (3.59)$$

If the fluctuations and the initial displacement of the particle are statistically independent the last term in (3.59) vanishes. If the fluctuations are delta-correlated in time with
strength D,

$$\langle \xi(t)\xi(t') \rangle_\xi = 2D\delta(t - t'), \quad (3.60)$$

we obtain from (3.59) the variance of the dynamical variable

$$\left\langle Q(t)^2 \right\rangle_\xi - \langle Q(t) \rangle_\xi^2 = \int_0^t dt_1 \int_0^t dt_2 \, 2D\delta(t_2 - t_1) = 2Dt. \quad (3.61)$$

Consequently, the variance of the displacement of an ensemble of Brownian particles
increases linearly in time. For an ergodic process, namely one for which the ensemble
average is equivalent to a long-time average, we can also interpret (3.61) as the variance
of the displacement of the trajectory of a single Brownian particle from its starting
position increasing linearly in time.

   The assumption of independence between the initial state of the dynamical variable
$Q(0)$ and the random force $\xi(t)$ holds when there exists a large time-scale separation between the dynamics of the *fast variable* ξ and the *slow variable* Q. However,
this hypothesis must be used with caution in applications, which often refer to situations in which the random force can have a slow long-time decay, thereby violating the
assumption of time-scale separation. For example, the autocorrelation function

$$\Phi_\xi(t) = \frac{\langle \xi(0)\xi(t) \rangle_\xi}{\langle \xi^2 \rangle_\xi} \quad (3.62)$$

can have a number of different forms. We have described the case of no memory, for
which the autocorrelation function is given by the Dirac delta function

$$\Phi_\xi(t) \propto \delta(t). \quad (3.63)$$

A memory in which the autocorrelation function is exponential,

$$\Phi_\xi(t) = e^{-\gamma t}, \quad (3.64)$$

and $1/\gamma$ is the decay time of the memory is often used. But in general the autocorrelation
function can have either a positive or a negative long-time tail. In molecular dynamics,
processes characterized by a long-time regime with an inverse power-law correlation
function

$$\Phi_\xi(t) \approx \pm \frac{1}{t^\beta} \quad (3.65)$$

have been called slow-decay processes ever since the pioneering work of Adler and
Wainwright [1]. Note that such an inverse power-law autocorrelation function implies
an exceptionally long memory and this is what we refer to as long-time memory. The
literature on this topic is quite extensive, but does not directly relate to the purpose
of our present discussion, so we just mention Dekeyser and Lee [15], who provided the
first microscopic derivation of an inverse power-law autocorrelation function in physics.

Implementing the assumption that the initial condition and the random force fluctuations are statistically independent of one another yields for the second moment of the web's dynamic response

$$\left\langle Q(t)^2 \right\rangle_\xi = \int_0^t dt_1 \int_0^t dt_2 \langle \xi(t_1)\xi(t_2) \rangle_\xi + Q(0)^2. \qquad (3.66)$$

The integral in (3.66) can be simplified by assuming that the ξ-process is stationary in time; that is, its moments are independent of the origin of time and consequently

$$\langle \xi(t_1)\xi(t_2) \rangle_\xi = \langle \xi(t_1 - t_2)\xi(0) \rangle_\xi. \qquad (3.67)$$

Inserting (3.67) into (3.66) and taking the time derivative leads to the differential equation of motion for the second moment

$$\frac{d\langle Q(t)^2 \rangle_\xi}{dt} = 2 \int_0^t dt' \langle \xi(t')\xi(0) \rangle_\xi. \qquad (3.68)$$

Clearly the second moment of the web response appearing on the lhs of (3.68) must be connected with the long-time diffusional regime described by

$$\left\langle Q(t)^2 \right\rangle_\xi = Ct^{2\alpha}, \qquad (3.69)$$

where C is a constant. It is evident that the physical bounds on the possible values of the scaling exponent in (3.69) are given by $0 \le \alpha \le 1$; $\alpha = 0$ defines the case of localization, which is the lower limit of any diffusion process. The value $\alpha = 1$ refers to the case of many uncorrelated deterministic trajectories with $Q(t) - Q(0)$ being linearly proportional to time for each of them. The constraint $\alpha \le 1$ is a consequence of the fact that a classical diffusion process cannot spread faster than a collection of deterministic ballistic trajectories!

It is worth pointing out that turbulent diffusion given by the motion of a passive scalar, such as smoke, in a turbulent flow field, such as wind a few meters above the ground, leads to a mean-square separation of the smoke particles that increases faster than ballistic; that is, $\alpha > 1$. In fact, Richardson [41] observed the width of smoke plumes from chimneys increasing with a power-law index slightly larger than $\alpha = 1.5$. Such behavior occurs because the flow field drives the diffusing scalar particles in a manner having to do with the scaling of vortices in turbulence. The first scaling model of fluid turbulence was that proposed by Kolmogorov [27] in which eddies are nested within one another with the spatially larger eddies rotating faster than the smaller ones. Consequently, when a local eddy pulls the scalar particles apart the effect is accelerated because an increasing separation distance produces a larger relative velocity. This shearing effect results in the growth of the mean-square separation of the passive scalar particles becoming faster than ballistic; it is a driven process. In this way the dynamics of the environment directly influences the measured properties of the stochastic web variable.

Finally, the condition $\alpha = 0.5$ is obtained for simple diffusion, where the variance increases linearly with time. We dwell on these physical models here because they are useful in interpreting phenomena outside the physical sciences that satisfy analogous equations. The deviation from the classical diffusion growth of the mean-square separation is particularly interesting because it may provide a probe into the dynamics of the environment in which the web of interest is embedded. But then again life is not always so straightforward and deviations from simple diffusion may also be caused by internal dynamics of the web rather than from the influence of the environment on the web.

One probe of the web's dynamics can be obtained by substituting the definition of the autocorrelation function given by (3.62) into (3.68), and thereby obtaining the time-dependent diffusion coefficient

$$2D(t) = \frac{d\langle Q(t)^2 \rangle_\xi}{dt} = 2\langle \xi^2 \rangle_\xi \int_0^t \Phi_\xi(t')dt'. \tag{3.70}$$

The deviation of the scaling index from the conventional diffusion value of $\alpha = 0.5$ can now be explained in terms of the long-term memory in the random force. This memory may be a consequence of internal web dynamics not explicitly included in the equations of motion or it can be produced by a modulation of the web dynamics by the motion of the environment. Part of the challenge is to determine the source of the memory from the context of the phenomenon being investigated. Inserting the inverse power-law form of the autocorrelation function given by (3.65) into (3.70) and taking another time derivative leads to the long-time prediction

$$\frac{d^2\langle Q(t)^2 \rangle_\xi}{dt^2} = 2\alpha(2\alpha - 1)Ct^{2\alpha-2} \approx \pm\frac{1}{t^\beta}. \tag{3.71}$$

The equalities in (3.71) are a consequence of having assumed that the long-time limit of the autocorrelation function is dominated by the inverse power law shown in (3.65). A realization of the positive sign in (3.71) is depicted by the solid line in Figure 3.3, whereas a realization of the negative sign is depicted by the dashed line in that figure. Note that the dashed curve goes negative near a lag time of 24 and then asymptotically approaches zero correlation from below.

The relation between the two exponents, the one for anomalous diffusion and the other for the random force autocorrelation function, is obtained using (3.71) to be

$$\alpha = 1 - \beta/2, \tag{3.72}$$

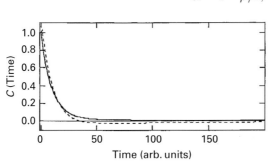

Figure 3.3. The solid curve is the positive inverse power-law autocorrelation function for the random force and remains positive for all time, approaching zero asymptotically from above. The dashed curve has the negative autocorrelation function at long times and approaches zero asymptotically from below [32]. Reproduced with permission.

where we have matched the time dependence on both sides of the equation. On comparing the exponents in (3.72) it is evident that $\alpha > 0.5$ implies a positive long-time correlation, whereas $\alpha < 0.5$ implies a negative long-time correlation. So let us summarize the result of this simple theoretical analysis with one eye on Figure 3.3:

solid curve: $1 > \alpha > 0.5$ and $1 > \beta > 0$;
dashed curve: $0 < \alpha < 0.5$ and $2 > \beta > 1$.

Thus, we see that the autocorrelation function depicted by the solid line in Figure 3.3 leads to superdiffusive behavior ranging from standard diffusion ($\alpha = 0.5$) to ballistic behavior ($\alpha = 1$). The autocorrelation function corresponding to the dashed curve leads to subdiffusive behavior ranging from standard diffusion to no motion at all.

It should be stressed that the autocorrelation function for a superdiffusive process diverges asymptotically, which is to say according to (3.70), $D(\infty) = \infty$; whereas, from the autocorrelation function in the subdiffusive case, the diffusion coefficient remains finite, $D(\infty) < \infty$. At early times in the diffusion process the mean-square value of the dynamic variable $Q(t)$ increases; then, when the negative part of the autocorrelation function becomes important, the rate of diffusion decreases. When the negative tail completely compensates for the positive part of the relaxation process, the rate of diffusion virtually vanishes. At this late stage further diffusion is rigorously prevented and the diffusing entity becomes localized. Processes of this kind have been discovered [13] and the theory presented here affords a remarkably straightforward explanation of them. It is interesting that such complex processes should admit such a straightforward interpretation.

3.2.2   Linear response theory (LRT)

Large aggregates of people, bugs or particles can arrange themselves into remarkably well-organized assemblies. The chemical interactions among particles can produce stripes, pinwheels or even oscillating colors in time. People fall into step with one another as they casually walk together and audiences can spontaneously clap their hands in rhythm. Fireflies can blink in unison even though there is no coordinating external leader. These kinds of synchronizations are the consequence of the internal dynamics of complex webs, which we have not taken into account in our discussion so far but take up in subsequent sections. For the moment we are interested in the simplest response of the web to a complex environment, both in terms of a stochastic force and a deterministic perturbation.

Let us return to the Langevin equation for a Brownian particle

$$\frac{dV(t)}{dt} = -\lambda V(t) + \xi(t) \tag{3.73}$$

which we wrote down earlier. The formal solution to this equation is

$$V(t) = \int_0^t e^{-\lambda(t-t')} \xi(t') dt' \tag{3.74}$$

when the initial velocity is zero. For each realization of the stochastic force (3.74) provides a trajectory solving the Langevin equation. As we learned in our discussion of the law of frequency of errors, the best representation of all these realizations is the average value, but we recall that the average of the ξ-fluctuations is zero, so the average velocity vanishes,

$$\langle V(t) \rangle = 0, \tag{3.75}$$

when the initial velocity is set to zero.

The lowest-order non-vanishing moment of the velocity of the Brownian particle is therefore the autocovariance function

$$\langle V(t_1)V(t_2) \rangle = \int_0^{t_1} e^{-\lambda(t_1 - t_1')} \int_0^{t_2} e^{-\lambda(t_2 - t_2')} \langle \xi(t_1')\xi(t_2') \rangle dt_1' \, dt_2', \tag{3.76}$$

which is averaged over an equilibrium ensemble of realizations of the fluctuations. Note that we no longer subscript the averaging bracket when there is little danger of confusion. The equilibrium autocorrelation function for the stochastic force is given by

$$\Phi_\xi(t_1, t_2) = \frac{\langle \xi(t_1)\xi(t_2) \rangle}{\langle \xi^2 \rangle}, \tag{3.77}$$

where $\langle \xi^2 \rangle$ is independent of time and, if we make the stationarity assumption,

$$\Phi_\xi(t_1, t_2) = \Phi_\xi(|t_1 - t_2|). \tag{3.78}$$

An analytic form of the autocorrelation function that is often used is the exponential

$$\Phi_\xi(|t_1 - t_2|) = e^{-\lambda|t_1 - t_2|}. \tag{3.79}$$

Note that the Laplace transform of the exponential $\Phi_\xi(t)$ is

$$\hat{\Phi}_\xi(u) = \mathcal{LT}[\Phi_\xi(t), u] \equiv \int_0^\infty e^{-ut} \Phi_\xi(t) dt = \frac{1}{u + \lambda}, \tag{3.80}$$

so that

$$\hat{\Phi}_\xi(0) = \frac{1}{\lambda} \equiv \tau_c. \tag{3.81}$$

We denote the correlation time of the noise by τ_c.

To simplify the calculations we assume that the random force is delta-correlated in time,

$$\langle \xi(t_1)\xi(t_2) \rangle = 2\langle \xi^2 \rangle \tau_c \delta(t_1 - t_2), \tag{3.82}$$

so that the stationary autocorrelation function reduces to

$$\Phi_\xi(t) = \frac{\langle \xi(t)\xi(0) \rangle}{\langle \xi^2 \rangle} = 2\tau_c \delta(t), \tag{3.83}$$

whose Laplace transform satisfies (3.81). Let us insert (3.82) into the double integral (3.76) and assume $t_1 > t_2$ to obtain

$$\langle V(t_1)V(t_2)\rangle = \int_0^{t_2} e^{-\lambda(t_1+t_2-2t_2')}\langle\xi^2\rangle\tau_c\,dt_2'$$

$$= \frac{\langle\xi^2\rangle\tau_c}{\lambda}\left[e^{-\lambda(t_1-t_2)} - e^{-\lambda(t_1+t_2)}\right]. \tag{3.84}$$

The equilibrium condition for the solution is obtained for the sum $t_1 + t_2 \to \infty$ and the difference $t_1 - t_2 < \infty$ so that (3.84) reduces to

$$\langle V(t_1)V(t_2)\rangle_{\text{eq}} = \frac{\langle\xi^2\rangle\tau_c}{\lambda}e^{-\lambda(t_1-t_2)}. \tag{3.85}$$

The equilibrium mean-square velocity obtained when $t_1 = t_2$ is, from (3.85),

$$\left\langle V^2\right\rangle_{\text{eq}} = \frac{\langle\xi^2\rangle\tau_c}{\lambda}, \tag{3.86}$$

and, on repeating the same calculation with $t_1 < t_2$, we finally obtain

$$\Phi_V(t_1, t_2) = e^{-\lambda|t_1-t_2|} \tag{3.87}$$

as the stationary velocity autocorrelation function.

Now let us apply a perturbation $E(t)$ to the web and replace the previous Langevin equation with

$$\frac{dV(t)}{dt} = -\lambda V(t) + \xi(t) + E(t), \tag{3.88}$$

whose exact solution when the initial velocity is set to zero is given by

$$V(t) = \int_0^t e^{-\lambda(t-t')}[\xi(t') + E(t')]dt'. \tag{3.89}$$

Again recalling that the average stochastic force vanishes, the average velocity of the Brownian particle is

$$\langle V(t)\rangle = \int_0^t e^{-\lambda(t-t')}E(t')dt', \tag{3.90}$$

where there is no averaging bracket on the perturbation because it is assumed to be independent of the environmental fluctuations. We can reexpress the average velocity in terms of the velocity autocorrelation function by using (3.87) to rewrite (3.90) as

$$\langle V(t)\rangle = \int_0^t \Phi_V(t - t')E(t')dt', \tag{3.91}$$

the well known Green–Kubo relation in which the kernel is the unperturbed velocity autocorrelation function.

It is important to stress that in the example discussed here the linear response theory (LRT) is an exact result, namely the response of the dynamic web variable to a perturbation is determined by the autocorrelation function of the web variable in the absence of the perturbation. In general this solution is not exact, but under certain conditions this prediction turns out to be correct for sufficiently weak perturbations. We take up the general theory of this approach in later chapters.

3.2.3 Double-well potential and stochastic resonance

The next step in increasing modeling complexity is to make the potential of order higher than quadratic. A nonlinear oscillator potential coupled to the environment is given by a dynamical equation of the form

$$\frac{d^2Q}{dt^2} = -\lambda\frac{dQ}{dt} - \frac{\partial U(Q)}{\partial Q} + \xi(t), \tag{3.92}$$

so, in addition to the potential, the particle experiences fluctuations and dissipation due to the coupling to the environment. Of most interest for present and future discussion is the double-well potential

$$U(Q) = \frac{A}{4}\left(Q^2 - a^2\right)^2, \tag{3.93}$$

which has minima at $Q = \pm a$ and a maximum at $Q = 0$. In the Smoluchosky approximation the inertial force, that is the second-derivative term in (3.92), is negligibly small so that (3.92) can be replaced with

$$\frac{dQ}{dt} = -\frac{\partial\Phi(Q)}{\partial Q} + \eta(t), \tag{3.94}$$

where $\eta(t) = \xi(t)/\lambda$ and the scaled potential is $\Phi(Q) = U(Q)/\lambda$ so that the dynamics are determined by

$$\frac{dQ}{dt} = -\frac{A}{\lambda}\left(Q^2 - a^2\right)Q + \eta(t), \tag{3.95}$$

where the fluctuations and dissipation terms model the coupling to the environment.

The Smoluchosky approximation is widely used in stochastic physics. One of the most popular applications is in the *synergetics* of Herman Haken [21]. An essential concept in synergetics is the order parameter, which was originally introduced to describe phase transitions in thermodynamics in the Ginzberg–Landau theory. The order-parameter concept was generalized by Haken to the *enslaving principle*, which states that the dynamics of fast-relaxing (stable) modes are completely determined by a handful of slow dynamics of the "order parameter" (unstable modes) [22]. The order dynamics can be interpreted as the amplitude of the unstable modes determining the macroscopic pattern. The enslaving principle is an extension of the Smoluchosky approximation from the physics of condensed matter to complex networks, including the applications made for understanding the workings of the brain [23, 24].

The problem of a thermally activated escape over a barrier was studied by Kramers in his seminal paper [28] and studies are continually revealing insights into such complex phenomena; see for example [33]. The theory provides a touchstone against which to assess more elaborate analyses, since it provides one of the few analytic results available. Consequently, we record here the evaluation of the rate of escape over a potential

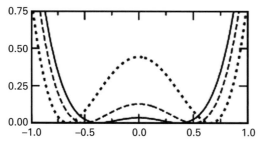

Figure 3.4.   A double-well potential is depicted for three different values of the parameter controlling the barrier height between the wells. The parameters have been adjusted so that the well minima are at ± 0.50 and the maximum is at zero.

barrier done by Kramers in the case of the double-well potential. Kramers applies the Smoluchosky approximation and adopts the Langevin equation given by (3.95), where $U(Q)$ is the potential sketched in Figure 3.4.

The rate at which a population of chemicals makes a transition from one well to the other due to thermal excitations is estimated here using Kramers' theory. We note that this is not an exact analytic calculation and that it includes an approximation of the shape of the double-well potential. However, numerical calculations of the rate of transition and of its dependence on the strength of the thermal fluctuations and the height of the barrier between the two wells compare very well with Kramers' rate. The Kramers theory is too specialized to explore here in detail, but the double-well potential has found numerous useful modeling applications. One such application involved climate change, for which a new physical concept was introduced by Benzi and Vulpiani [7]. They were trying to understand whether a relatively small periodic force can be amplified by internal nonlinear interactions in the presence of noise. The mechanism they identified as stochastic resonance (SR) applied to the understanding of the Milankovitch cycle, whose period of 10^5 years is observed in climate records. This mechanism was independently identified and applied to the same problem by Nicolis and Nicolis [37]. However, the potential importance of the SR mechanism did not become apparent until after the article by Moss and Wiesenfeld [35] on the neurologic applications of SR.

It is possible to begin the analysis of SR from a Hamiltonian having a double-well potential and a periodic forcing term to change (3.95) into

$$\frac{dQ}{dt} = -\frac{A}{\lambda}(Q^2 - a^2)Q + \frac{A_{\mathrm{p}}}{\lambda}\cos(\omega_{\mathrm{p}}t) + \eta(t) \tag{3.96}$$

and here too we have included the coupling to the environment. Note that (3.96) describes a particle in a time-dependent potential

$$\Phi(Q) = \frac{A}{4\lambda}(Q^2 - a^2)^2 - Q\frac{A_{\mathrm{p}}}{\lambda}\cos(\omega_{\mathrm{p}}t). \tag{3.97}$$

We leave the description of how the potential behaves as Problem 3.3. From (3.96) we see that the particle oscillates in the vicinity of one of the minima and the potential rhythmically changes the relative position of the minima. If the change in minima matches the fluctuations, a particle can be propelled over the potential barrier. This is essentially a synchronization of the thermally activated hopping event between the potential minima and the weak periodic forcing. Kramers calculated this rate of transition between the minima of such potential wells to be

$$r = \frac{\omega_a \omega_0}{4\pi\lambda} \exp\left[-\frac{\lambda \, \Delta U}{D}\right], \tag{3.98}$$

where the frequency of the linear oscillations in the neighborhood of the stable minima is $\omega_a = |U''(a)|^{1/2}$; the frequency at the unstable maximum is $\omega_0 = |U''(0)|^{1/2}$ and ΔU is the height of the potential barrier separating the two wells.

The average long-time response of the nonlinear oscillator does not look very different from that obtained in the case of linear resonance. However, here, because the potential is nonlinear, we use LRT to obtain the solution for a small-amplitude periodic forcing function

$$\langle Q(t) \rangle_\xi = \frac{A_p \langle Q^2 \rangle_0}{D} \frac{2r}{\sqrt{4r^2 + \omega_p^2}} \cos[\omega_p t + \phi_p] \tag{3.99}$$

with phase shift

$$\phi_p = -\tan^{-1}\left[\frac{\omega_p}{2r}\right] \tag{3.100}$$

and Kramers' rate r replaces the frequency difference obtained in the case of linear resonance. Inserting the expression for the rate of transition into the asymptotic solution for the average signal allows us to express the average signal strength as a function of the intensity of the fluctuations. By extracting the D-dependent function from the resulting equation,

$$G(D) = \frac{\exp[-\lambda \, \Delta U/D]}{D\sqrt{4\exp[-2\lambda \, \Delta U/D] + \omega_p^2}}, \tag{3.101}$$

and choosing the parameters for the potential to be $A_p = a = 1$ and the frequency of the driver to be unity we arrive at the curve given as Problem 3.3.

The average signal strength determined by the solution to the weakly periodically driven stochastic nonlinear oscillator is proportional to $G(D)$. It is clear from this function, or the curve drawn from it in Problem 3.3, that the average signal increases with increasing D, the level of the random force, implying that weak noise enhances the average signal level. At some intermediate level of fluctuation, determined by the parameters of the dynamic web, the average signal reaches a maximum response level. For more intense fluctuations the average signal decreases, asymptotically approaching an inverse power-law response in the noise intensity D. This non-monotonic response of the periodically driven web to noise is characteristic of SR.

SR is a statistical mechanism whereby noise influences the transmission of information in both natural and artificial networks such that the signal-to-noise ratio is no longer monotonic. There has been an avalanche of papers on the SR concept across a landscape of applications, starting with the original paleoclimatology studies and going all the way to the influence of noise on the information flow in sensory neurons, ion-channel gating and visual perception shown by Moss *et al.* [36]. Most of the theoretical analyses have relied on the one-dimensional bistable double-well potential, or a reduced version given by two-state models, both of which are now classified as dynamic SR. Another version of SR, which is most commonly found in biological webs, involves the concurrence of a threshold with a subthreshold signal and noise and is called threshold SR [36].

Visual perception is one area in which there have been a great many experiments done to demonstrate the existence of a threshold. Moss *et al.* [36] point out that through the addition of noise the SR mechanism has been found to be operative in the perception of gratings, ambiguous figures and letters and can be used to improve the observer's sensitivity to weak visual signals. The nonlinear auditory network of human hearing also manifests threshold SR behavior. The absolute threshold for the detection and discrimination of pure tones was shown by Zeng *et al.* [49] to be lowered in people with normal hearing by the addition of noise. They also showed that the same effect could be achieved for individuals with cholera or brain-stem implants by the addition of an optimal amount of broad-band noise.

The greatest success of SR has been associated with modeling and simulation of webs of neurons, where the effects of noise on sensory function match those of experiment. Consequently, the indications are that the theory of threshold SR explains one of the fundamental mechanisms by which neurons operate. This does not carry over to the brain, however, since the SR mechanism cannot be related to a specific function. This theme will be taken up again later.

3.3　　More on classical diffusion

We discussed classical diffusion from the point of view of the trajectory of the Brownian particle being buffeted by the lighter particles of the background fluid. We again take up this fundamental physical process but now from a less restrictive vantage point and base our discussion on two physical concepts: particle transport and the continuity of fluid motion. The net transport of material across a unit surface is proportional to the gradient of the material density in a direction perpendicular to the unit area. Particle transport is modeled by the spatially isotropic form of Fick's law,

$$\mathbf{j}(\mathbf{r}, t) \equiv -D \nabla n(\mathbf{r}, t), \tag{3.102}$$

where $\mathbf{j}(\mathbf{r}, t)$ is the particle flux (current) across the unit area, D is the diffusion coefficient, which may be a function of position $\mathbf{r}$ and time t, and $n(\mathbf{r}, t)$ is the particle number density at the location $\mathbf{r}$ and time t. The particle current is defined in terms of the particle velocity $\mathbf{v}(\mathbf{r}, t)$ by

$$\mathbf{j}(\mathbf{r}, t) = n(\mathbf{r}, t)\mathbf{v}(\mathbf{r}, t). \tag{3.103}$$

In free space material is conserved so that during movement the diffusing particles are neither created nor destroyed. This principle is expressed by the continuity equation

$$\frac{\partial n(\mathbf{r}, t)}{\partial t} + \nabla \cdot \mathbf{j}(\mathbf{r}, t) = 0. \tag{3.104}$$

From these three equations one can derive two basic differential equations of motion for the physical or appropriate biological network, one for the particle density and the other for the total particle velocity. Traditional theory emphasizes the diffusion equation, which is obtained by substituting (3.102) into (3.104) to obtain

$$\frac{\partial n(\mathbf{r}, t)}{\partial t} = \nabla \cdot [D \nabla n(\mathbf{r}, t)]. \tag{3.105}$$

When the diffusion coefficient is independent of both space and time (3.105) simplifies to

$$\frac{\partial n(\mathbf{r}, t)}{\partial t} = D \nabla^2 n(\mathbf{r}, t), \tag{3.106}$$

which in one spatial dimension becomes

$$\frac{\partial n(q, t)}{\partial t} = D \frac{\partial^2 n(q, t)}{\partial q^2}. \tag{3.107}$$

Once the particle density has been determined from the diffusion equation, the current $\mathbf{j}(\mathbf{r}, t)$ can be determined from (3.102).

An alternative way to describe diffusion is through the velocity field $\mathbf{v}(\mathbf{r}, t)$. Once the velocity is known, the particle density can be obtained by combining (3.102) and (3.103). In the case of a constant diffusion coefficient we can write

$$\mathbf{v}(\mathbf{r}, t) = -D \nabla \log n(\mathbf{r}, t). \tag{3.108}$$

A differential equation for $\mathbf{v}(\mathbf{r}, t)$ is obtained by noting from (3.103) and (3.104) that

$$\frac{\partial \mathbf{v}(\mathbf{r}, t)}{\partial t} = -D \nabla \frac{\partial}{\partial t} \log n(\mathbf{r}, t) = D[n(\mathbf{r}, t)^{-1} \nabla \cdot n(\mathbf{r}, t)\mathbf{v}(\mathbf{r}, t)]$$
$$= D \nabla [\nabla \cdot \mathbf{v}(\mathbf{r}, t) + \mathbf{v}(\mathbf{r}, t) \cdot \nabla \log n(\mathbf{r}, t)], \tag{3.109}$$

so that inserting (3.108) into (3.109) yields

$$\frac{\partial \mathbf{v}(\mathbf{r}, t)}{\partial t} = \nabla [D \nabla \cdot \mathbf{v}(\mathbf{r}, t) - \mathbf{v}(\mathbf{r}, t) \cdot \mathbf{v}(\mathbf{r}, t)]. \tag{3.110}$$

Therefore, even elementary diffusion theory is nonlinear if one formulates the theory in terms of the "wrong" variable. Here the equation for the velocity is quadratic in the velocity. From this construction it is obvious that the solution to the quadratically nonlinear partial differential equation for the velocity field is proportional to the gradient of the logarithm of the solution of the classical linear diffusion equation given by (3.108). The relationships between the diffusion equation and a number of other nonlinear equations have been recognized over the years and were discussed in this context over thirty years ago by Montroll and West [34].

Much of our previous attention focused on the distribution of a quantity of interest, whether it is the number or magnitude of earthquakes, how wealth is distributed in a population, how influence is concentrated, or the number of papers published by scientists. We are now in a position to calculate the probability density associated with these phenomena from first principles. The first principles are physically based and not all the webs of interest satisfy the conditions laid out above, but let us see where the present line of argument takes us.

The manner in which problems are commonly formulated in diffusion theory is to specify the initial distribution of the quantity of interest; here we continue to refer to that quantity as particles, and use the diffusion equation to determine the evolution of the web away from that initial state. An efficient way to solve the partial differential equation is to use Fourier transforms. The Fourier transform of the number density in one dimension is $\tilde{n}(k, t)$, so from (3.107) we obtain

$$\frac{\partial \tilde{n}(k, t)}{\partial t} = D \int_{-\infty}^{\infty} e^{ikq} \frac{\partial^2 n(q, t)}{\partial q^2} dq. \tag{3.111}$$

Assuming that the particle concentration and its gradient vanish asymptotically at all times, that is, $\partial n(q, t)/\partial q|_{q=\pm\infty} = n(q = \pm\infty, t) = 0$, we can integrate the equation by parts twice to obtain

$$\frac{\partial \tilde{n}(k, t)}{\partial t} = -Dk^2 \tilde{n}(k, t). \tag{3.112}$$

The solution to this equation in terms of the initial value of the particle concentration is

$$\tilde{n}(k, t) = \tilde{n}(k, 0)e^{-Dk^2t}. \tag{3.113}$$

Hence by inverting the Fourier transform we obtain

$$n(q, t) = \frac{1}{2\pi} \int_{-\infty}^{\infty} e^{-ikq} e^{-Dk^2t} dk \int_{-\infty}^{\infty} e^{iky} n(y, 0)dy. \tag{3.114}$$

A particularly important initial distribution has all the mass locally concentrated and is given by a delta-function weight at a point $q = q_0$,

$$n(q, t = 0) = \delta(q - q_0), \tag{3.115}$$

which, when substituted into (3.114), yields

$$n(q, t) = \frac{1}{2\pi} \int_{-\infty}^{\infty} e^{-ik(q-q_0)} e^{-Dk^2t} dk$$

$$= \frac{1}{\sqrt{4\pi Dt}} e^{-\frac{(q-q_0)^2}{4Dt}}$$

$$\equiv P(q - q_0; t). \tag{3.116}$$

The function $P(q - q_0; t)$ is the probability that a particle initially at q_0 diffuses to q at time t and is a Gaussian propagator with a variance that grows linearly in time. Here we have assumed that the mass distribution is normalized to unity, thereby allowing us to adopt a probabilistic interpretation. Consequently, we can write for the general solution to the one-dimensional diffusion equation

$$n(q, t) = \int_{-\infty}^{\infty} P(q - q_0; t)n(q_0, 0)dq_0, \tag{3.117}$$

which can be obtained formally by relabeling terms in (3.114).

On the other hand, we know that the Gaussian distribution can be written in normalized form as

$$P(q) = \frac{1}{\sqrt{2\pi \langle q^2 \rangle}} e^{-\frac{q^2}{2\langle q^2 \rangle}}.$$ (3.118)

However, multiplying (3.107) by q^2 and integrating the rhs by parts yields

$$\frac{d \langle q^2 \rangle}{dt} = 2D,$$ (3.119)

which integrates to

$$\langle q^2; t \rangle = 2Dt,$$ (3.120)

and therefore the Gaussian can be written

$$P(q, t) = \frac{1}{\sqrt{2\pi \langle q^2; t \rangle}} e^{-\frac{q^2}{2\langle q^2; t \rangle}},$$ (3.121)

with the second moment given by (3.120).

It is also interesting to note that (3.121) can be derived from the scaled form of the probability density

$$P(q, t) = \frac{1}{t^\delta} F\left(\frac{q}{t^\delta}\right)$$ (3.122)

by setting $\delta = 1/2$, in which case, with $y = q/\sqrt{t}$,

$$F(y) = \frac{1}{\sqrt{4\pi D}} e^{-\frac{y^2}{4D}}.$$ (3.123)

This is the property of scaling in the context of the probability density; in fact, when a stochastic variable is said to scale, this is what is meant. It is not literally true that $Q(\lambda t) = \lambda^H Q(t)$, but rather the scaling occurs from the probability density of the form (3.122) and the relation between the two scaling parameters δ and H depends on the functional form of $F(\cdot)$.

Thus, when the physical conditions for simple diffusion are satisfied we obtain the distribution of Gauss. From this we can infer that the complex web phenomena that do not have Gauss statistics do not satisfy these physical conditions and we must look elsewhere for unifying principles. So let us generalize our considerations beyond the simple diffusion equation and investigate the behavior of the probability densities for more general dynamical networks.

3.3.1   The Fokker–Planck equation (FPE)

The stochastic differential equations distinguish between the deterministic forces acting on a particle through a potential and the random forces produced by interactions with the environment. In the present discussion of a single-variable equation of motion the dynamical variable is denoted by $Q(t)$, the Langevin equation is

$$\frac{dQ(t)}{dt} = -\frac{\partial \Phi(Q)}{\partial Q} + \eta(t), \tag{3.124}$$

the associated phase-space variable is denoted q and the dynamics of the variable are given by the potential $U(q)$. The path from the Langevin equation to the phase-space equation for the probability density can be very long and technically complicated. It depends on whether the random fluctuations are assumed to be delta-correlated in time or to have memory; whether the statistics of the fluctuations are Gaussian or have a more exotic nature; and whether the change in the probability density is local in space and time or is dependent on a heterogeneous, non-isotropic phase space. Each condition further restricts the phase-space equation for zero-centered, delta-correlated-in-time fluctuations with Gaussian statistics

$$\langle \eta(t) \rangle = 0 \qquad \text{and} \qquad \langle \eta(t) \eta(t') \rangle = \frac{2D}{\lambda^2} \delta(t - t'), \tag{3.125}$$

with the final result being the form of the Fokker–Planck equation (FPE)

$$\frac{\partial P(q, t|q_0)}{\partial t} = \frac{\partial}{\partial q} \left\{ \frac{\partial \Phi(q)}{\partial q} + \frac{D}{\lambda^2} \frac{\partial}{\partial q} \right\} P(q, t|q_0). \tag{3.126}$$

The FPE describes the phase-space evolution of complex webs having individual trajectories described by (3.124) and an ensemble of realizations described by $P(q, t|q_0)$.

The steady-state solution to the FPE is independent of the initial condition,

$$\frac{\partial P_{ss}(q, t|q_0)}{\partial t} = \frac{\partial P_{ss}(q)}{\partial t} = 0, \tag{3.127}$$

and is given by the probability density

$$P_{ss}(q) = Z \exp \left[-\frac{\lambda U(q)}{D} \right] = Z \exp \left[-\frac{U(q)}{k_B T} \right], \tag{3.128}$$

where we have used the Einstein fluctuation–dissipation relation (FDR)

$$\frac{D}{\lambda} = k_B T \tag{3.129}$$

to express the strength of the fluctuations D in terms of the bath temperature T; Z is the partition function

$$Z = \int \exp \left[-\frac{U(q)}{k_B T} \right] dq \tag{3.130}$$

and the distribution has the canonical form given in equilibrium statistical mechanics.

The FPE can also be expressed as

$$\frac{\partial P(q, t|q_0)}{\partial t} = \mathcal{L}_{FP} P(q, t|q_0) \tag{3.131}$$

in terms of the Fokker–Planck operator

$$\mathcal{L}_{FP} \equiv \frac{\partial}{\partial q} \left\{ \frac{\partial \Phi(q)}{\partial q} + \frac{D}{\lambda^2} \frac{\partial}{\partial q} \right\}. \tag{3.132}$$

The formal solution to (3.131) can be obtained by factoring the solution as

$$P(q, t|q_0) = W(t)F(q, q_0)$$

to obtain the eigenvalue equation

$$\frac{1}{W}\frac{dW}{dt} = \frac{1}{F}\mathcal{L}_{\mathrm{FP}}F = -\gamma. \tag{3.133}$$

Upon indexing the eigenvalues in order of increasing magnitude $|\gamma_0| < |\gamma_1| < \ldots$ we obtain the solution as the expansion in eigenfunctions $\phi_j(q, q_0)$ of the Fokker–Planck operator

$$P(q, t|q_0) = \sum_j e^{-\gamma_j t}\phi_j(q, q_0). \tag{3.134}$$

The FPE has been the work horse of statistical physics for nearly a century and some excellent texts have been written describing the phenomena to which it can be applied. All the simple random walks subsequently discussed can equally well be expressed in terms of FPEs. However, when the web is complex, or the environment has memory, or when any of a number of other complications occur, the FPE must be generalized. Even in the simple case of a dichotomous process the statistics of the web's response can be non-trivial.

Multiplicative fluctuations

The above form of the FPE assumes that the underlying dynamical web is thermo-dynamically closed; that is, the fluctuations and dissipation have the same source and are therefore related by the FDR. An open stochastic web, in which the sources of the fluctuations and any dissipation in the web may be different, might have a dynamical equation of the form

$$\frac{dQ(t)}{dt} = G(Q) + g(Q)\xi(t), \tag{3.135}$$

where $G(\cdot)$ and $g(\cdot)$ are analytic functions of their arguments. If we again assume that the fluctuations have Gaussian statistics and are delta-correlated in time then the Stratonovitch version of the FPE is given by

$$\frac{\partial P(q, t|q_0)}{\partial t} = \frac{\partial}{\partial q}\left\{-G(q) + Dg(q)\frac{\partial}{\partial q}g(q)\right\}P(q, t|q_0). \tag{3.136}$$

There is a second probability calculus, due to Itô, which is completely equivalent to that of Stratonovitch, but we do not discuss that further here. Solving (3.136) in general is notoriously difficult, but it does have the appealing feature of yielding an exact steady-state solution.

The steady-state solution to (3.136) is obtained from the condition

$$\frac{\partial P_{\mathrm{ss}}(q)}{\partial t} = 0 = \frac{\partial}{\partial q}\left\{-G(q) + Dg(q)\frac{\partial}{\partial q}g(q)\right\}P_{\mathrm{ss}}(q), \tag{3.137}$$

for which the distribution is independent of time and the initial state of the web. Imposing a zero-flux boundary condition yields

$$-G(q)P_{ss}(q) + Dg(q)\frac{\partial}{\partial q}\left[g(q)P_{ss}(q)\right] = 0,$$

so that

$$\frac{1}{g(q)P_{ss}(q)}\frac{\partial}{\partial q}[g(q)P_{ss}(q)] = \frac{G(q)}{Dg(q)^2}. \tag{3.138}$$

Integrating (3.138) yields the exact steady-state probability density

$$P_{ss}(q) = \frac{Z}{g(q)}\exp\left[\frac{1}{D}\int\frac{G(q')dq'}{g(q')^2}\right], \tag{3.139}$$

where Z is the normalization constant. It will prove useful to have this solution available to us for later analysis of complex webs.

An example of multiplicative fluctuations is given by an open network whose connections are determined by a stochastic rate. Assume that the growth of a network, that is, the number of connections of a node to other nodes, is determined by a combination of a feedback that keeps the growth from becoming unstable and a preferential attachment having a random strength such that

$$\frac{dk(t)}{dt} = -\lambda k(t) + \xi(t)\,k(t). \tag{3.140}$$

Here $k(t)$ is the number of connections of the node of interest to other nodes in the web at time t. We assume that the fluctuating rate is zero-centered, delta-correlated in time and has Gaussian statistics. With these assumptions regarding the fluctuations we can construct a FPE of the form (3.136) with the functions

$$G(k) = -\lambda k \tag{3.141}$$

and

$$g(k) = k. \tag{3.142}$$

The steady-state solution to this FPE is given by (3.139) and substituting (3.141) and (3.142) into this expression yields the analytic solution

$$P_{ss}(k) = \frac{A}{k^\mu}, \tag{3.143}$$

with the power-law index, given in terms of the parameters of this example,

$$\mu = 1 + \frac{\lambda}{D}. \tag{3.144}$$

This solution and others from nonlinear multiplicative Langevin equations are discussed by Lindenberg and West [30] as well as the conditions for normalization for $k > 0$.

Note that the inverse power-law solution to the FPE, when suitably restricted, agrees with the preferential-attachment model of Barabási and Albert (BA) [5], which we discuss later, if we take the parameters to be in the ratio

$$\frac{\lambda}{D} = 2, \tag{3.145}$$

indicating that in a balanced steady state the feedback has twice the strength of the fluctuations. In the present discussion there is no need to adopt this particular ratio for the solution; any positive-definite ratio for $k > 0$ is sufficient. However, it is interesting to note that when the strength of the fluctuations becomes sufficiently large that they overwhelm the feedback control,

$$\frac{\lambda}{D} < 1, \tag{3.146}$$

the first moment

$$\langle k \rangle = \int_1^\infty k' P_{\mathrm{ss}}(k')dk' = \frac{A}{\mu - 2} \tag{3.147}$$

diverges and the condition

$$\mu < 2 \tag{3.148}$$

is realized. The condition for non-ergodicity (3.148) is realized when the fluctuations dominate dissipation in the multiplicative Langevin equation. This is another interesting alternative to the perspective of the BA model of network growth [5].

3.3.2   The drift-diffusion model (DDM) of decision making

Suppose that an individual is asked to make a decision between two alternatives with limited information and only one of which has a positive outcome. Such a situation is called a two-alternative forced-choice (TAFC) task by psychologists. One of the most often used models for deciding between two such choices is the drift-diffusion model (DDM), which is based on the stochastic differential equation of the previous sections. However, other important concepts enter into this model, such as the idea of optimality, meaning that a decision of a specified accuracy is made in the shortest possible time. The speed and accuracy of the decisions affect the cumulative reward obtained and collectively are called the TAFC paradigm in the psychology literature.

As discussed in the excellent review by Bogacz *et al.* [11], TAFC task models typically make three assumptions: (1) the evidence for each choice is accumulated over time; (2) the uncertainty in the evidence is modeled as random fluctuations in time; and (3) a decision is triggered when the accumulated evidence for one of the alternatives first exceeds a specified threshold. The accumulation of information for each choice can be aggregated in a number of ways, most of which are found to be equivalent [11], but it is often the difference in evidence that is integrated and it is this difference that triggers the choice of one alternative over the other. This integrated difference is modeled by reducing the number of variables to describe the dynamics from two, the evidence or information for each of the alternatives separately, to one, which is the difference between the two alternatives. It is often the case that this difference can be related to neuronal models of inhibition and excitation, as we subsequently discuss. Another model based on the difference in probability of making a given decision is discussed later.

The simplest form of the stochastic mathematical model in a decision-making context has individual accumulated data to support the two hypotheses and is given by (3.56). The integrated value of the evidence at time t is denoted by the dynamic variable $Q(t)$ and we assume that the initial state is unbiased so that $Q(0) = 0$. The DDM with unbiased evidence is given by the dynamical equation

$$\frac{dQ(t)}{dt} = C + \xi(t) \tag{3.149}$$

in a notation consistent with the physics literature. Here the small change in net evidence dQ given in a time increment dt is the constant $C\,dt$, which is the deterministic increase in evidence of the correct decision per unit time due to a constant drift. As Bogacz et al. explain, we have $C > 0$ if hypothesis H_+ is correct for the realization in question and $C < 0$ if hypothesis H_- is correct and the random-force term is usually chosen to be white noise as it is for physical processes. We emphasize here that the choice of statistical distribution for the random fluctuations is made for convenience, with little or no justification in terms of experimental evidence. In fact the evidence we discuss in subsequent chapters suggests that statistical distributions other than those usually used in the literature for the random force might be more appropriate.

The formal solution to the dynamical equation, when $Q(0)=0$, is given by

$$Q(t) = Ct + \int_0^t \xi(t')dt'. \tag{3.150}$$

This solution can be used to construct the characteristic function for the decision-making process. The characteristic function is given by the Fourier transform of the probability density $p(q, t)$, namely

$$\phi(k, t) = \mathcal{F}T[p(q, t); k] \equiv \int_{-\infty}^{\infty} e^{ikq} p(q, t)dq = \left\langle e^{ikQ(t)} \right\rangle_\xi. \tag{3.151}$$

Here it should be clear that we are evaluating the average value of the exponential function and for this we can use the solution to the Langevin equation and thereby obtain

$$\phi(k, t) = e^{ikCt} \left\langle e^{ik \int_0^t \xi(t')dt'} \right\rangle_\xi. \tag{3.152}$$

The average in (3.152) can be evaluated by expanding the exponential and using the assumed statistical properties of the random force to evaluate the individual moments in the series expansion. In particular, on treating the moments of a Gaussian random variable factor as

$$\left\langle x^{2n} \right\rangle = \left\langle x^2 \right\rangle^n$$

and using the delta-correlated character of the random force, the series can be resummed to yield for the characteristic function

$$\phi(k, t) = e^{ikCt - 2Dtk^2}. \tag{3.153}$$

The probability density for the decision-making process is then given by the inverse Fourier transform of the characteristic function

$$p(q, t) = \mathcal{F}\mathcal{T}^{-1}[\phi(k, t); q] \equiv \frac{1}{2\pi} \int_{-\infty}^{\infty} e^{-ikq} \phi(k, t) dk, \qquad (3.154)$$

so that inserting (3.153) into (3.154) and carrying out the indicated integration yields

$$p(q, t) = \frac{1}{\sqrt{4\pi Dt}} \exp\left[-\frac{(q - Ct)^2}{2Dt}\right]. \qquad (3.155)$$

It is clear from this form of the probability density that the average value of the dynamical variable is given by

$$\langle Q(t) \rangle = Ct, \qquad (3.156)$$

so the mode of the distribution moves linearly in the direction of the drift. The variance of the dynamic variable is given by

$$\sigma^2(t) \equiv \left\langle [Q(t) - \langle Q(t) \rangle]^2 \right\rangle = 2Dt, \qquad (3.157)$$

just as in classical diffusion.

Bogacz *et al.* [11] model the interrogation paradigm by asking whether, at the interrogation time T, the value of the dynamic variable lies above or below zero. If the hypothesis H_+ is appropriate, a correct decision occurs if $Q(T) > 0$ and an incorrect one if $Q(T) < 0$. The average error rate (ER) can therefore be determined from the probability that the solution $Q(T)$ lies below zero, so that, using the probability density (3.155), the average error rate is given by the expression

$$\text{ER} = \frac{1}{\sqrt{2\pi}} \int_{-\infty}^{-Ct/\sqrt{Dt}} e^{-u^2/2} du \qquad (3.158)$$

and the integral on the rhs is the error function

$$\text{erf}(x) \equiv \frac{1}{\sqrt{2\pi}} \int_{-\infty}^{x} e^{-u^2/2} du. \qquad (3.159)$$

In the DDM the decision is made when $Q(t)$ reaches one of two fixed thresholds, usually selected symmetrically at the levels $+z$ and $-z$. Figure 3.5 depicts some sample paths of the solution to the Langevin equation with a constant drift; these are separate realizations of the trajectory $Q(t)$. From Figure 3.5 it is evident that the short-time average of the trajectory does not necessarily move in the direction determined by the sign of the constant drift C. The fluctuations produce erratic motion, sometimes resulting in the trajectory crossing the wrong threshold, indicating how uncertainty can lead to the wrong decision. The mathematics required to formally determine the probability of crossing one of the two thresholds indicated by the dark horizontal lines for the first time comes under the heading of first-passage-time problems. Consequently, in order to determine the first-passage-time probability density we must develop the calculus for probabilities, which we do later. For the time being we examine how to model decision making with increasing levels of complexity.

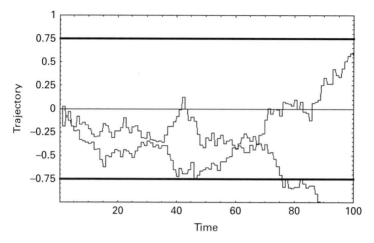

Figure 3.5. Two trajectories with the same initial condition are depicted. The random force is seen to drive one to the lower decision and the other apparently to the upper decision, both from the same initial state.

It should be pointed out that this kind of modeling has been used in neural science, and is called the integrate-and-fire model. In that application it concerns the firing of various neurons that contribute to the neuron of interest; when the voltage at the specified neuron exceeds a prescribed value that neuron, as well, fires and in turn contributes to the firing of neurons to which it is connected.

One way to generalize the DDM for decision making is by assuming that the drift is a random variable. In this situation the "constant" drift would not be a stochastic variable in time like the random force, but would vary randomly from one realization of the trajectory to another. It is often assumed that the constant drift is subject to the law of frequency of errors so that its variation from realization to realization would have normal statistics with a finite mean and variance.

Another generalization of the DDM is to introduce a control process that reduces the deviation from the equilibrium state. This is a Langevin equation with dissipation, called an Ornstein–Uhlenbeck (OU) process [43], which can be written as

$$\frac{dQ(t)}{dt} = C - \lambda Q(t) + \xi(t). \tag{3.160}$$

In physics the sign of the parameter λ is fixed because it is the physical dissipation in the web of interest and therefore is positive definite. In a communications context λ is a feedback control parameter and may be of either sign. In the absence of the random force the fixed point for (3.160) is given by

$$Q_f = \frac{C}{\lambda}, \tag{3.161}$$

where the velocity vanishes. The deviation of the trajectory from the fixed point can be determined from $Q = Q_f + \delta Q$ such that

$$\frac{d\,\delta Q(t)}{dt} = -\lambda\,\delta Q(t), \tag{3.162}$$

whose solution exponentially decays to zero for $\lambda > 0$, and consequently the fixed point is an attractor for positive λ. For $\lambda < 0$ the fixed point is unstable and the solution to (3.162) diverges to infinity.

The solution of the OU Langevin equation for $Q(0) = 0$ is given by

$$Q(t) = \frac{C}{\lambda}[1 - e^{-\lambda t}] + \int_0^t e^{-\lambda(t-t')}\xi(t')dt', \qquad (3.163)$$

yielding the average value of the accumulated evidence

$$\langle Q(t)\rangle_\xi = \frac{C}{\lambda}[1 - e^{-\lambda t}] \qquad (3.164)$$

with variance

$$\sigma^2(t) = \frac{D}{\lambda}[1 - e^{-2\lambda t}] \qquad (3.165)$$

the same as that given by (3.157). The variance remains unchanged in functional form because it is a measure of how much the trajectories diverge from one another over time. This divergence is influenced not only by the random force but also by the dissipation, so asymptotically the variance becomes constant. What is interesting is that at early times, $t \ll 1/\lambda$, the exponential in the average evidence (3.164) can be expanded to obtain

$$\lim_{t \to 0} \langle Q(t)\rangle_\xi = Ct \qquad (3.166)$$

in agreement with the constant-drift model. At the other extreme the exponential in the average evidence vanishes, giving rise to

$$\lim_{t \to \infty} \langle Q(t)\rangle_\xi = \frac{C}{\lambda}, \qquad (3.167)$$

resulting in a time-independent average. The dissipation or negative feedback therefore prevents the average evidence from reaching an unrealistically high value, thereby increasing the likelihood that an incorrect decision can be made under more realistic circumstances.

The probability density can be obtained for the OU process using the characteristic function as we did for the constant-drift case. In the OU case the characteristic function can be written

$$\phi(k, t) = e^{ik\langle Q(t)\rangle_\xi} \left\langle e^{ik\int_0^t e^{-\lambda(t-t')}\xi(t')dt'} \right\rangle_\xi \qquad (3.168)$$

and again the properties of white noise can be used to evaluate the average. The inverse Fourier transform of the characteristic function yields

$$p(q, t) = \frac{1}{\sqrt{4\pi\sigma^2(t)}} \exp\left[-\frac{(q - \langle Q(t)\rangle_\xi)^2}{2\sigma^2(t)} \right], \qquad (3.169)$$

where the variance is given by (3.165) and the mean is given by (3.164).

It is worth reviewing that the sign of the control parameter λ determines whether the trajectory is attracted towards the fixed point as it would be for negative feedback

($\lambda > 0$) or repelled away from the fixed point as it would be for positive feedback ($\lambda < 0$). Bogacz *et al.* [11] interpret this behavior in terms of decision making as follows:

> In the stable case, all solutions approach and tend to remain near the fixed point, which lies nearer the correct threshold, so they typically slow down before crossing it, corresponding to conservative behavior. The unstable case corresponds to riskier behavior. Solutions on the *correct* side of the fixed point accelerate toward the correct threshold, giving faster responses, but solutions on the *incorrect* side accelerate toward the incorrect threshold, possibly producing more errors.

From this interpretation of the trajectories it is possible to propose that rewards for correct responses should increase the magnitude of the control parameter, whereas punishments for errors should decrease it, as suggested by Busemeyer and Townsend [12]. These authors noted the decay effect of the control parameter for $\lambda > 0$, incorporating the fact that earlier input does not influence the trajectory (accumulated evidence) as much as does later input. By the same token, there is a primacy effect of the control parameter for $\lambda < 0$, namely that the earlier input influences the trajectory more than later inputs. This earlier influence is manifest in the divergence of the trajectory.

3.3.3   First-passage-time distributions

Consider a web characterized by a continuous state variable q and assume that it is initially in the state q_0. There exist numerous situations such that when a specific state q_1 is achieved for the first time the web changes its nature or perhaps activates a device to change the character of another web that is controlled by the first. An example would be the climate-control system in your home that has two temperature settings. When the temperature falls below a certain temperature T_l the heating system is turned on, warming the rooms, and when the temperature reaches T_l from below the heating system is turned off. If the temperature continues to climb to the level T_u then the air conditioning comes on, cooling the rooms, and when the temperature falls to T_u from above the air conditioning is turned off. In the statistics literature the distribution that describes this behavior is called the first-passage time. The earliest work on the subject dates back to Daniel Bernoulli (1700–1782) in his discussion of the gambler's ruin. Ruin was identified as the consequence of the vanishing of the gambler's fortune "for the first time" in a succession of gambling trials.

Much of our previous discussion was concerned with determining the statistical properties of the stochastic variable $Q(t)$. Now we are concerned with a more restricted question associated with when the dynamical variable takes on certain values for the first time. The DDM model just discussed is of this form. The dynamical variable was the accumulated evidence for making a decision, either positive or negative, and had the form of determining when a diffusive trajectory crossed symmetric barriers. Consider the case when the variable of interest is in the interval $q_1 < q < q_2$ and we consider a class of trajectories connecting q_1 with q_2 that must proceed by going through the point q; that is, the process cannot jump over any point. This transition can be decomposed into two independent stages, namely the first taking the trajectory from q_1 to the intermediate point q for the first time in the interval τ, and the second taking the trajectory from q to q_2 in the time remaining $(t - \tau)$, not necessarily for the first time. Then we can write the integral equation for the pdfs

$$P(q_2 - q_1, t) = \int_0^t P(q_2 - q, t - \tau)W(q - q_1, \tau)d\tau \qquad (3.170)$$

and the function $W(q, \tau)$ is the first-passage-time pdf for the transition $0 \to q$ in time τ.

The convolution form of (3.170) allows use of Laplace transforms to obtain the product form

$$\widehat{P}(q_2 - q_1, u) = \widehat{P}(q_2 - q, u)\widehat{W}(q - q_1, u), \qquad (3.171)$$

so that the Laplace transform of the first-passage-time pdf is

$$\widehat{W}(q - q_1, u) = \frac{\widehat{P}(q_2 - q_1, u)}{\widehat{P}(q_2 - q, u)}. \qquad (3.172)$$

What is now necessary is to take the inverse Laplace transform of the ratio of the two point pdfs.

As an example of the application of this general formula let $P(q, t)$ be the solution to the FPE for a diffusive process, that is, the unbiased Gaussian transition probability

$$P(q, t) = \frac{1}{\sqrt{4\pi Dt}} \exp\left[-\frac{q^2}{4Dt}\right] = \frac{1}{2\pi} \int_{-\infty}^{\infty} e^{-ikq} e^{-Dtk^2} dk. \qquad (3.173)$$

The next step is to take the Laplace transform of this pdf to obtain

$$\widehat{P}(q, u) = \frac{1}{2\pi} \int_{-\infty}^{\infty} e^{-ikq} (u + Dk^2)^{-1} dk$$

$$= \frac{1}{2\sqrt{Du}} \exp\left[-|q| \left|\frac{u}{D}\right|^{1/2}\right]. \qquad (3.174)$$

Finally, by taking into account the ordering of the variables in (3.172) we can write

$$\widehat{W}(q - q_1, u) = \exp\left[-(q - q_1) \left|\frac{u}{D}\right|^{1/2}\right], \qquad (3.175)$$

so that the inverse Laplace transform of (3.175) yields

$$W(q - q_1, t) = \frac{q - q_1}{\sqrt{4\pi D}} \frac{1}{t^{3/2}} \exp\left[-\frac{(q - q_1)^2}{4Dt}\right], \quad \text{for } t > 0. \qquad (3.176)$$

Note that this pdf does not have a finite first moment, as can be shown using

$$\langle t \rangle = \int_0^{\infty} t W(q, t) dt = -\lim_{u \to 0} \frac{\partial}{\partial u} \int_0^{\infty} e^{-ut} W(q, t) dt$$

$$= -\lim_{u \to 0} \frac{\partial \widehat{W}(q - q_1, u)}{\partial u}, \qquad (3.177)$$

so that, upon inserting (3.175) into (3.177), the first-passage-time moment diverges as $1/\sqrt{u}$ as $u \to 0$.

An alternative procedure can be developed for determining the first-passage-time pdf through the application of the theory of a stochastic process on a line bounded on one side by an absorbing barrier. Assuming that the absorbing barrier is located at the point q_a, the probability that a particle survives until time t without being absorbed is

$$F(q_a, t) = \int_{-\infty}^{q_a} \mathcal{P}(q, t) dq. \tag{3.178}$$

The quantity $\mathcal{P}(q, t)dq$ is the probability of the particle being in the interval $(q, q + dq)$ at time t in the presence of the absorbing barrier. The probability of its being absorbed in the time interval $(t, t + \delta t)$ is

$$F(q_a, t) - F(q_a, t + \delta t) = -\frac{dF(q_a, t)}{dt} \delta t \tag{3.179}$$

and consequently the first-passage-time pdf can be expressed as

$$W(q_a, t) = -\frac{dF(q_a, t)}{dt} = -\frac{d}{dt} \int_{-\infty}^{q_a} \mathcal{P}(q, t) dq. \tag{3.180}$$

It is a straightforward matter to rederive (3.176) starting from (3.180) using the unbiased Gaussian distribution.

The first-passage-time pdf for a homogeneous diffusion process can be determined using the Kolmogorov backward equation [16]

$$\frac{\partial P(q, t; q_0)}{\partial t} = a(q_0) \frac{\partial P(q, t; q_0)}{\partial q_0} + b(q_0) \frac{\partial^2 P(q, t; q_0)}{\partial q_0^2}, \tag{3.181}$$

where the derivatives are being taken with respect to the initial condition $Q(0) = q_0$. The Laplace transform of this equation with $q_a > q > q_0$ gives

$$u \widehat{P}(q, u; q_0) = a(q_0) \frac{\partial \widehat{P}(q, u; q_0)}{\partial q_0} + b(q_0) \frac{\partial^2 \widehat{P}(q, u; q_0)}{\partial q_0^2}, \tag{3.182}$$

so that using the ratio of probabilities in (3.172) yields the following second-order differential equation for the Laplace transform of the first-passage-time distribution:

$$b(q_0) \frac{\partial^2 \widehat{W}(q, u; q_0)}{\partial q_0^2} + a(q_0) \frac{\partial \widehat{W}(q, u; q_0)}{\partial q_0} - u \widehat{W}(q, u; q_0) = 0. \tag{3.183}$$

Feller [16] proved that the condition (3.176) and the requirement that

$$\widehat{W}(q_a, u; q_0) < \infty \, \forall q_0 \tag{3.184}$$

together determine that the solution to (3.183) is unique.

3.4   More on statistics

The increase in complexity of a web can be achieved by extending the Hamiltonian to include nonlinear interactions, as we do later. Another way to increase complexity is to put the web in contact with the environment, which in the simplest physical case provides interrelated fluctuations and dissipation. In the motion of a Brownian particle the force fluctuations were found to have Gaussian statistics and to be delta-correlated in time. The response of the Brownian particle at equilibrium to these fluctuations is determined by the canonical distribution. However, the statistics of the heat bath need not be either Gaussian or delta-correlated in character, particularly for non-physical networks. So another way complexity can enter network models is through modification of the properties of the fluctuating force in the Langevin equation.

We advocate an approach to complexity that is based on a phase-transition perspective. Complexity emerges from the interaction among many elements located at the nodes of a web, and the complex structure of the web makes it possible to realize this phase transition without using the frequently made approximation that every node of the network interacts with all the other nodes (all-to-all coupling). We return to this point in Chapter 5. Nevertheless, the web will not leave a given state for all time, but undergoes abrupt transitions from one state to another. These transitions may be visible or they may be hidden by the emergence of macroscopic coherence. In order to establish a satisfactory command of this complexity condition it is necessary to become familiar with the notion of renewal and non-renewal statistics.

3.4.1 Poisson statistics

Let us consider the simplest statistical process that can be generated by dynamics, namely an ordinary Poisson process. Following Akin *et al.* [2], we use a dynamical model based on a particle moving in the positive direction along the Q-axis and confined to the unit interval $I \equiv [0, 1]$. The equation of motion is chosen to be the rate equation

$$\frac{dQ(t)}{dt} = \gamma Q(t),$$
(3.185)

where $\gamma \ll 1$. Statistics are inserted into the deterministic equation (3.185) through the boundary condition; whenever the particle reaches the border $Q = 1$, it is injected back into the unit interval I to a random position having uniform probability. Throughout this discussion we refer to this back injection as an event, an event that disconnects what happens in one sojourn on the interval from what happens during another.

With this assumption of randomness for back injection the initial value for the solution $Q(t)$ after an event is a random variable

$$Q(0) = \xi$$
(3.186)

with ξ standing for a random number in the unit interval I. Once the initial time has been set, the sojourn time of the particle in the unit interval is determined by the solution to (3.185),

$$Q(t) = Q(0)e^{\gamma t} = \xi e^{\gamma t}.$$
(3.187)

We denote by t the total time it takes for the particle to reach the border $Q = 1$ from a random initial point ξ. Using (3.187) and setting $Q(\tau) = 1$, we obtain

$$\xi = e^{-\gamma \tau},$$
(3.188)

so that the length of time spent within the unit interval is

$$\tau = -\frac{1}{\gamma} \ln \xi, \tag{3.189}$$

and, since ξ is a random variable, the sojourn time τ is also a random variable. The waiting- or sojourn-time distribution function $\psi(\tau)$ is consequently determined by

$$\psi(\tau)d\tau = p(\xi)d\xi \tag{3.190}$$

and by assumption the random distribution of initial conditions determined by back injection is uniform, so that $p(\xi) = 1$. Thus, (3.190) can be solved to determine the waiting-time distribution

$$\psi(\tau) = \left|\frac{d\xi}{d\tau}\right| = \gamma e^{-\gamma \tau}, \tag{3.191}$$

which is an exponential probability distribution.

It is worth spending a little time discussing the properties of Poisson processes since they are used so often in modeling statistical phenomena in the physical, social and life sciences. Moreover, it is by modifying these properties that complexity is made manifest in a variety of the webs we discuss. Consider the notion of renewal defined by the requirement that the variables $\{\tau_1, \tau_2, \ldots\}$ are independent identically distributed (iid) random variables, all with the same probability density function $\psi(\tau)$. The characteristic function corresponding to the occurrence of n events in the time interval $(0, t)$ is given by the product of the probabilities of n independent events,

$$\hat{\psi}_n(u) = \langle \exp[-u(\tau_1 + \tau_2 + \cdots + \tau_n)] \rangle_\tau, \tag{3.192}$$

where the brackets with a subscript τ denote an average over the distribution times for the n variables. Note that this technique is used to generate the time series $t_1 = \tau_1, t_2 = \tau_1 + \tau_2$ and so on. The iid statistical assumption implies that the probability density factors into a product of n identical terms, and from (3.192) we obtain

$$\hat{\psi}_n(u) = \prod_{k=1}^{n} \langle e^{-u\tau_k} \rangle_\tau = \left[\hat{\psi}(u)\right]^n. \tag{3.193}$$

Assume that the statistics for a single event is exponential so that the Laplace transform of the waiting-time distribution given by (3.191) is

$$\hat{\psi}(u) = \frac{\gamma}{\gamma + u}. \tag{3.194}$$

Inserting this into (3.193) and taking the inverse Laplace transform yields

$$\psi_n(t) = \frac{\gamma(\gamma t)^{n-1}}{\Gamma(n)} e^{-\gamma t}. \tag{3.195}$$

In the renewal-theory [14] and queueing-theory [20] literature (3.195) goes by the name of the Erlang distribution. Note that it is *not* the Poisson distribution.

Consider the occurrence of events in our dynamical equation to be a counting process $\{N(t), t \geq 0\}$, where $N(t)$ is the total number of events up to and including time t, with $N(0) = 0$. The probability that exactly n events occur in a time interval t can be determined from

$$p_n(t) = \Pr\{N(t) = n\} = F_n(t) - F_{n-1}(t),$$ (3.196)

where the functions on the rhs of (3.196) are the probabilities of the sum of $(n + 1)$ events at times exceeding t, given by

$$F_n(t) = \Pr\{\tau_1 + \tau_2 + \cdots + \tau_{n+1} > t\} = \int_t^\infty \psi_{n+1}(t')\,dt'.$$ (3.197)

Inserting the waiting-time distribution function (3.195) into the integral in (3.197) allows us to do a straightforward integration and obtain

$$F_n(t) = \sum_{k=1}^n \frac{(\gamma t)^k}{k!} e^{-\gamma t},$$ (3.198)

which is the cumulative distribution for a Poisson process. Consequently, inserting (3.198) into (3.196) and taking the indicated difference yields the Poisson distribution

$$p_n(t) = \frac{(\gamma t)^n}{n!} e^{-\gamma t}.$$ (3.199)

Thus, an exponential distribution of waiting times implies a Poisson distribution for the number of events occurring in a given time interval, indicating that the statistics of the time intervals and the statistics of the number of time intervals are not the same, but are related.

3.4.2 Statistics with memory

The dynamical process generated using (3.185) is renewal because the initial condition ξ of (3.186) is completely independent of earlier initial conditions. Consequently the sojourn time τ of (3.189) is independent of earlier sojourn times as well, the time series $\{\tau_j\}$ has no memory and the order of the times τ_j adopted to produce the memoryless time series $\{t_j\}$ is not important. So now let us introduce some complexity into this renewal process by perturbing the rate in the dynamical generator (3.185) to obtain

$$\frac{dQ(t)}{dt} = \gamma(t)Q(t).$$ (3.200)

The solution to (3.200) after a time t' is given by

$$Q(t + t') = \xi \exp\left[\int_{t'}^{t+t'} \gamma(t'')dt''\right]$$ (3.201)

and, using the boundary condition $Q = 1$, (3.188) is replaced with

$$\xi = \exp\left[-\int_{t'}^{t+t'} \gamma(t'')dt''\right].$$ (3.202)

Upon implementing the uniform-back-injection assumption again we obtain the conditional waiting-time distribution density

$$\psi(t \mid t') = \gamma(t + t') \exp\left[-\int_{t'}^{t+t'} \gamma(t'')dt''\right], \qquad (3.203)$$

from which it is clear that the memory in the waiting-time statistics is determined by the time dependence of the perturbation. The function $\psi(t|t')$ is the conditional probability that an event occurs at time $t = t' + \tau$, given the fact that the sojourn in the interval I began at time t'. On the other hand, $t' = \tau_1 + \tau_2 + \cdots + \tau_n$. Thus, the event that occurred at time t' depends on an earlier event that occurred at time $\tau_1 + \tau_2 + \cdots + \tau_{n-1}$, and so on, so that by following this argument back the dependence on the initial time $t = 0$ is established. This buried dependence on the initial state of the system is the sense in which we mean that the perturbation induces memory into the stochastic process. We discuss this more completely in subsequent sections.

The new waiting-time distribution function (3.203) is no longer exponential in time and consequently the underlying process is no longer Poisson. The precise non-Poisson form of the process depends on the nature of the time dependence of the perturbation. The time dependence of the rate in the dynamical generator (3.200) has been intentionally left arbitrary and we explore certain choices for the perturbation later. There are various ways in which perturbations of the rate can be realized: (1) by externally exciting the web with a time-varying force; (2) by introducing nonlinear interactions into the web dynamics; (3) by nonlinearly coupling the web to the environment; and (4) by coupling the web to a second nonlinear dynamic web, to name a few.

Herein we use the non-Poisson nature of the statistical process as a working definition of one type of complexity. We know full well that this definition is not acceptable to all investigators, but we take solace in the fact that the properties of non-Poisson statistics are sufficiently rich that such distributions can be used to explain the characteristics of a large number of complex phenomena in the physical, social and life sciences.

3.4.3   Inverse power-law correlations

The covariance function for non-stationary time series is given by (3.77), which is interesting as a mathematical expression but does not provide much guidance regarding how to interpret the underlying data set. Suppose we have N discrete measurements given by $\{Q_j\}$, $j = 1, 2, \ldots, N$; how might we go about determining whether there is a pattern in the data? One measure is the correlation coefficient defined by

$$r_n = \frac{\mathrm{Cov}_n \, Q}{\mathrm{Var} \, Q} = \frac{[1/(N-n)] \sum_{j=1}^{N-n} (Q_j - \overline{Q})(Q_{j+n} - \overline{Q})}{(1/N) \sum_{j=1}^{N-n} (Q_j - \overline{Q})^2}, \qquad (3.204)$$

the ratio of the covariance between elements in the discrete time series separated by n nearest neighbors and the variance of the time series. This is a standard definition applied to raw data and the correlation coefficient is in the interval $-1 \leq r_n \leq 1$ for every n.

We want to construct a simple analytic form for (3.204) that captures any long-time correlation pattern in the data. One way to determine the correlations in the time series, aside from the brute-force method of directly evaluating (3.204) is by aggregating the data into groups of increasing size. In this way the scaling behavior changes as a function of the number of data points we aggregate and it is this dependence we seek to exploit. Segment the sequence of data points into adjacent groups of n elements each so that the n nearest neighbors of the jth data point aggregate to form

$$Q_j^{(n)} = Q_{nj} + Q_{nj-1} + \cdots + Q_{nj-n-1}. \tag{3.205}$$

The average over the aggregated sets yields

$$\overline{Q^{(n)}} = \frac{1}{\left[\frac{N}{n}\right]} \sum_{j=1}^{\left[\frac{N}{n}\right]} Q_j^{(n)} = n\overline{Q}. \tag{3.206}$$

We have used the fact that the sum yields $N\overline{Q}$ and the brackets $\left[\frac{N}{n}\right]$ denote the closest integer determined by N/n. The ratio of the standard deviation to the mean value is denoted as the relative dispersion and is given by [4]

$$R^{(n)} = \frac{\sqrt{\text{Var } Q^{(n)}}}{\overline{Q^{(n)}}} = \frac{\sqrt{\text{Var } Q^{(n)}}}{n\overline{Q}}. \tag{3.207}$$

If the time series scales the relative dispersion satisfies the relation

$$R^{(n)} = n^{H-1} R^{(1)}, \tag{3.208}$$

which results from the standard deviation scaling as n^H and the mean increasing linearly with lag time. The variance of the coarse-grained data set is determined in analogy to the nearest-neighbor averaging to be

$$\text{Var } Q^{(n)} = \frac{1}{\left[\frac{N}{n}\right]} \sum_{j=1}^{\left[\frac{N}{n}\right]} \left(Q_j^{(n)} - \overline{Q_j^{(n)}} \right)^2$$

$$= n \text{ Var } Q + 2 \sum_{j=1}^{n-1} (n-j) \text{Cov}_n Q. \tag{3.209}$$

Substituting (3.208) for the relative dispersion into (3.209) allows us, after a little algebra, to write

$$n^{2H-1} = n + 2 \sum_{j=1}^{n-1} (n-j) r_n. \tag{3.210}$$

Equation (3.210) provides a recursion relation for the correlation coefficients from which we construct the hierarchy of relations

$$r_1 = (2^{2H} - 2)/2,$$

$$2r_1 + r_2 = (3^{2H} - 2)/2,$$

$$\vdots$$

$$nr_1 + (n-1)r_2 + \cdots + 2r_{n-1} + r_n = [(n+1)^{2H} - (n+1)]/2. \tag{3.211}$$

If we select the two equations from (3.211) with $n = m$ and $n = m - 2$ and subtract from their sum twice the expression with $n = m - 1$ we obtain

$$r_m = \frac{1}{2} \left[(m+1)^{2H} - 2m^{2H} + (m-1)^{2H} \right], \tag{3.212}$$

which is valid for all $m \geq 1$. Consider the case of large m so that we can Taylor expand the correlation coefficient

$$r_m = \frac{m^{2H}}{2} \left[\left(1 + \frac{1}{m}\right)^{2H} - 2 + \left(1 - \frac{1}{m}\right)^{2H} \right]$$

and, by combining terms, obtain to lowest order in inverse powers of the lag time [4]

$$r_m \approx H(2H - 1)m^{2H-2} \tag{3.213}$$

for $0 \leq H \leq 1$.

This correlation coefficient decays very slowly with lag time, much more slowly than does an exponential correlation, for example. In Figure 3.6 this correlation coefficient is graphed as a function of lag time for various values of the scaling exponent. We can see from Figure 3.6 that scaling time series with H less than 0.7 have values of the correlation coefficient for $m > 2$ that are so low, less than 0.2, that it is easy to understand why fractal signals would historically be confused with uncorrelated noise, even though uncorrelated noise is nearly two orders of magnitude smaller in amplitude.

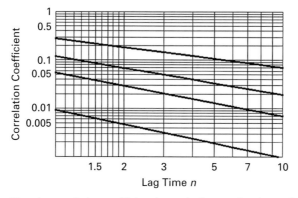

Figure 3.6.   Here the correlation coefficient is graphed versus lag time n for $H = 0.509, 0.55, 0.60$ and 0.70 from the bottom up.

The spectrum corresponding to the correlation coefficient (3.213) can be determined from a Tauberian theorem to be

$$S(\omega) \propto \frac{1}{\omega^{2H-1}}, \tag{3.214}$$

where it is clear that for $H < 1/2$ the spectrum increases with frequency. Are there natural complex webs that behave in this way? The answer to this question is yes. In fact, one of the first complex webs to be systematically studied was the beating of the human heart and the statistics of the change in the time intervals between beats were found to behave in this way.

Heart-rate time series are constructed by recording the time interval between adjacent beats as data, for example, let $f(n)$ be the interval between beats n and $n + 1$, measured as the distance between maxima on an electrocardiogram. In direct contradiction to the notion of "normal sinus rhythm" the interval between beats of the human heart has a great deal of variability, as shown in Figure 3.7. The mechanism for this variability apparently arises from a number of sources as discussed in Chapter 2.

The power spectrum, the square of the Fourier transform of the increments of the time intervals $I(n) = f(n + 1) - f(n)$, was determined by Peng *et al.* [38] to be given

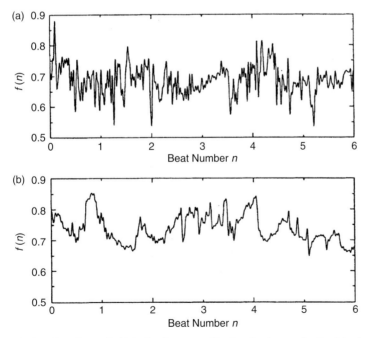

Figure 3.7. The interbeat interval $f(n)$ after low-pass filtering the time series for (a) a healthy subject and (b) a patient with severe cardiac disease (dilated cardiomyopathy). The healthy-heartbeat time series shows more complex fluctuations than those in the diseased case [38]. Reproduced with permission.

by (3.214). They observed for data sets of patients with severe heart failure that the spectrum is essentially constant in the low-frequency regime, indicating that the $I(n)$ are not correlated over long times; see Figure 3.8. The interpretation of this conclusion in terms of random walks will be taken up in later chapters. For the time being we note that for healthy individuals the data set yields $2H - 1 = -0.93$ so that $H \approx 0.04$, indicating a long-time correlation in the interbeat interval differences. On comparing this with classical diffusion we see that the time exponent is nearly zero. Peng *et al.* use this result to point out that $0 < H \le 1/2$ implies that the interbeat intervals are anticorrelated, which is consistent with a nonlinear feedback web that "kicks" the heart rate away from extremes. This tendency operates statistically on a wide range of time scales, not on a beat-to-beat basis [48].

Note that the naturally evolved control mechanism for the heart is one that imposes the inverse power-law on the long-time memory, anticipating when the next long delay is going to occur and suppressing it. Consequently, this control web operates to reduce the influence of extreme excursions and narrows the otherwise even broader range of

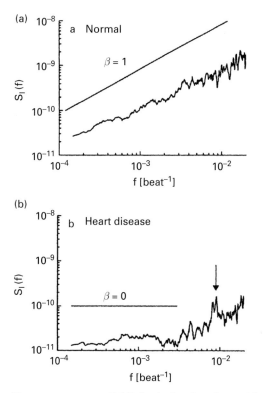

Figure 3.8.   The power spectrum $S_I(f)$ for the interbeat interval increment sequences over approximately 24 hours for the same subjects as in Figure 3.7. (a) The best-line fit for the healthy adult has a slope of $\beta = 0.93$. (b) The best-line fit for the patient with severe heart failure has an overall slope of 0.14 [38]. Reproduced with permission.

cardiac beat intervals. Here again we have a competition between the extreme variability of the heart rate and the long-time correlation necessary to keep the longest excursions, those that could produce death, from occurring. Consequently, even though the control cannot predict when the largest excursions will occur, it acts so as to statistically suppress them, thereby anticipating the improbably long intervals in a healthy heart.

3.5 Moving from Hamiltonians to events

In this section we offer some general advice on how to use events to describe web dynamics as an alternative to the continuous Hamiltonian prescription. In the literature on renewal processes such events correspond to failure processes. The renewal condition is established imagining that immediately after a failure a machine is brought back to the condition of being brand new; it is reconstructed. Thus, an event occurrence can be interpreted also as an instantaneous rejuvenation. Note that the dynamical description of such a renewal process would be strongly nonlinear. The event description has the nonlinear dynamical structure built into it.

One area of investigation in which it is important to characterize the statistical properties of things is operations research, which is a branch of engineering that quantifies the behavior of everyday activities such as decision making. Suppose that you have a car and are considering replacing it with a new one; this is not a decision to make lightly. It is necessary to balance the cost of maintenance, the number of remaining payments and the reliability of the present car against those properties of the replacement vehicle. There are well-established procedures for doing such calculations that often involve probabilities, since we do not have complete information about the future and we have a need to quantify the properties of interest. One such property is the probability that the car will just fail tomorrow or the day after. So let us examine the probability of failure, or its complement, the survival probability of your car.

3.5.1 Age-specific failure rates

According to the renewal-theory literature [14] the *age-specific failure rate* $g(t)$ is defined by

$$g(t) = \lim_{\Delta t \to 0} \frac{\Pr(t < \tau \leq t + \Delta t | t < \tau)}{\Delta t}, \tag{3.215}$$

where the numerator is the probability that the time of a failure τ lies in the interval $(t, t + \Delta t)$ conditional on no failure occurring before time t. Thus, using the calculus of conditional probabilities, we factor the joint probability of two events A and B occurring as the product of the probability of event A occurring conditional on event B occurring and the probability of event B occurring:

$$Pr(A \cup B) = Pr(A|B)Pr(B). \tag{3.216}$$

Now, if we identify A with a failure in the specified interval and B with no failure up to time t, we have

$$Pr(A \cup B) = Pr(t < \tau \le t + \Delta t) = \psi(\tau)\Delta t. \tag{3.217}$$

Here $\psi(\tau)$ is the probability distribution density for the first failure to occur in the time interval $(t, t + \Delta t)$. We also introduce the survival probability $\Psi(t)$ through the integral

$$\Psi(t) = \int_t^\infty \psi(\tau)d\tau, \tag{3.218}$$

so that the probability that a failure has not occurred up to time t is given by

$$Pr(B) = Pr(t < \tau) = \Psi(t). \tag{3.219}$$

Consequently the age-specific failure rate (3.215) is given by the ratio

$$g(t) = \frac{\psi(t)}{\Psi(t)}. \tag{3.220}$$

As a consequence of the integral relation (3.218) we can also write

$$\psi(t) = -\frac{d\Psi(t)}{dt}, \tag{3.221}$$

which, when inserted into (3.220), yields the equation for the rate

$$g(t) = -\frac{d \ln[\Psi(t)]}{dt}, $$

which is integrated to yield the exponential survival probability

$$\Psi(t) = \exp\left[-\int_0^t g(\tau)d\tau\right]. \tag{3.222}$$

Note that the probability density $\psi(t)$ is properly normalized,

$$\int_0^\infty \psi(\tau)d\tau = 1, \tag{3.223}$$

since it is certain that a failure will occur somewhere between the time extremes of zero and infinity. The normalization condition is consistent with the survival probability having the value

$$\Psi(t = 0) = 1; \tag{3.224}$$

that is, no failures occur at time $t = 0$.

3.5.2   Possible forms of failure rates

The simplest possible choice for the age-specific failure rate is a time-independent constant

$$g(t) = \gamma. \tag{3.225}$$

A constant failure rate implies that the rate of failure does not depend on the age of the process and consequently the survival probability is the simple exponential

$$\Psi(t) = e^{-\gamma t} \tag{3.226}$$

and, from the derivative relation, the probability density is

$$\psi(t) = \gamma e^{-\gamma t}. \tag{3.227}$$

Here we see that the exponential probability density implies that the rate of failure for, say, a machine is the same for each time interval. One can also conclude from this argument that the number of failures in a given time interval has Poisson statistics.

It should be reemphasized that the concept of an age-specific failure rate has been developed having in mind engineering problems [9] and maintenance [19]. The concept is not limited to this class of applications but can also be applied to living networks, including human beings, under the name of mortality risk [18]. One of the more popular formulas for the mortality risk was proposed by Gompertz [17]:

$$g(t) = Ae^{\alpha t}. \tag{3.228}$$

The exponential risk generates a survival probability that may look strange,

$$\Psi(t) = \exp\left[\frac{A}{\alpha}(1 - e^{\alpha t})\right], \tag{3.229}$$

so that the rate of decrease of the survival probability is doubly exponentially fast. This peculiar formula captures the rise in mortality, the loss of survivability, in a great variety of species [17].

Another well-known proposal for the rate was given by Weibull [10]:

$$g(t) = \alpha\beta t^{\beta-1} \tag{3.230}$$

with $\alpha, \beta > 0$. In this case of an algebraic age-specific failure rate we obtain

$$\Psi(t) = \exp[-\alpha t^\beta], \tag{3.231}$$

so that the Weibull probability density is

$$\psi(t) = \alpha\beta t^{\beta-1} \exp[-\alpha t^\beta]. \tag{3.232}$$

It is important to notice that, for $\beta < 1$, the mortality risk decreases rather than increases as a function of time. This may be a proper description of infant mortality risk [10]. There are cases in which $g(t)$ may decrease at short times and then increase at long times.

In the materials-science literature the Weibull distribution characterizes the influence of cracks and imperfections in a piece of material on the overall response to stress and the ultimate failure of the material to the applied stress. Up to this point the distributions have been phenomenological in that the failure rates have been chosen to accurately describe a given data set. Subsequently we show a dynamical basis for the Weibull distribution as well as others.

It is worth studying the general case of the rate

$$g(t) = r_0[1 + r_1 t]^\eta \tag{3.233}$$

in terms of three parameters. In the long-time limit $t \gg 1/r_1$ the general rate becomes equivalent to the mortality risk (3.230) with

$$\eta = \beta - 1. \tag{3.234}$$

By expanding the general form of the rate (3.233) in a Taylor series,

$$g(t) = r_0(r_1 t)^\eta \left[1 + \frac{1}{r_1 t} \right]^\eta \approx r_0(r_1 t)^\eta,$$

and comparing it with (3.230), we obtain for the lumped parameter

$$\alpha = \frac{r_0 r_1^\eta}{\eta + 1}. \tag{3.235}$$

The border between where the Weibull distribution has a maximum at $t > 0$ (for $\eta > 0$) and the region where it diverges at the origin is at $\eta = 0$. The emergence of a maximum at $t > 0$ is clearly depicted in Figure 3.9, which illustrates some cases with $\beta > 1$. The adoption of (3.233) serves the purpose of turning this divergent behavior into a distribution density with a finite value at $t = 0$:

$$\Psi(t) = \exp\left[-\frac{r_0}{r_1} \frac{1}{1 + \eta} \left([1 + r_1 t]^{1+\eta} - 1 \right) \right]. \tag{3.236}$$

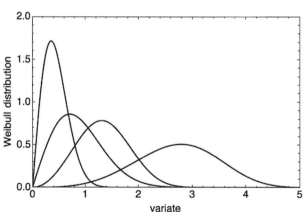

Figure 3.9. The four curves depicted have values of $\beta = 2$ and $\lambda = 1/\alpha^{1/\beta}$, where the lumped parameter is given by (3.235) and from left to right λ is 0.5, 1.0, 1.5 and 3.0, respectively.

There is a singularity in (3.236) at $\eta = -1$ that is vitiated by direct integration of (3.222) to yield

$$\Psi(t) = \exp\left[-\frac{r_0}{r_1}\ln(1 + r_1 t)\right]$$

$$= \exp\left[-\frac{r_0}{r_1}\ln\left(\frac{1}{r_1} + t\right) + \frac{r_0}{r_1}\ln(r_1)\right],$$

which after some algebra simplifies to the hyperbolic distribution

$$\Psi(t) = \left(\frac{T}{T+t}\right)^{\mu-1}. \tag{3.237}$$

The new parameters are expressed in terms of the old as

$$\mu = 1 + \frac{r_0}{r_1} \quad \text{and} \quad T = \frac{1}{r_1}. \tag{3.238}$$

The corresponding distribution density, $\psi(t)$, is easily obtained from the derivative to be given by the hyperbolic distribution density

$$\psi(t) = (\mu - 1)\frac{T^{\mu-1}}{(T+t)^{\mu}}. \tag{3.239}$$

The singularity at $\eta = -1$ is the origin of the hyperbolic distribution density. It is important to notice that the conventional Poisson case can be derived from the general age-specific failure rate by setting $r_1 = 0$. When $\eta = -1$, and we run an algorithm corresponding to the prescription of (3.233), we probably obtain first the times $t \ll T$, which have a larger probability, and only later do the times $t \gg T$ appear. However, the renewal character of the process requires that a very long time may appear after a very short one, and the illusion of Poisson statistics is due to the truncation produced by adopting a short data sequence.

The properties of the hyperbolic distribution density are in themselves very interesting. For example, the power spectrum $S(f)$ of such hyperbolic sequences has been shown to be

$$S(f) \propto \frac{1}{f^{\eta}}, \tag{3.240}$$

where

$$\eta = 3 - \mu. \tag{3.241}$$

Consequently, this spectrum diverges for $f \to 0$ when μ falls in the interval

$$1 < \mu < 3. \tag{3.242}$$

For this reason, the events described by (3.239) with μ in the interval (3.242) are called *critical events*. Another remarkable property of these events is determined by the central moments $\langle t^n \rangle$ defined by

$$\langle t^n \rangle = \int_0^{\infty} t^n \psi(t)dt. \tag{3.243}$$

The first and second moments are given by

$$\langle t \rangle = \frac{T}{\mu - 2} \tag{3.244}$$

and

$$\langle t^2 \rangle = \frac{2T^2}{(\mu - 2)(\mu - 3)}, \tag{3.245}$$

so the variance of the hyperbolic distribution is

$$\Delta \tau^2 \equiv \langle t^2 \rangle - \langle t \rangle^2 = \frac{\mu - 1}{(\mu - 2)^2 (\mu - 3)} T^2. \tag{3.246}$$

Equation (3.246) implies that the width of the distribution density $\psi(t)$, denoted by the standard deviation $\Delta \tau$, is divergent when μ is in the interval (3.242).

Thus, we can define the critical events as those whose waiting-time distribution density has a divergent width. In addition, the crucial events with $\mu < 2$ are characterized by the divergent mean value $\langle t \rangle = \infty$. We shall see that these events do not satisfy the ergodicity condition and consequently are the least ordinary in the more general class of crucial events.

3.5.3   Various Poisson processes

Up to this point we have adopted the jargon of the engineering literature and called $g(t)$ a failure rate, but in keeping with our intention of making the presentation as discipline-rich as possible we replace the term *failure* with the term *event*. We will have occasion to return to the nomenclature of failures, but the general discussion is much broader. In fact much of what we focus our attention on is a special class of events called *crucial events*.

Imagine that, when an event occurs, its occurrence time is recorded and immediately afterward we are ready to record the occurrence time of a subsequent event. The engineering perspective of failure and the process of repair immediately after failure is especially convenient when we want to design a procedure to generate a time series. Imagine that the repair process is instantaneous and that the *restoration* or *repair effectiveness* is given by a quantity q in the interval $0 \le q \le 1$. Renewal theory [14] is based on the assumption that the repair is instantaneous and perfect. This means that after repair the machine is "as good as new." Kijma and Sumita [26] adopted a *generalized-renewal-process* (GRP) perspective based on the notion of *virtual age*:

$$A_n = q S_n, \tag{3.247}$$

where A_n is the web's age before the nth repair, S_n is the web's age after the nth repair and q is the restoration factor.

Ordinary renewal theory is recovered by setting the restoration factor to zero $q = 0$. In this case the virtual age of the machine immediately after a repair is zero. The machine is "as good as new" or *brand new*. This condition is illustrated in Figure 3.10,

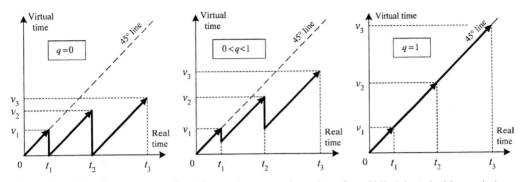

Figure 3.10. Virtual age versus real age for varying restoration values, from [44]. Adopted with permission.

from which we see that the virtual age tends to increase with time but jumps to the vanishing value when an event occurs. With increasing time the virtual age may become much smaller and in some cases infinitesimally smaller than the real time. If the repair is not perfect, see the condition $0 < q < 1$ of Figure 3.10, the virtual age tends to increase with time. Finally, if the repair is very bad, the restoration factor is $q = 1$, and the network is restored to the "same as old" condition.

In order to describe this virtual age mathematically let's go back to the general rate given by (3.233). The condition $q = 0$ corresponds to assuming that at time time τ_1, the time at which the first event occurs, the repair process generates a new machine, whose failure rate $g(t)$, is given by

$$g(t) = r_0[1 + r_1(t - \tau_1)]^\beta \qquad (3.248)$$

for $\tau_1 \leq t < \tau_1 + \tau_2$. The next event occurs at time $\tau_1 + \tau_2$ and after this second repair the rate $g(t)$ ages according to

$$g(t) = r_0[1 + r_1(t - \tau_1 - \tau_2)]^\beta \qquad (3.249)$$

and so on. In the limiting case $q = 1$ in (3.247), the time evolution of $g(t)$ is independent of the event occurrence, and the virtual age of the machine coincides with real time.

In general the condition $q = 1$ is indistinguishable from a *non-homogeneous Poisson process* (NHPP). In the literature of anomalous statistical physics an attractive example of NHPP is given by the superstatistics of Beck [6]. To generate superstatistics, imagine that the probability of the time-independent rate g is given by a distribution function $\Pi(g)$. In this case we can express the distribution density $\psi(t)$ for time-independent g as

$$\psi(t) = \int_0^\infty \Pi(g)ge^{-gt}\,dg. \qquad (3.250)$$

Let us assign to $\Pi(g)$ the analytic form

$$\Pi(g) \equiv \frac{T^{\mu-1}}{\Gamma(\mu - 1)}g^{\mu-2}e^{-gT}. \qquad (3.251)$$

By inserting the distribution $\Pi(g)$ of (3.251) into (3.250), replacing the integration variable g with $x = g(t + T)$, using the definition of the gamma function and the well-known property $\Gamma(z + 1) = z\Gamma(z)$ it is possible to prove that the distribution density resulting from the integration in (3.250) becomes identical to the hyperbolic distribution density (3.239).

The interpretation of this result is that the emergence of a hyperbolic law can be caused by the superposition of infinitely many exponential relaxation functions. However, this approach to complexity has caused some confusion in the literature. In fact, as pointed out by Allegrini *et al.* [3], for this approach to make physical sense we have to imagine that the web remains in a state in the neighborhood of a given rate g for an extended period of time. For example, it is possible that the distribution of rates yields

$$\Pi(g) \propto \frac{1}{dg/dt}, \tag{3.252}$$

in which case the renewal condition for the statistics is lost. This is especially evident in the case where $g(t)$ is a monotonically decreasing function of time. The time interval between two consecutive events tends to become longer and longer, whereas in a renewal process a very long time interval between two consecutive events may appear on any time scale.

The Weibull distribution function for $\beta = 0$ corresponds to a singularity that is responsible for the emergence of the hyperbolic behavior so ubiquitous in the field of complexity. This is reminiscent of the myth of Pandora's box. After Prometheus' theft of the secret of fire, Zeus ordered Hephaestus to create the woman Pandora as part of the punishment for mankind. Pandora was given many seductive gifts from Aphrodite, Hermes, Charites and Horae (according to *Works and Days* [25]). For fear of additional reprisals, Prometheus warned his brother Epimetheus not to accept any gifts from Zeus, but Epimetheus did not listen and married Pandora. Pandora had been given a large jar and instructions by Zeus to keep it closed, but she had also been given the gift of curiosity, and ultimately opened it. When she opened it, all the evils, ills, diseases and burdensome labor that mankind had not known previously escaped from the jar, but the one thing that was good that came out of the jar was hope. Our box is given by (3.236). In this book, under the influence of curiosity we explore the singularity in (3.236). If we move from this representation to the conventional Weibull picture, the condition $\beta \ll 1$ apparently generates an extremely slow exponential relaxation $\Psi(t) = e^{-\alpha t}$. This is equivalent to not opening Pandora's box. If we do open it, we see the emergence of the complex world of the inverse power law. Of course, we are giving an optimistic interpretation of the myth of Pandora, as the discoverer of complexity, namely of the phenomena of organized webs, including diseases and the spreading of epidemics. Understanding their origin may contribute to improving the conditions of the world.

What is the origin of this inverse power-law behavior? We have seen two distinct approaches to this condition. The superstatistics approach [6], in which an ensemble of exponentials with different rates is assumed to govern the process, implies a deviation from the renewal condition. In our discussion we focus more on the renewal approach, corresponding to $q = 0$. However, we have to keep in mind that reality can be more complex and that it may be a mixture of renewal and non-renewal properties.

3.6 Overview

In this chapter we have introduced a variety of concepts that are usually discussed in the context of several non-overlapping phenomena. The mechanical basis of Newton's laws, namely the conservation of energy, forms the content of a number of undergraduate and graduate courses in physics and engineering. The dynamical equations for mechanical webs establish a paradigm for understanding the motion of matter from planets to elementary particles. The solutions to such equations make prediction possible and, although our private lives do not lend themselves to such prescriptions, the technology of our society is predictable in this way. However, being able to predict the operation of inanimate objects and not being able to do the same for animate objects sets up a tension in our intellectual map of the world. We can understand the operation of the devices cluttering up our world, but we do not understand the relationships between these devices and human beings. A simple example is the personal computer, the engineering of which we understand, but the way it has transformed society over the past decade was not anticipated by anyone, neither scientist nor social commentator.

To model our inability to predict the behavior of the more complex phenomena in our world, such as human beings, we introduced uncertainty in the form of random variables. The randomness shifted the emphasis from a prediction of the single trajectory solving the deterministic equation of motion to an ensemble of such trajectories. This change replaces the prediction of a definite outcome stemming from an exact initial condition by a distribution of possible futures with varying degrees of probability. Certain phenomena, such as the spreading of a drop of ink in water or a rumor in the population of a tightly coupled community, cannot be described by deterministic trajectories but are well represented by probability densities. The dynamical description therefore shifts to how the probability density changes over time, and this unfolding of the probability suggests the possible futures for the phenomena being studied.

It is natural that such human activities as decision making lend themselves more readily to the uncertainty associated with the probability description than to the predictability of the trajectory description. The tools from statistical physics that describe classical diffusion lend themselves to phenomena that are a combination of deterministic and random behavior. The deterministic behavior is expressed in terms of the average of the dynamical variable and the random behavior is in terms of fluctuations. We understand the world, at least in part, by smoothing over these fluctuations. Imagine looking down from the top of a twenty-story building in New York City at the movement of people. In the aggregated flow below you do not see the individuals moving to avoid collisions, to let someone by, to dodge a young child; all these personal choices of moving faster or slower and changing direction are lost in the smoothed view from the top. In the case of people the uncertainty associated with fluctuations may be the variability in choices that psychologists try to understand, but these variations might also be the fluctuations in the numbers of people making a particular decision, a vote, say; and that concerns the sociologist. So we have tools for handling fluctuations, but to understand the phenomenon we must be able to associate the properties of the fluctuations with the web's complexity.

The simplicity of physical webs even when they are not deterministic is described by equally simple statistical processes, those of Poisson and Gauss. The increasing complexity of phenomena of interest to us is reflected in the increasing complexity of the random fluctuations in such properties as long-term memory and the diverging of central moments. Two ways we have discussed in which these properties are manifest is through inverse power-law correlations and hyperbolic probability densities that asymptotically become inverse power-law probabilities. This discussion established the groundwork for introducing the crucial events described by inverse power laws, which we take up in more detail in the next chapter.

3.7   Problems

3.1 Slowly modulated oscillators

Consider a set of independent linear harmonic oscillators coupled to a slowly varying external harmonic perturbation. The Hamiltonian for this web is

$$H'(\mathbf{a}, \mathbf{a}^*) = \sum_{j=1}^{N} \omega_j a_j a_j^* + \sum_{j=1}^{N} \Gamma_j (a_j + a_j^*)\cos(\Omega t),$$

where the natural frequencies of the oscillators are much higher than that of the perturbation $\omega_j \gg \Omega$. The notion of perturbation implies that the strength of the external driver is small, $\Gamma_j < 1$, for all j. Construct the solution to Hamilton's equations and discuss the frequencies that naturally enter into the web's dynamics. Also discuss the case of resonance when the external frequency falls within the spectral range of the oscillators.

3.2 A Hamiltonian heat bath

The analysis of linear dynamical equations performed in this section can be simplified to that of a single harmonic oscillator coupled to a heat bath, which in turn is modeled as a network of harmonic oscillators. Consider the linear oscillator Hamiltonian in terms of mode amplitudes $H_0 = \omega_0 a^* a$, the Hamiltonian for the environment, also in the mode-amplitude representation

$$H_{\mathrm{B}} = \sum_{k=1}^{\infty} \omega_k b_k^* b_k,$$

and the interaction Hamiltonian

$$H_{\mathrm{int}} = \sum_{k=1}^{\infty} \Gamma_k \left(b_k^* + b_k\right)\left(a^* + a\right).$$

The mode amplitudes and their complex conjugates constitute sets of canonical variables, so their dynamics are determined by Hamilton's equations. Put the equations of motion into the form (3.32) both for the web and for the bath. Solve the bath equations formally and substitute that solution into the web equation of motion as was done in the text and obtain an equation of the form

$$\frac{da}{dt} + i\omega_0 a = f(t) + \int_0^t dt'\, K(t - t')[a(t') + a^*(t')].$$

Find explicit expressions for $f(t)$ and $K(t)$ in terms of the bath parameters; explain how the former is noise, whereas the latter is a dissipative memory kernel, and how the two are related to one another.

3.3 Stochastic resonance

In Section 3.2.3 we discussed SR and presented the equation of motion. Use linear perturbation theory to solve the equation of motion for the average signal strength given by (3.99). Graph the signal strength versus the intensity of the fluctuations D and explain what is significant about the resulting curve.

3.4 Complete derivation

In the text we did not include all the details in the derivation of the probability densities (3.155) and (3.169). The key step in both cases is evaluating the characteristic function. Go back and derive these two distributions, including all the deleted steps in the process, and review all the important properties of Gaussian distributions.

3.5 The inverse power law

Consider the dynamical equation defined on the unit interval [0, 1]

$$\frac{dQ}{dt} = aQ^z$$

and apply the same reasoning as was used at the beginning of this section to obtain its solution. Each time the dynamical variable encounters the unit boundary $Q(\tau) = 1$ record the time and reinject the particle to a random location on the unit interval. The initial condition is then a random variable $\xi = Q(0)$, as is the time at which the reinjection occurs. Show that for a uniform distribution of initial conditions the corresponding times are distributed as an inverse power law.

References

[1] B. J. Adler and T. E. Wainwright, "Velocity autocorrelation function of hard spheres," *Phys. Rev. Lett.* **18**, 968 (1967); "Decay of the velocity autocorrelation function," *Phys. Rev. A* **1**, 18 (1970).

[2] O. C. Akin, P. Paradisi and P. Grigolini, *Physica A* **371**, 157–170 (2006).

[3] P. Allegrini, F. Barbi, P. Grigolini and P. Paradisi, "Renewal, modulation and superstatistics," *Phys. Rev. E* **73**, 046136 (2006).

[4] J. B. Bassingthwaighte, L. S. Liebovitch and B. J. West, *Fractal Physiology*, New York: Oxford University Press (1994).

[5] A. L. Barabási and R. Albert, "Emergence of scaling in random networks," *Science* **286**, 509 (1999).

[6] C. Beck, "Dynamical foundations of nonextensive statisical mechanics," *Phys. Rev. Lett.* **87**, 180601 (2001).

[7] R. Benzi, A. Sutera and A. Vulpiani, *J. Phys. A* **14**, L453 (1981); R. Benzi, G. Parisi, A. Sutera and A. Vulpiani, *Tellus* **34**, 10 (1982).

[8] M. Bianucci, R. Mannella, B. J. West and P. Grigolini, "Chaos and linear response: analysis of the short-, intermediate-, and long-time regime," *Phys. Rev. E* **50**, 2630 (1994).

[9] B. S. Blanchard, *Logistics Engineering and Management*, 4th edn., Englewood Cliffs, NJ: Prentice Hall (1992), pp. 26–32.

[10] N. Bousquet, H. Bertholon and G. Cedeux, "An alternative competing risk model to the Weibull distribution for modeling aging in lifetime data analysis," *Lifetime Data Anal.* **12**, 481 (2006).

[11] R. Bogacz, E. Brown, J. Moehlis, P. Holmes and J.D. Cohen, "The physics of optimal decision making: a formal analysis of models of performance in two-alternative forced-choice tasks," *Psych. Rev.* **113**, 700 (2006).

[12] J. R. Busemeyer and J. T. Townsend, "Decision field theory: a dynamic–cognitive approach to decision making in an uncertain environment," *Psych. Rev.* **100**, 432 (1993).

[13] S. W. Chiu, J. A. Novotny and E. Jakobsson, "The nature of ion and water barrier crossings in a simulated ion channel," *Biophys. J.* **64**, 98 (1993).

[14] D. R. Cox, *Renewal Theory*, London: Metheun & Co. (1967); first printed in 1962.

[15] R. Dekeyser and H. H. Lee, "Nonequilibrium statistical mechanics of the spin-1/2 van der Walls model. II. Autocorrelation function of a single spin and long-time tails," *Phys. Rev. B* **43**, 81238147 (1991).

[16] W. Feller, *An Introduction to Probability Theory and Its Applications*, Vol. II, New York: John Wiley & Sons (1971).

[17] C. E. Finch, *Longevity, Senescence and the Genome*, Chicago, IL: University of Chicago Press (1990).

[18] M. Finkelstein, "Aging: damage accumulation versus increasing mortality rate," *Math. Biosci.* **207**, 104 (2007).

[19] M. Finkelstein, "Virtual age of non-reparable objects," *Reliability Eng. and System Safety* **94**, 666 (2009).

[20] D. Gross and C. M. Harris, *Fundamentals of Queueing Theory*, 3rd edn., New York: Wiley-Interscience (1998).

[21] H. Haken, *Synergetics, an Introduction: Nonequilibrium Phase Transitions and Self-Organization in Physics, Chemistry and Biology*, 3rd rev. edn., New York: Springer (1983).

[22] H. Haken, "Slaving principle revisited," *Physica D* **97**, 97 (1996).

[23] H. Haken, "Synergetics of brain function," *Int. J. Psychophys.* **60**, 110 (2006).

[24] H. Haken, *Brain Dynamics, Synchronization and Activity Patterns in Pulse-Coupled Neural Nets with Delays and Noise*, Berlin: Springer (2007).

[25] Hesoid, *Works and Days*, translated from the Greek, London (1728).

[26] M. Kijma and A. Sumita, "A useful generalization of renewal theory: counting processes governed by non-negative Markov increments," *J. Appl. Prob.* **23**, 71 (1986).

[27] A. N. Kolmogorov, "Local structure of turbulence in an incompressible liquid for very large Reynolds numbers," *Comptes Rendus (Dokl.) Acad. Sci. URSS (N.S.)* **26**, 115–118 (1941); "A refinement of previous hypotheses concerning the local structure of turbulence in a viscous incompressible fluid at high Reynolds number," *J. Fluid Mech.* **13**, 82–85 (1962).

[28] H. A. Kramers, "Brownian motion in a field of force and the diffusion model of chemical reactions," *Physica* **7**, 284 (1940).

[29] P. Langevin, *Comptes Rendus Acad. Sci. Paris* **146**, 530 (1908).

[30] K. Lindenberg and B. J. West, *The Nonequilibrium Statistical Mechanics of Open and Closed Systems*, New York: VCH Publishers (1990).

[31] M. Longuet-Higgins, "The effect of non-linearities on statistical distributions in the theory of sea waves," *J. Fluid Mech.* **17**, 459 (1963); "Statistical properties of an isotropic random surface," *Phil. Trans. Roy. Soc. London A* **250**, 158 (1957).

[32] R. Mannella, P. Grigolini and B. J. West, "A dynamical approach to fractional Brownian motion," *Fractals* **2**, 81–124 (1994).

[33] G. Margolin and E. Barkai, "Single-molecule chemical reactions: reexamination of the Kramers approach," *Phys. Rev. E* **72**, 025101 (2005).

[34] E. W. Montroll and B. J. West, "On an enriched collection of stochastic processes," in *Fluctuations Phenomena*, 2nd edn., eds. E. W. Montroll and J. L. Lebowitz, Amsterdam: North-Holland (1987); first published in 1979.

[35] F. Moss and K. Wiesenfeld, "The benefits of background noise," *Sci. Am.* **273**, 66–69 (1995).

[36] F. Moss, L. M. Ward and W. G. Sannita, "Stochastic resonance and sensory information processing," *Clin. Neurophysiol.* **115**, 267–281 (2004).

[37] C. Nicolis and G. Nicolis, "Is there a climate attractor?," *Nature* **311**, 529–532 (1984).

[38] C. K. Peng, J. Mietus, J. M. Hausdorff *et al.*, "Long-range autocorrelations and non-Gaussian behavior in the heartbeat," *Phys. Rev. Lett.* **70**, 1343 (1993).

[39] J. Perrin, *Brownian Movement and Molecular Reality*, London: Taylor and Francis (1910).

[40] J. W. Strutt (Lord Rayleigh), *Phil. Mag.* **X**, 73–78 (1880); in *Scientific Papers*, Vol. I, New York: Dover Publications (1880), pp. 491–496.

[41] L. F. Richardson, "Atmospheric diffusion shown on a distance–neighbor graph," *Proc. Roy. Soc. London Ser. A* **110**, 709–725 (1926).

[42] D. Ruelle, *Chaotic Evolution and Strange Attractors*, Cambridge: Cambridge University Press (1989).

[43] G. E. Uhlenbeck and L. S. Ornstein, "On the theory of Brownian motion," *Phys. Rev.* **36**, 823 (1930).

[44] B. Veber, M. Nagode and M. Fajdiga, "Generalized renewal process for repairable systems based on finite Weibull mixture," *Reliability Eng. and System Safety* **93**, 1461 (2008).

[45] J. N. Weiss, A. Garfinkel, M. L. Spano and W. L. Ditto, "Chaos control for cardiac arrhythmias," in *Proc. 2nd Experimental Chaos Conference*, eds. W. Ditto, L. Pecora, M. Shlesinger, M. Spano and S. Vohra, Singapore: World Scientific (1995).

[46] B. J. West and A. L. Goldberger, "Physiology in fractal dimensions," *Am. Sci.* **75**, 354–364 (1987).

[47] B. J. West, "Physiology in fractal dimensions: error tolerance," *Ann. Biomed. Eng.* **18**, 135–149 (1990).

[48] B. J. West, *Where Medicine Went Wrong*, Singapore: World Scientific (2006).

[49] F. G. Zeng, Q. Fu and R. Morse, *Brain Res. Inter.* **869**, 251 (2000).

4 Random walks and chaos

In the late nineteenth century, it was believed that a continuous function such as those describing physical processes must have a derivative "almost everywhere." At the same time some mathematicians wondered whether there existed functions that were everywhere continuous, but which did not have a derivative at any point (continuous everywhere but differentiable nowhere). Perhaps you remember our discussion of such strange things from the first chapter. The motivation for considering such pathological functions was initiated by curiosity within mathematics, not in the physical or biological sciences where one might have expected it. In 1872, Karl Weierstrass (1815–1897) gave a lecture to the Berlin Academy in which he presented functions that had the remarkable properties of continuity and non-differentiability. Twenty-six years later, Ludwig Boltzmann (1844–1906), who connected the macroscopic concept of entropy with microscopic dynamics, pointed out that physicists could have invented such functions in order to treat collisions among molecules in gases and fluids. Boltzmann had a great deal of experience thinking about such things as discontinuous changes of particle velocities that occur in kinetic theory and in wondering about their proper mathematical representation. He had spent many years trying to develop a microscopic theory of gases and he was successful in developing such a theory, only to have his colleagues reject his contributions. Although kinetic theory led to acceptable results (and provided a suitable microscopic definition of entropy), it was based on the time-reversible dynamical equations of Newton. That is a fundamental problem because entropy distinguishes the past from the future, whereas the equations of classical mechanics do not. This basic inconsistency between analytic dynamics and thermodynamics remains unresolved today, although there are indications that the resolution of this old chestnut lies in microscopic chaos.

It was assumed in the kinetic theory of gases that molecules are not materially changed as a result of scattering by other molecules, and that collisions are instantaneous events as would occur if the molecules were impenetrable and perfectly elastic. As a result, it seemed quite natural that the trajectories of molecules would sometimes undergo discontinuous changes. As mentioned in the discussion on classical diffusion, Robert Brown observed the random motion of a speck of pollen immersed in a water droplet. The discontinuous changes in the speed and direction of the motion of the pollen mote observed are shown in Figure 3.2. Jean Baptiste Perrin, of the University of Paris, experimentally verified Einstein's predictions on the classical diffusion of Brown's pollen mote, and received the Nobel Prize for his work leading to the experimental determination of Avogadro's number in 1926. Perrin, giving a physicist's view of mathematics in 1913, stated that curves without tangents (derivatives) are more common than those special, but interesting ones, like the circle, that have tangents. In fact he was quite adamant in his arguments emphasizing the importance of non-analytic functions for describing complex physical phenomena, such as Brownian motion. Thus, there are valid physical reasons for looking for these types of functions, but the scientific reasons became evident to the general scientific community only long after the mathematical discoveries made by Weierstrass and other mathematicians.

A phenomenological realization of the effect of these discrete processes is given by random-walk models in which the moving entity, the random walker, is assumed to have a prescribed probability of taking certain kinds of steps. We show that after N steps, where N is a large number, the probability of being a given distance from the starting point can be calculated. This strategy for modeling random phenomenon has a long lineage and is very useful for understanding the properties of certain simple stochastic processes such as the diffusion of ink in water, the spreading of rumors, the flow of heat and certain kinds of genetic transfer, to name just a few. A simple random walk is then generalized to capture the inverse power-law behavior that we have observed in so many complex phenomena.

Finally, the influence of memory on the dynamics of linear complex webs is discussed using the fractional calculus. This type of complexity is a manifestation of the non-locality of the web's dynamics, so that the complete history of the web's behavior determines its present dynamics. This history-dependence became even more significant when the web was again embedded in an environment that introduces a random force. The resulting fractional random walk is a relatively new area of investigation in science [44].

A number of the analyses pursued in foregoing chapters had to be cut short because some of the technical concepts associated with random variables, such as the definition of a probability density, had not been properly introduced. As we know, it is difficult to proceed beyond superficial pleasantries without a proper introduction. To proceed beyond the superficial, we now present the notion of probabilities, probability densities, characteristic functions and other such things and replace the dynamics of web observables with the dynamics of web distribution functions. Here again we strive more for practicality than for mathematical elegance. We begin with the simplest of probability arguments having to do with random walks, which form the intuitive basis of probability theory in the physical sciences. From there we proceed to the more mathematically sophisticated concept of chaos as another way of introducing uncertainty into the modeling.

The concept of chaos, as mentioned, was developed as a response to scientists' inability to obtain closed-form solutions in any but a handful of nonlinear ordinary differential equations. This inability to obtain analytic solutions required the invention of a totally different way of examining how the dynamics of a complex web unfolds. The methods for analyzing general nonlinear dynamic webs are ultimately based on mapping techniques to provide visualization of the underlying trajectories. Consequently, we begin the discussion of dynamics with maps and see how even the simplest nonlinear map has chaotic solutions.

In previous chapters the dynamics of the variables thought to describe complex webs were addressed from a variety of perspectives. We examined the conservative dynamics of classical mechanics and Hamilton's equations of motion. In this chapter we show how linear interactions give way to nonlinear interactions resulting in the dynamics of chaos as web interactions become more complicated. In this replacement of the linear with the nonlinear it also becomes clear that the precise and certain predictions for which scientists strive are not available when the web is truly complex. Chaos, with its sensitive dependence on initial conditions, provides predictions for only very short times, since infinitesimally distant trajectories exponentially separate from one another in time. So the chaos of unstable Hamiltonian trajectories, as well as the strange attractors of dissipative networks, both lead to complexity that defies long-time predictability in real webs.

Part of our purpose is to seek out and/or develop analysis techniques that are not restricted to describing the behavior of physical webs. More specifically we are looking for universal principles such as the variational principle used in the Hamiltonian formalism; that is, the variation of the total energy vanishes and yields the equations of motion. Here we find that linear response theory (LRT) in conjunction with chaos theory provides a dynamical foundation for the fluctuation–dissipation relation. This marriage of the two concepts suggests that LRT may be universal; this is because it does not refer to a single trajectory, but rather to the manifold in phase space on which all the trajectories in an ensemble of initial conditions unfold. Thus, although the manifold is linearly distorted by a perturbation, the underlying dynamics of the process can be arbitrarily deformed and consequently the dynamic response of the web is not linear. However, unlike the response of a single trajectory, the average dynamical response of the web to the perturbation *is* linear. The average is taken over the chaotic fluctuations generated by the microscopic variables.

A Hamiltonian coupling a macroscopic variable to microscopic chaos is used to motivate the universality of LRT to model the average web response to perturbation. This argument sets the stage for the more rigorous presentation in later chapters. But for the time being the formalism establishes computational consistency between a microscopic nonlinear dynamic model coupled to a macroscopic variable and Brownian motion; microscopic chaos is manifest as fluctuations in the equation for the macroscopic variable.

4.1   Random walks

The disruption of web predictability was investigated at the turn of the last century by introducing random forces into dynamical equations. In physics these stochastic differential equations are called Langevin equations for historical reasons and were discussed in the previous chapter. In Langevin equations the random force is intended to mimic the influence of the environment on the dynamics of the web of interest. In this way the complexity of the context in which even simple webs are embedded results in a degradation of a scientist's ability to make predictions. The exact deterministic future predicted by simple dynamical equations is replaced with a distribution of alternative futures. In our examination of the DDM model we obtained analytic forms for the probable decisions an individual can make, given uncertain and incomplete information. In this way we do not predict a person's decision, but rather we provide a functional form for the likelihood that the person will choose one future over another. This is a rather pompous way of saying that we do not know what a person's decision will be, but some decisions are more likely than others. We determine the odds and anticipate a given choice, giving one possibility more weight than another, but we do predict not the choice, but only its probability.

4.1.1 Simple random walks

Random walks date from two 1905 articles in the magazine *Nature*. In the first article the biostatistician Karl Pearson posed a question that probably would not be published today. The question frames a mathematics problem in a form whose solution would provide insight into a variety of scientific phenomena. The question he asked was the following:

A man starts from a point 0 and walks l yards in a straight line: he then turns through any angle whatever and walks another l yards in a second straight line. He repeats this process n times. I require the probability that after these n stretches he is at a distance between r and $r + dr$ from his starting point 0. The problem is one of considerable interest, but I have only succeeded in obtaining an integrated solution for two stretches. I think, however, that a solution ought to be found, if only in the form of a series of powers of $1/n$ where n is large.

Lord Rayleigh immediately saw the generic form of Pearson's question and responded in the same issue of *Nature* with the following answer:

This problem, proposed by Prof. Karl Pearson in the current number of *Nature*, is the same as that of the composition of n iso-periodic vibrations of unit amplitude and of phase distributed at random, considered in *Philosophical Magazine*, X. p. 73, 1880, XLVII. p. 246, 1899 (*Scientific Papers*, I. p. 491; IV. p. 370). If n be very great, the probability sought is

$$2n^{-1}e^{-r^2/n}r\,dr$$

Probably methods similar to those employed in the papers referred to would avail for the development of an approximate expression applicable when n is only moderately great.

The ability to strip away irrelevant detail and focus on the important aspects of a problem is the hallmark of a good modeler and being right is the sign of a great modeler. In this regard Lord Rayleigh's explanation of the random walk is reminiscent of Newton's thinking about the speed of sound in air, at least in terms of the clarity of the approach.

The Bernoulli sequence

Of course, history has endorsed Lord Rayleigh's intuition concerning the equivalence of a random walk to the superposition of acoustic modes with random phases. A simple argument that captures Lord Rayleigh's approach is now given. Let us consider a variable Q indexed by the discrete time $j = 1, 2, \ldots, N$ which defines the set of values $\{Q_j\}$ determined by the equation

$$Q_{j+1} = Q_j + U_j, \tag{4.1}$$

where U is a random variable taking on the value of $+1$ or -1. The solution to the iterative equation (4.1) is given by

$$Q_N = \sum_{j=1}^{N} U_j, \tag{4.2}$$

which is the sum of N identically distributed random variables. The set of values taken on by the random variable $\{U_j\} = \{+1, +1, -1, \ldots\}$ is called a Bernoulli sequence after the scientist Daniel Bernoulli. This is the same Bernoulli who introduced the utility function into science and who first studied how people make decisions under uncertainty. The curly brackets denote a particular realization of such a sequence, each term of which can be obtained by flipping a coin to decide whether a $+1$ or a -1 is put in a specific location. A Bernoulli sequence is a completely random process, by which we mean that no element in the sequence is dependent upon any other element in the sequence; each toss of the coin is independent of every other toss and therefore so too is each element in the sequence independent of every other element. Here we visualize this process by considering a walker on a one-dimensional lattice taking a step in the indicated direction, to the right for $+1$ and to the left for -1, Figure 4.1.

Consequently, summing over the elements of a Bernoulli sequence yields a simple random-walk process. Pearson's question regarding the probability of being at a specific location after taking N steps can now be formally addressed. We emphasize that the conclusions we are able to draw from this simple random-walk argument are independent of the values of the constant step length put into the random sequence of steps and we use the Bernoulli sequence both because of its conceptual simplicity and on account of its usefulness later when we develop some rather formidable formalisms.

It is useful to briefly explore the properties of the Bernoulli sequence. Every computer has a program to generate a random number between 0 and 1 without bias, meaning that any value on the interval [0, 1] is equally likely. If we take the interval and divide it into two equal parts, [0, 0.5] and [0.5, 1], then a random number falling on the left interval is given the value -1 and a random number falling on the right interval is given the value $+1$; the probability of falling on either side of 0.5 is equal to one half, just as in a coin toss. In this way a Bernoulli sequence can be generated on the computer. Suppose that we generate 10,000 data points $\{U_j\}$ and we want to determine whether in fact the computer-generated sequence is a completely random process. Consider the discrete autocorrelation function

$$-3\ -2\ -1\ \ 0\ \ 1\ \ 2\ \ 3$$

Figure 4.1.   Homogeneous lattice sites for a one-dimensional random walk.

$$r_k \equiv \frac{\left[1/(N-k)\right] \sum\limits_{j=1}^{N-k} \left[U_j - \overline{U}\right]\left[U_{j+k} - \overline{U}\right]}{(1/N) \sum\limits_{j=1}^{N} \left[U_j - \overline{U}\right]^2}; \quad N \gg k, \tag{4.3}$$

where the sums of the sequences in (4.3) are calculated using the average value

$$\overline{U} = \frac{1}{N} \sum_{j=1}^{N} U_j \tag{4.4}$$

for a sequence of N computer-generated coin tosses. The autocorrelation coefficient determines whether the value of the dynamical process at one instant of time influences its value at subsequent times. The autocorrelation coefficient is normalized to one such that when $k = 0$, the absence of time lag, there is complete self-determination of the data. The magnitude of this coefficient as a function of k determines how much the present value is determined by history and how much the present will determine the future. The autocorrelation function is graphed versus the lag time in Figure 4.2 for the Bernoulli sequence. The correlation is seen to plummet to zero in one step and to fluctuate around this no-influence value of zero for the remainder of the time.

There are many phenomena in the physical, social and biological sciences for which we use random-walk models to describe their evolution over time. For example, just as Bachelier had foreseen [4], the stock market with its responses to inflation and the level of employment, its discount of taxes, its reaction to wars in various parts of the world, all of which are uncontrollable, does in fact buffet the price of stock in a manner not unlike Igen Hauz's charcoal or Brown's pollen motes. This simple reasoning has also been applied to a bacterium's search for food. The bacterium moves in a straight line for a while and then tumbles to reorient itself for another straight-line push in a random direction. In each of these applications of the random walk the rationale is that the best model is the simplest one that can explain all the available experimental data with the fewest assumptions. Alternative models are those that make predictions

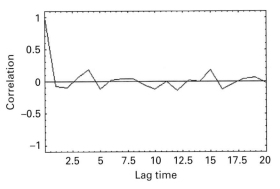

Figure 4.2. The discrete autocorrelation function is calculated for a Bernoulli sequence. The coefficient has an initial value of one and drops to zero within one unit of delay time, the time it takes to carry out one step. The correlation subsequently fluctuates around zero for arbitrarily long delay times.

and can assist in formulating new experiments that can discriminate between different hypotheses. Simple random walks were, in fact, among the most quoted models of speculation in the stock market [32] until fairly recently, when the fractal market hypothesis was introduced; see Peters [37] or Mandelbrot [29] for a more complete discussion and the relatively new discipline of econophysics [30]. This example is taken up again subsequently, but for the moment we follow the random-walk strategy to gain an understanding of how to construct a stochastic process as we increase the complexity of the phenomenon being modeled.

In this simple random-walk model we assume that the walker takes one step in each unit of time, so that the "time" variable can be defined by $t = N \Delta t$, where Δt is the time interval between steps and is usually taken to be unity between elements in the Bernoulli time sequence. The time becomes continuous as the time increment becomes a differential, $\Delta t \to 0$, and the number of steps becomes very large, $N \to \infty$, in such a way that their product remains constant, that being the continuous time t. Physically, time can be considered a continuous variable even though it is treated as discrete in the walk as long as the sampling time of the experiment is sufficiently small and the sample size is sufficiently large.

We refer to (4.2) as a diffusion trajectory since it signifies the location of the walker after each step. This nomenclature stems from the use of single-particle trajectories to denote the instantaneous position of a diffusing particle. The mean-square value of the diffusion trajectory is given by

$$\left\langle Q_N^2 \right\rangle = \sum_{j=1}^{N} \sum_{k=1}^{N} \langle U_j U_k \rangle = \sum_{j=1}^{N} \sum_{k=1}^{N} \delta_{jk} = N. \tag{4.5}$$

The angle brackets denote an average over an ensemble of realizations of Bernoulli sequences and, since the elements in the Bernoulli sequence are independent of one another, the average vanishes except when the indices are equal, as formally denoted by the Kronecker delta $\delta_{jk} = 1$ if $j = k$; $\delta_{jk} = 0$ if $j \neq k$. Consequently the random diffusion variable, as measured by its root-mean-square (rms) value, increases as the square root of the number of steps taken by the walker. This increase in the rms value of the random variable suggests that the normalized quantity $Q_N / \sqrt{N}$ will remain of order unity as N becomes arbitrarily large. Interpreting the number of steps as the time (4.5) shows that the second moment increases linearly with time, giving rise to a value $\alpha = H = 0.5$ for the scaling index introduced in the first chapter.

This analysis can be extended beyond the Bernoulli sequence by introducing the probability of making a transition between two sites arbitrarily far apart. For this more general random walk the equation for the probability of being at a lattice site s after $n + 1$ steps is given by

$$P_{n+1}(q) = \sum_{q'} p(q - q') P_n(q'), \tag{4.6}$$

where the probability of stepping from site q' to site q is $p(q - q')$. In the simple case of the Bernoulli sequence q' is only one lattice space to the right or to the left of

q and $p(q \pm 1) = 1/2$ with all other transition probabilities being zero, so that (4.6) reduces to

$$P_{n+1}(q) = \frac{1}{2}[P_n(q + 1) + P_n(q - 1)]. \tag{4.7}$$

The solution to the doubly discrete probability equation (4.7) is [13]

$$P_N(m) = \frac{N!}{[(N + m)/2]! \, [(N - m)/2]!} \frac{1}{2^N} \tag{4.8}$$

where m is the integer label of the lattice site and N is the total number of steps taken. We leave it as a problem to prove that when $N \gg m$ (4.8) is well approximated by a Gaussian distribution.

This random-walk model can be generalized to the case where the second moment of the diffusion variable is given by

$$\langle Q(t)^2 \rangle \propto t^{2H} \tag{4.9}$$

and H is the Hurst exponent confined to the interval $0 \le H \le 1$. The case $H = 0.5$ corresponds to the simple random-walk outline above, in which the probability of moving to the right or the left on a regular one-dimensional lattice is the same, resulting in ordinary diffusion. The case $H > 0.5$ corresponds to a random walk in which the probability of a walker stepping in the direction of her previous step is greater than that of reversing directions. The walk is called *persistent* and yields a mean-square displacement that increases more rapidly in time than does the displacement in ordinary diffusion, so the process is called *superdiffusive*. The case $H < 0.5$ corresponds to a random walk in which the probability of a walker stepping in the direction of her previous step is less than that of reversing directions. This walk is called *anti-persistent* and yields a mean-square displacement that increases more slowly in time than does the displacement in ordinary diffusion, so the process is *subdiffusive*. Thus, one kind of anomalous diffusion has to do with memory in time for the random steps taken on a regular lattice. This model has been used extensively in the interpretation of the fluctuations in physiologic time series, as well as in physical and social phenomena [45]. Anomalous diffusion will be discussed in more detail in the following section.

The random-walk model discussed above assumes a spatial lattice upon which a walker takes randomly directed steps in equal time intervals. Each step is completed instantaneously and in the limit of a great many such steps the central limit theorem (CLT) assures us that the resulting distribution of locations of the random walker is normal or Gaussian. However, there are a great many phenomena for which this kind of diffusion assumption does not apply, either because the medium in which the walker moves has structure, structure that might inhibit or facilitate the walker's steps, or because the properties of the walker change over time. In either case a more general kind of random walk is required, but before we turn our attention to that let us consider an alternative limit of (4.8).

4.1.2   Poisson webs

The binary nature of the above process naturally leads to the binomial distribution, even if it has a somewhat different form from what we are accustomed to seeing. Consider the asymmetric binary process where p is the probability of realizing outcome 1 and $1 - p$ is the probability of realizing outcome 2. In this case the probability of realizing the first outcome m times out of N trials can be obtained from the binomial expansion

$$(1 - p + p)^N = \sum_{m=0}^{N} P_N(m), \tag{4.10}$$

where the probability is determined by the binomial expansion to be

$$P_N(m) = \frac{N!}{(N - m)!m!} p^m (1 - p)^{N-m}. \tag{4.11}$$

When the number of events is very large and the probability of a particular outcome is very small the explicit form (4.11) is awkward to deal with and approximations to the exact expression are more useful. This was the case in the last section when the probability of each outcome was the same for very large N and the Gaussian distribution was the approximate representation of the binomial distribution.

Suppose that we have N disagreements among various groups in the world, such as between nations, unions and management, etc. Suppose further that the probability of any one of these disagreements undergoing a transition into a "deadly quarrel" in a given interval of time is p. What is the probability that m of the disagreements undergo transitions to deadly quarrels in the time interval? The Quaker scientist Richardson [39] determined that the distribution describing the number of "deadly quarrels" is the Poisson distribution, so that from the conditions of his argument we can extract the Poisson distribution as another approximation to the above binomial distribution.

The first factor on the rhs of (4.11) has the explicit form

$$\frac{N!}{(N - m)!} = N(N - 1)...(N - m + 1), \tag{4.12}$$

which has m factors, each of which differs from N by terms of order $m/N \ll 1$ and can therefore be neglected, yielding

$$\frac{N!}{(N - m)!} \approx N^m. \tag{4.13}$$

Replacing the ratio of factorials in the binomial distribution with this expression yields

$$P_N(m) \approx \frac{N^m}{m!} p^m (1 - p)^{N-m} \tag{4.14}$$

and in the limit of very large N and very small p, but such that their product is constant,

$$\lambda = Np, \tag{4.15}$$

the approximate binomial (4.14) becomes

$$P_N(m) \approx \frac{\lambda^m}{m!} (1 - p)^{N-m}. \tag{4.16}$$

Table 4.1. The occurrence of wars. The rows are x, the number of outbreaks of war per year; y, the frequency of years with number of outbreaks x; and z, the frequency from a Poisson distribution with mean $\lambda \approx 0.69$ years. The table is reproduced with permission from Richardson [39] using the data collected by Wright [46].

				x			
	0	1	2	3	4	>4	Total
y	223	142	48	15	4	0	432
z	216.2	149.7	51.8	12	2.1	0.3	432.1

Finally, the last factor in this expression can be written as an exponential,

$$(1-p)^{N-m} \approx (1-p)^{\lambda/p} = \left[(1-p)^{1/p}\right]^{\lambda} \approx e^{-\lambda}, \tag{4.17}$$

and the resulting form is the Poisson distribution

$$p(m; \lambda) = \lim_{N \to \infty, p \to 0} P_N(m) = \frac{\lambda^m}{m!} e^{-\lambda}. \tag{4.18}$$

The mean number of "deadly quarrels" can be calculated using this expression to be

$$\langle m \rangle = \sum_{m=0}^{N} m p(m; \lambda) = \lambda, \tag{4.19}$$

so that the Poisson distribution is parameterized in terms of the average value.

In Table 4.1 the "deadly quarrels" over the past 500 years are recorded in a way that emphasizes their historical pattern. Richardson [39] was interested in the causes of wars between nations, but he found that, independently of the moral position of a country, its economy or any of the other reasons that are generally offered to justify war, the number of wars initiated globally in any given interval of time follows a Poisson distribution.

Wright [46] published information on wars extending from 1482 to 1940 including 278 wars, with the start and end dates, the names of treaties of peace, participating states and so on. Richardson [39] took these data and counted the numbers of years in which no war began, one war began, two wars began, and so on, as listed in the second row of the table. In the final row is the fit to the data using a Poisson distribution. It is clear from Table 4.1 that this quantitative discussion captures an essential aspect of the nature of war; not that we can predict the onset of a particular event, but that we can determine a useful property of the aggregate, that is, of the collection of events.

The Poisson distribution also emerges in the elementary discussion of the number of connections formed among the N elements of a web. The elements could represent the airports distributed in cities around the world, the power stations located outside urban areas in the United States, neurons in the human brain, scientific collaborators, or the elements in any of a growing number of complex webs. For the purpose of the present discussion it does not matter what the elements are, only that they are to be connected in some specified way in order to form a web. In the sociology literature an element is also called a node or a vertex and the connecting line or arc is called an edge, a terminology developed in the mathematics of graph theory that dates back 300 years to the polymath Leonard Euler (1707–1784). The mathematicians Erdös and Rényi [20] in the 1950s and 1960s applied this theory to the investigation of the properties of random webs and determined that the probability that m vertices terminate on a given vertex is given by a Poisson distribution. There are some excellent popular reviews [5, 12, 42, 43] of the Erdös–Rényi ground-breaking work on random networks and we review some of that work in Chapter 6.

An interesting property of the Poisson distribution is that it has a single peak located at the average value; in this regard it is very similar to the distribution of Gauss. In fact, similarly to processes described by the normal distribution, those random webs described by the Poisson distribution are dominated by averages. In a Poisson world, just like in a Gaussian world, as Barabási [5] pointed out, most people would have roughly the same number of acquaintances; most neurons would connect to the same typical number of other neurons; most companies would trade with approximately the same number of other companies; most web sites would be visited by the same number of visitors, more or less. But the real world is not like that of Gauss or Poisson. Barabási [5] and Watts [43] showed that average measures are not appropriate for characterizing real-world networks, such as the Internet and the World Wide Web. West [45] showed that average measures in medicine are equally inappropriate and a number of the physiologic webs he considered are discussed in more detail later.

Figure 4.3 depicts the cumulative distribution in the number of connections to web sites with data obtained from the number of links to web sites found in a 1997 web crawl of about 200 million pages [11]. Clauset *et al.* [15] point out that this is a limited sample of the much larger entire web with an inverse power-law index of approximately 2.34. The exact value of the inverse power-law index should not be given much credence because of the large fluctuations in the data, a property common in the tails of such distributions. The best we can do here, without going into the details of data processing, is to recognize the inverse power-law behavior of the phenomenon.

A related graph is given in Figure 4.4, where the cumulative distribution in the number of hits received by web sites, to servers not to pages, from customers of the America

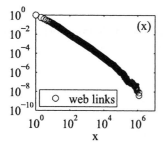

Figure 4.3.   The cumulative distribution in the number of links to web sites is shown by the circles, and is approximately described by an inverse power law with slope −2.34 [15]. Reproduced with permission.

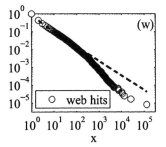

$$10^0$$
$$10^{-1}$$ (w)
$$10^{-2}$$
$$10^{-3}$$
$$10^{-4}$$
$$10^{-5}$$ | ○ web hits |
$$10^0 \ 10^1 \ 10^2 \ 10^3 \ 10^4 \ 10^5$$
x

Figure 4.4. The cumulative distribution in the number of hits to web sites from AOL users is shown by the circles and an inverse power law with slope -1.81 is depicted as a dashed-line segment [15]. Reproduced with permission.

On Line (AOL) Internet service in a single day is depicted [2]. Here we see a deviation of the data from a strict inverse power law with index 1.81. The deviation might be modeled as a second straight line, but the best fit to the data is found to be an inverse power law truncated by an exponential [15]. Here again we do not explore the data-processing techniques, since this would lead us too far afield. The intent here is to reinforce the notion that simple random walks with Gaussian statistics are unable to capture the richness of complex web data.

4.2 Fractal random walks

The underlying assumption for classical diffusion is that the process changes smoothly and that a given data point in a time-ordered sequence is most strongly influenced by adjacent data points. This was evident in the simple random-walk models where the probability of a displacement increased or decreased at each point in time in a spatially symmetric way. The basic assumption used in the argument is that the steps are local and symmetric on the lattice and depend only on the distance stepped, not on where the walker is located on the lattice at the time of the step. Assuming a finite mean-square displacement has been a mainstay of all modelers since de Moivre (1732) [18] first derived the distribution which now bears the name of Gauss.

Classical physicists have argued that certain general properties of complex physical webs are independent of the detailed microscopic dynamics. The phenomenological giant, thermodynamics, with its three laws was built from this perspective. The discipline of statistical physics concerns itself with understanding how microscopic dynamics, which is deterministic classically, manifests stochastic behavior in macroscopic measurements and gives rise to the observed thermodynamic relations. Past discussions have centered on equilibrium properties of processes, so the traditional wisdom of equilibrium thermodynamics could be applied and much of the microscopic behavior could be ignored. The condition necessary for the validity of these assertions is a separation of scales between the microscopic and macroscopic dynamics. In our subsequent discussion of the fractal calculus we find that this is not always the case. In the present random-walk context this separation of scales requires that the jump probability density $p(q)$ has a finite second moment and is symmetric; the random walk is otherwise rather insensitive to its precise form.

We now investigate some consequences of having a single-step probability density that has diverging central moments and find that it leads us to the fractional calculus.

4.2.1   The Weierstrass walk

The assumption that the mean-square step length diverges as

$$\langle q^2 \rangle = \sum_q q^2 p(q) = \infty \tag{4.20}$$

is not sufficient to specify the form of the random walk, since the transition probability leading to this divergence is not unique. Consider a jump distribution that leads to non-diffusive behavior,

$$p(q) = \frac{a-1}{2a} \sum_{n=0}^{\infty} \frac{1}{a^n} \left[\delta_{q,b^n} + \delta_{q,-b^n} \right], \tag{4.21}$$

which was introduced by Hughes *et al.* [25], where b and a are dimensionless constants greater than one. The mean-square jump length is then given by

$$\langle q^2 \rangle = \frac{a-1}{a} \sum_{n=0}^{\infty} \left(\frac{b^2}{a} \right)^n, \tag{4.22}$$

which for $b^2 > a$ is infinite since the series diverges. The lattice structure function is the discrete Fourier transform of the jump probability and is the analog of a characteristic function

$$\tilde{p}_k = \sum_{s=-\infty}^{\infty} e^{iks} p_s = \frac{a-1}{a} \sum_{n=0}^{\infty} \frac{1}{a^n} \cos(b^n k) \tag{4.23}$$

and for obvious reasons was called the Weierstrass walk by Hughes *et al.* [25]. The divergence of the second moment is a consequence of the non-analytic properties of the Weierstrass function.

Consider the physical interpretation of the process defined by the jump distribution (4.21) wherein the probability of taking a step of unit size is $1/a$, the probability of taking a step a factor b larger is a factor $1/a$ smaller and so on, with increasingly larger steps occurring with increasingly smaller probabilities. This is a Bernoulli scaling process that emerged in the resolution of the St. Petersburg paradox, where the player tries to determine a characteristic scale from a process that does not possess such a scale. If a billion people play the game, then, on average, half will win one coin, a quarter will win two coins, an eighth will win four coins, and so on. Winnings occur on all scales, just as do the wiggles in the Weierstrass function, which is the paradigm of fractals.

So how does this winning an order of magnitude more money (in base b) occurring with an order-of-magnitude-lower probability (in base a) translate into a random walk? As the Weierstrass walk progresses the set of sites visited by the walker consists of localized clusters of sites, interspersed by gaps, followed by clusters of clusters over a

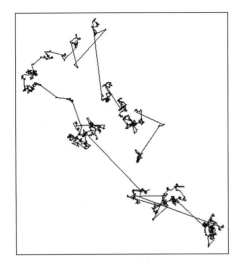

Figure 4.5.   The landing sites for the Weierstrass walk are depicted and the islands of clusters are readily seen.

larger spatial scale. This walk generates a hierarchy of clusters, which are statistically self-similar; see for example Figure 4.5. The parameter a determines the number of subclusters within a cluster and the parameter b determines the spatial scale size between clusters.

The renormalization-group analysis done previously can be used to determine the properties of the present fractal walk. Consider the structure-function argument in (4.23) scaled by the parameter b,

$$\widetilde{p}_{bk} = \frac{a-1}{a} \sum_{n=0}^{\infty} \frac{1}{a^n} \cos(b^{n+1}k),$$ (4.24)

so that, on shifting the summation index by unity, we have the renormalization equation

$$\widetilde{p}_{bk} = a\widetilde{p}_k - \frac{a-1}{a} \cos k.$$ (4.25)

The solution to the scaling equation consists of a homogeneous part $\widetilde{p}^h$ and a singular part $\widetilde{p}^s$ and the details for the derivation of the homogeneous part are given in Hughes et al. [25]. The singular part relies on the arguments presented earlier in the analysis of the Weierstrass function,

$$\widetilde{p}^s_{bk} = a\widetilde{p}^s_k,$$ (4.26)

and yields the solution

$$\widetilde{p}^s_k = \sum_{n=-\infty}^{\infty} A_n |k|^{H_n}$$ (4.27)

with the complex power-law indices

$$H_n = \frac{\ln a}{\ln b} + i\frac{n2\pi}{\ln b} = \mu + i\frac{n2\pi}{\ln b}.$$ (4.28)

The analytic forms for the coefficients in (4.27) are also given in Hughes *et al.* [25].

To simplify the analysis somewhat we go to the continuum and determine the form of the jump probability density by inverse Fourier transforming the structure function. Here again we duck some of the more complicated algebra and assume that the dominant behavior of the structure function is given by the scaling of the lowest-order singular piece, so that

$$\widetilde{p}_k = \widetilde{p}_k^h + \widetilde{p}_k^s \approx A_0 |k|^\mu . \tag{4.29}$$

Then the jump probability in one dimension is

$$p(q) = \frac{1}{2\pi} \int_{-\infty}^{\infty} e^{-iqk} \widetilde{p}_k \, dk \approx \frac{A_0}{2\pi} \int_{-\infty}^{\infty} e^{-iqk} |k|^\mu dk \tag{4.30}$$

and introducing the new variable $z = kq$ into the integral yields

$$p(q) \approx \frac{A_0}{\pi |q|^{\mu+1}} \int_0^\infty z^\mu \cos z \, dz,$$

which simply integrates to

$$p(q) \approx \frac{1}{|q|^{\mu+1}} \frac{A_0 \Gamma(\mu + 1)\cos[(\mu + 1)\pi]}{\pi}. \tag{4.31}$$

Thus, we have that the singular part of the Weierstrass walk produces an inverse power-law step distribution. It should be clear that this behavior also dominates the stepping in the random walk and consequently justifies throwing away all the other terms as we did in the above argument.

Substituting the discrete form of (4.31) into (4.6) yields

$$P_{(n+1)}(q) = \sum_{q'} \frac{A}{|q - q'|^{\mu+1}} p_n(q'), \tag{4.32}$$

which we find in the next chapter suggests a fractional integral in space. That is the direction in which this analysis is taken subsequently when a more general form of the random-walk approach is used to prove this most interesting conjecture. The discrete version of (4.32) was analyzed in 1970 by Gillis and Weiss [22], who found that the solution yields a type of general limit distribution studied by Paul Lévy and called α-stable distributions or Lévy distributions [33]. Such distributions are often found to replace the Gaussian in characterizing complex stochastic webs and are discussed later.

4.2.2   Lévy random flights

The previous section contained a demonstration of one way the second moment of the step length in a random walk can diverge. It is worth emphasizing that when the central moments of the transition probability $p(q)$ are not finite then one no longer has a simple random walk and consequently the walk no longer describes Brownian motion. The mathematician Lévy studied the general properties of such processes prior to World War II and generalized the CLT, to which the law of frequency of errors is the forerunner, to include phenomena with diverging central moments. He uncovered a class of limit distributions having properties very different from normal statistics, which were subsequently referred to as Lévy processes and Lévy distributions. In the present context the random-walker nomenclature is replaced by that of the random flyer to indicate that the moving entity does not touch all the intermediate lattice points but flies above them to touch down some distance from launch. The Lévy flight produces the kind of clustering so prominent in Figure 4.5. This clustering arises from the separation of time scales as discussed in the previous section and to which we return below. It is worth pointing out here that asymptotically the Lévy distribution has an inverse power-law form and therefore is often closely related to phenomena having those statistical properties. Let us now consider what happens when the steps in the random flight can become arbitrarily long, resulting in the divergence of the second moment.

For simplicity consider the transition probability used in the last section to have the inverse power-law form

$$p(q) = \frac{C}{|q|^{\mu+1}}, \quad 0 < \mu < 2, \tag{4.33}$$

where C is the normalization constant. The characteristic function of the transition probability is given by the Fourier transform to be

$$\tilde{p}(k) = \mathcal{FT}\left[p(q); k\right] \approx 1 - c|k|^{\mu} \tag{4.34}$$

for small k, whose inverse Fourier transform gives the probability per unit time of making long flights. The probability density for N flights is an N-fold convolution of the single-flight transition probabilities, which implies that the characteristic function for the N-flight process is given by the product of the characteristic functions for the single-flight process N times,

$$\tilde{P}_N(k) = \left[\tilde{p}(k)\right]^N \approx \left[1 - c|k|^{\mu}\right]^N \approx \exp\left[-Nc|k|^{\mu}\right]. \tag{4.35}$$

The exponential approximation used here can be made rigorous, but that does not contribute to the argument and therefore is not presented. In the continuum limit where we replace the total number of jumps N by the continuum time t, we replace (4.35) with

$$\tilde{P}(k, t) = \exp\left[-\gamma t|k|^{\mu}\right] \tag{4.36}$$

and the probability density is given by the inverse Fourier transform

$$P_{\mathrm{L}}(q, t) = \frac{1}{2\pi} \int_{-\infty}^{\infty} \exp\left[-ikq - \gamma t|k|^{\mu}\right] dk; \quad 0 < \mu \leq 2. \tag{4.37}$$

This is the equation for the centrosymmetric Lévy distribution in one dimension and consequently we use the subscript indicated.

Montroll and West [33] showed that for large values of the random variable the Lévy distribution becomes an infinite algebraic series. The lowest-order term in this series for large values of the variate is the inverse power law

$$\lim_{q \to \infty} P_{\mathrm{L}}(q, t) \propto \frac{1}{|q|^{\mu+1}}; \quad 0 < \mu \leq 2. \tag{4.38}$$

Note that (4.38) has the same form as the transition probabilities (4.33) and therefore satisfies our intuition about fractals, that being that the parts and the whole share a common property. The scaling property of the Lévy distribution is obtained by scaling the displacement q with a constant λ and the time t with a constant ζ to obtain

$$P_L(\lambda q, \zeta t) = \lambda^{-\mu} P_L(q, t) \tag{4.39}$$

as long as $\lambda = \zeta^{1/\mu}$. It should also be observed that (4.39) has the same scaling form as fractional Brownian motion if the Lévy index μ is the same as $1/H$. Thus, the self-affine scaling results from Lévy statistics with $0 < \mu < 2$ or from Gaussian statistics with a power-law spectrum having $0.5 < H \leq 1$. The scaling relation is the same in both cases and therefore cannot be used to distinguish between the two.

A further property of Lévy processes can be observed by taking the derivative of the characteristic function (4.36) to obtain

$$\frac{\partial \widetilde{P}(k, t)}{\partial t} = -\gamma |k|^{\mu} \widetilde{P}(k, t). \tag{4.40}$$

In the case $\mu = 2$ the inverse Fourier transform of this equation yields a second derivative in the displacement. The resulting equation has the form of the FPE discussed in the previous chapter. However, in general there is no explicit form for the inverse of the Fourier transform of the Lévy characteristic function and consequently there is no differential representation for the dynamics of the probability density. We shall subsequently see that the resulting equation for the Lévy probability density is a fractional equation in the spatial variable.

The lack of a differential representation for Lévy statistics can perhaps be best interpreted using the random-walk picture. Unlike Brownian motion, for a Lévy process, the random walk has steps at each point in time that can be of arbitrary length. Those steps adjacent in time are not necessarily nearby in space, and the best we can do in the continuum limit is obtain an integro-differential equation to describe the evolution of the probability density. Consider a random flight defined by a jump distribution such that the probability of taking a jump of unit size is $1/a$, and the probability of taking a jump a factor b larger is a factor $1/a$ smaller. This argument is repeated again and again as we did for the Weierstrass walk in the last section. As the flight progresses, the set of sites visited consists of localized clusters of sites, interspersed by gaps, followed by clusters of clusters over a larger spatial scale. This random flight generates a hierarchy of clusters, the smallest having about a members in a cluster, the next largest being of size b with approximately a^2 members in the larger cluster and so on. The parameters a and b determine the number of subclusters in a cluster and the spatial scale size between clusters, respectively. The fractal random flight is characterized by the parameter

$$\mu = \frac{\ln a}{\ln b}, \tag{4.41}$$

which is the fractal dimension of the set of points visited during the flight. It is worth emphasizing that the inverse power-law index is given by the parameters that determine how the spatial scales are separated in the random flight.

4.2.3   Fractional operator walk

The simple random walk has provided a way to understand the Gaussian distribution with its aggregation of a large number of identically distributed random variables having finite variance. This is the background for the law of frequency of errors discussed in the first chapter. More general statistical phenomena can be modeled by means of fractals, and we considered the case in which the jump-size distribution function is an inverse power law. The spatial heterogeneity modeled by inverse power-law jump probabilities gives rise to the class of Lévy distributions of which the Gaussian is the unique member having a finite variance. Another way to model the complexity of dynamic webs is through non-integer differential and integral operators, which we now develop in the random-walk context.

The idea of a fractional derivative or integral was strange when it was first introduced; perhaps it is still strange, but at least it is well defined and even useful. In an analogous way the notion of a fractional difference can be developed to incorporate long-term memory into the random walks we have discussed. To develop fractional differences we introduce the shift operator B defined by the operation

$$BQ_{j+1} = Q_j \tag{4.42}$$

which shifts the data from element $j + 1$ to element j; that is, B shifts the process Q_j from its present value to that of one time step earlier. Using this operator a simple random walk can be formally written as

$$(1 - B)Q_{j+1} = \xi_j, \tag{4.43}$$

where ξ_j is the discrete random force. Note that this is a discrete analog of the Langevin equation for a free particle with the time step set to unity, with a solution given by

$$Q_N = \sum_{j=1}^{N} \xi_j, \tag{4.44}$$

where the equal-magnitude contributions for the Bernoulli walk are replaced by independent identically distributed random variables. If the second moment of the random fluctuations is finite the CLT assures us that the statistical distribution of the sum will be Gaussian.

Hosking [24] generalized (4.43) to the fractional difference equation

$$(1 - B)^{\alpha} Q_{j+1} = \xi_j, \tag{4.45}$$

where the power-law index α need not be an integer. We assert here without proof that equations of the fractional random-walk form are the direct analog of the fractional Langevin equation discussed in Chapter 5 in the same way that the traditional random walks are direct analogs of the ordinary Langevin equations discussed in Chapter 3. The question remains that of how to operationally interpret the formal expression (4.45).

Equation (4.45) is made usable by interpreting the operator $(1 - B)^\alpha$ in terms of the binomial theorem when the power-law index is an integer,

$$(1 - B)^\alpha = 1 - \alpha B - \frac{1}{2!}\alpha(1 - \alpha)B^2 - \frac{1}{3!}\alpha(1 - \alpha)(2 - \alpha)B^3 - \cdots$$

$$= \sum_{n=0}^{\infty} \binom{\alpha}{n} (-B)^n , \tag{4.46}$$

where, of course, this procedure is defined only when operating on an appropriate function. The binomial coefficient for a positive integer α is formally given by

$$\binom{\alpha}{n} = \frac{\Gamma(\alpha + 1)}{\Gamma(n + 1)\Gamma(\alpha + 1 - n)}, \tag{4.47}$$

where $\Gamma(\cdot)$ denotes a gamma function. Note that for negative integers the gamma function has poles so that the binomial coefficient is zero if $n > \alpha$ in (4.46) and α is integer-valued. In the case of arbitrary complex α and β with the restriction $\alpha \neq -1, -2, \ldots$, we can write

$$\binom{\alpha}{\beta} = \frac{\Gamma(\alpha + 1)}{\Gamma(\beta + 1)\Gamma(\alpha + 1 - \beta)} = \frac{\sin[(\beta - \alpha)\pi]}{\pi} \frac{\Gamma(\alpha + 1)\Gamma(\beta - \alpha)}{\Gamma(\beta + 1)}, \tag{4.48}$$

which is readily established using the Euler integral of the second kind

$$\Gamma(z) = \int_0^\infty x^{z-1}e^{-x}\,dx; \quad \mathrm{Re}\,z > 0. \tag{4.49}$$

Using this extension to the complex domain for the gamma function, the formal expression (4.46) is equally valid; however, the series no longer cuts off for $n > \alpha$ and is an infinite series for non-integer values of the power-law index.

The fractional difference equation has the formal solution

$$Q_{n+1} = (1 - B)^{-\alpha}\xi_j, \tag{4.50}$$

where we define the inverse operator

$$Q_{j+1} = \sum_{k=0}^{\infty} \theta_k B^k \xi_j = \sum_{k=0}^{\infty} \theta_k \xi_{j-k}. \tag{4.51}$$

Here we see that the solution to the fractional difference equation at time $j + 1$ is determined by random fluctuations that extend infinitely far back into the past with the strength of that influence being determined by the functional dependence of the coefficients θ_k on k. These coefficients are determined from the convergent expansion of the function $\theta(z) = (1 - z)^{-\alpha}$ for $|\alpha| < 1/2$,

$$(1 - z)^{-\alpha} = \sum_{n=0}^{\infty} \binom{-\alpha}{n} (-z)^n , \tag{4.52}$$

and using some identities among gamma functions,

$$\binom{-\alpha}{n} = \frac{\Gamma(1 - \alpha)}{\Gamma(n + 1)\Gamma(1 - \alpha - n)} = \frac{(-1)^n \Gamma(\alpha + n)}{\Gamma(n + 1)\Gamma(\alpha)},$$

so that

$$\theta(z) = \sum_{n=0}^{\infty} \frac{\Gamma(\alpha + n)}{\Gamma(n + 1)\Gamma(\alpha)} z^n \tag{4.53}$$

for $|z| \leq 1$. Therefore the coefficients in the solution (4.51) are given by

$$\theta_k = \frac{\Gamma(\alpha + k)}{\Gamma(k + 1)\Gamma(\alpha)}, \tag{4.54}$$

from which the influence of the distant past is determined by the asymptotic form of the coupling coefficient. Using the analytic properties of gamma functions it is straightforward to show that for $k \to \infty$

$$\theta_k \approx \frac{k^{\alpha-1}}{\Gamma(\alpha)}, \tag{4.55}$$

so the strength of the contribution to the solution decreases with increasing time lag as an inverse power law asymptotically as long as $\alpha < 1/2$.

Financial time series

The fractional walk was originally developed for the study of long-time memory in financial time series. The observable most often used to characterize the memory is the slope of the spectrum obtained from the average square of the Fourier amplitude of the time series. Given a set of measurements defined by (4.51) the discrete Fourier amplitudes of the observed time series can be written as the product

$$\tilde{Q}_\omega = \tilde{\theta}_\omega \tilde{\xi}_\omega \tag{4.56}$$

due to the convolution form of the solution to the fractional difference equation. The spectrum of the solution is

$$S(\omega) = \left\langle \left| \tilde{Q}_\omega \right|^2 \right\rangle_\xi, \tag{4.57}$$

where the subscripted angle brackets denote an average over an ensemble of realizations of the fluctuations driving the financial network. Inserting (4.56) into (4.57) yields

$$S(\omega) = \left| \tilde{\theta}_\omega \right|^2 \left\langle \left| \tilde{\xi}_\omega \right|^2 \right\rangle_\xi, \tag{4.58}$$

where for white-noise fluctuations the noise spectrum is the constant D_ξ.

The Fourier coefficient $\tilde{\theta}_\omega$ is calculated by taking the discrete Fourier transform of θ_k. Thus, for $0 < \omega \leq \pi$,

$$\tilde{\theta}_\omega = \sum_{k=0}^{\infty} \theta_k e^{-ik\omega}$$

$$= \sum_{k=0}^{\infty} \frac{(k + \alpha - 1)!}{k!(\alpha - 1)!} \left(e^{-i\omega} \right)^k = \frac{1}{\left(1 - e^{-i\omega} \right)^\alpha}, \tag{4.59}$$

which, when inserted into (4.58) and on rearranging terms, yields

$$S_Q(\omega) = \frac{D_\xi}{(2\sin[\omega/2])^{2\alpha}} = \frac{D_\xi}{\omega^{2\alpha}} \text{ for } \omega \to 0. \tag{4.60}$$

Equation (4.60) is the inverse power-law spectrum for the fractional-differenced white noise in the low-frequency limit. This inverse power-law behavior in the spectrum is often called $1/f$ noise and the underlying process is described as a $1/f$ phenomenon.

From this analysis we see that the infinite moving-average representation of the fractional-differenced white-noise process shows that the statistics of Q_j are the same as those of ξ_j since (4.51) is a linear equation. Thus, because the statistics of ξ_j are Gaussian, so too are the statistics of the observed process. The spectrum of the random force is flat, since it is white noise, and the spectrum of the observed time series is an inverse power law. From these analytic results we conclude that Q_j is analogous to fractional Brownian motion. The analogy is complete if we set $\alpha = H - 1/2$ so the spectrum can be written as

$$S_Q(\omega) = \frac{D_\xi}{\omega^{2H-1}} \text{ for } \omega \to 0. \tag{4.61}$$

We now consider a random-walk process driven by fluctuations with long-time memory using

$$X_{j+1} = X_j + Q_j \tag{4.62}$$

with the solution given by the sum of measurements, each with a long-time memory,

$$X_j = \sum_{k=0}^{j} Q_k. \tag{4.63}$$

The autocorrelation function for this process is

$$C(t) = \langle X_j X_{j-t} \rangle \propto t^{2H}, \tag{4.64}$$

where we again use $\alpha = H - 1/2$. The corresponding spectrum is given by

$$S_X(\omega) \propto \frac{D_\xi}{\omega^{2H+1}}, \tag{4.65}$$

which can be seen intuitively from the fact that the time series X_j is essentially the time integral of Q_j and consequently the spectra are related by

$$S_X(\omega) = \frac{S_Q(\omega)}{\omega^2}. \tag{4.66}$$

The long-time memory of financial time-series data has been well established starting from the application of fractional difference equations by Hosking [24] in 1981. One of the more recent applications of these ideas to financial time-series data was given by Pesee [36], who studied the daily changes in the values of various currencies against the US Dollar (USD), including the French Franc (FRF), the Deutsche Mark (DM), the Euro (EUR) and the Japanese Yen (JPY). Each of the time series has a variability

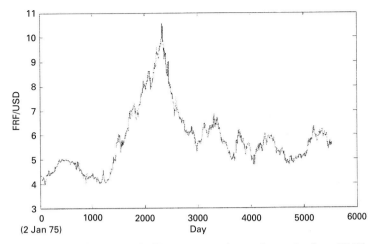

Figure 4.6. The time series for the FRF/USD currency exchange time series from 1/2/75 to 1/15/99 with a total of 5,529 data points [36]. Reproduced with permission.

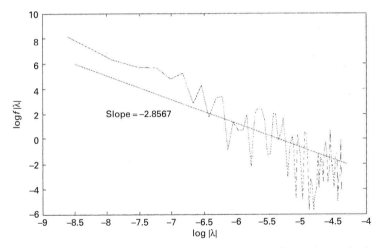

Figure 4.7. The power spectral density for the FRF/USD currency exchange time series from 1/2/75 to 1/15/99 with a slope of –2.86 and a corresponding Hurst exponent of $H = 0.93$ [36]. Reproduced with permission.

such as that shown in Figure 4.6, which is a combination of short-time variations and long-time excursions.

The typical financial time series in Figure 4.6 gives rise to the inverse power-law spectral density shown in Figure 4.7. The best fit to the individual slopes of the spectra lies in the interval $-2.86 \leq$ slope ≤ -2.11. The individual slopes are associated with $-(2H + 1)$ in the spectral density, so the Hurst parameter lies in the interval $0.55 \leq H \leq 0.93$. Pesee argues that since the Hurst indices for JPY/USD and EUR/USD approach 0.5 there is no memory in these cases and these financial time series can be represented by standard random walks, similarly to the observation made by Bachelier. On the other hand, there is clear evidence of the long-time memory for the FRF/USD and the DM/USD time series because the Hurst index in these cases approaches the deterministic value of one.

Anomalous diffusion

If we restrict the discussion to random walks then a second moment of a time series that increases algebraically in time, $\langle Q^2(t) \rangle \propto t^\alpha$, describes a phenomenon that can potentially be separated into three parts: classical diffusion with $\alpha = 1$, superdiffusion with $2 \geq \alpha > 1$ and subdiffusion with $1 \geq \alpha > 0$. Any algebraic time behavior other than $\alpha = 1$ is referred to as anomalous. In the simple random-walk context it is assumed in the literature that $\alpha = 2H$ and the Hurst exponent is restricted to the range $1 \geq H > 0$.

However, in the above discussion a random force with long-term memory was shown to induce a power-law exponent $\alpha > 2$ or equivalently $H > 1$, a value not allowed in the simple random-walk picture but which is reasonable when memory is included in the description. Consequently, such general random walks might also be used to "explain" the inverse power laws observed in the connections between elements in certain kinds of complex webs. The question of whether this modeling of the second moment is unique arises and, if it is not, is there a test for the non-uniqueness?

The interpretation in terms of long-time memory is a consequence of the ordering of the data in the time series. If the data points in the time series are randomly shuffled the memory disappears; however, since no data points are added or subtracted in the shuffling process, the statistics of the time series remains unchanged. A direct test of whether the algebraic form of the second moment results from memory or from some exotic statistical distribution is to randomly rearrange the data points and recalculate the second moment. In this way a time series with a second moment with $\alpha \neq 1$ before shuffling and $\alpha = 1$ after shuffling is shown to have long-time memory. On the other hand, if the value of the power-law index is not changed by shuffling the data, the anomaly in the second moment is due to non-Gaussian statistics.

We subsequently determine that the observed scaling of the second moment can guide us in the formulation of the proper model of the $1/f$ phenomenon. However, phenomenological random walks are not the only way in which statistics can be introduced into the dynamics of a process. Another way in which the variability can appear random is through the chaotic solutions of nonlinear equations of motion, as mentioned earlier. So now let us turn our attention to maps, chaos and strange attractors.

4.3   Maps, chaos and strange attractors

As late as the end of the 1960s, when the authors were graduate students in physics, it was not general knowledge that the traditional techniques for solving the equations of motion for complex mechanical networks were fundamentally flawed. Most physicists were ignorant of the defects in the canonical perturbation methods developed for celestial mechanics. These techniques were taught by them unabashedly in most graduate courses on analytic dynamics. The story of the documentation of the breakdown of canonical perturbation theory begins in 1887 when the king of Sweden, Oscar II, offered a prize of 2,500 crowns for the answer to the question of whether or not the solar system is stable. One of the respondents (and the eventual winner of this contest) was Henri Poincaré (1854–1912). The King wanted to know whether the planets would remain in their present orbits for eternity, or whether the Moon would escape from the Earth or crash into it, or whether the Earth would meander into the Sun. This was and still is one of the open questions in astronomy, not so much for its practical importance, but because we ought to be able to answer the question. It is a testament to his genius for analysis that Poincaré was able to win the prize without directly answering the question.

In 1890 Poincaré submitted a monograph establishing a new branch of mathematics, or more properly a new way of analyzing an established branch of mathematics. He used topological ideas to determine the general properties of dynamical equations. This new tool was applied to the problem in celestial mechanics of many bodies moving under mutual Newtonian gravitational attraction. He was able to show that the two-body problem had periodic solutions in general, but that if a third body is introduced into this universe no simple analytic solution could be obtained. The three-body problem could not be solved by the methods of celestial mechanics! He was able to show that if the third body had a lighter mass than the other two its orbit would have a very complex structure and could *not* be described by simple analytic functions. One hundred years later it was determined that this complicated orbit is fractal, but to Poincaré the complexity of the three-body problem indicated the unanswerability of the King's question regarding the stability of the solar system. For an excellent discussion see either Stewart [41] or Ekeland [19], or, better yet, read both. Thus, even for the epitome of clockwork motion given by the motion of the planets, unpredictability intrudes.

Somewhat later Poincaré [38] was able to articulate the general perspective resulting from the failure to solve the three-body problem:

A very small cause which escapes our notice determines a considerable effect that we cannot fail to see, and then we say that the effect is due to chance. If we knew exactly the laws of nature and situation of the universe at the initial moment, we could predict exactly the situation of that same universe at a succeeding moment. But even if it were the case that the natural laws had no longer any secret for us, we could still only know the initial situation approximately. If that enables us to predict the succeeding situation with the same approximation, that is all we require, and we should say that the phenomenon had been predicted, that it is governed by laws. But it is not always so: it may happen that small differences in the initial conditions produce very great ones in the final phenomena. A small error in the former will produce an enormous error in the latter. Prediction becomes impossible, and we have the fortuitous phenomenon.

Poincaré sees an intrinsic inability to make long-range predictions due to a sensitive dependence of the evolution of a network on its initial state, even though the network is deterministic. This was a departure from the view of Laplace, who believed in strict determinism, and to his mind this implied absolute predictability. Uncertainty for Laplace was a consequence of imprecise knowledge, so that probability theory was necessitated by incomplete and imperfect observations.

It is Poincaré's "modern" view of mechanics, a view that took over 100 years for the scientific community to adopt, if only in part, that we make use of in this chapter. Moreover, we develop this view keeping in mind that our end goal extends far beyond the mechanical behavior of physical networks, to the generic dynamics of other complex webs. For the moment we quote Poincaré's great theorem:

The canonical equations of celestial mechanics do not admit (except for some exceptional cases ...) any analytic and uniform integral besides the energy integral.

Of course, deterministic equations are only one way to describe how a complex web changes over time and chaos is only one form of uncertainty. But it is worth pointing out that there is more than one kind of chaos. One type of chaos that is not widely known outside physics is associated with the breakdown of nonlinear conservative dynamic webs. Another sort of chaos has to do with strange attractors, those being nonlinear webs with dissipation. We only briefly touch on the latter here, but enough to trace the source of unpredictability to the dynamics. The other source of unpredictability has to do with the statistics of webs arising from their contact with infinitely unknowable environments. Linear webs of this kind are modeled by linear stochastic differential equations as we saw in the last chapter, and those whose complexity is influenced by memory of the past are accounted for using fractional stochastic differential equations as discussed in the previous section.

The oldest scientific discipline dealing with dynamics is classical mechanics dating from the attempts of Galileo and Newton to describe the dynamics of our physical world. Newton's formalism developed into the equations of motion of Lagrange and then into those of Hamilton. In the middle of the last century it became well known that classical mechanics suffered from certain fundamental flaws and the theory of Kolmogorov [27], Arnol'd [3] and Moser [34] (KAM) laid out the boundaries of the classical description. KAM theory introduces the theory of chaos into non-integrable Hamiltonian systems. The pendulum provides the simplest example of the continuous nonlinear phenomena with which KAM theory is concerned. The breakdown of trajectories heralded by KAM theory is the first proof that the traditional tools of physics might not be adequate for describing general complex webs.

The KAM theory for conservative dynamical systems describes how the continuous trajectories of particles determined by Hamiltonians break up into a chaotic sea of randomly disconnected points. Furthermore, the strange attractors of dissipative dynamical systems have a fractal dimension in phase space. Both these developments in classical dynamics, KAM theory and strange attractors, emphasize the importance of non-analytic functions in the description of the evolution of deterministic nonlinear dynamical networks. We briefly discuss such dynamical webs herein, but refer the reader to a number of excellent books on the subject, ranging from the mathematically rigorous, but readable [35], to provocative picture books [1] and texts with extensive applications [44].

4.3.1   The logistic equation

The size of the human population grew gradually for about sixteen centuries and then at an increasing rate through the nineteenth century. Records over the period 1200 to 1700, while scanty, show some fluctuations up and down, but indicate no significant trend.

By the end of the eighteenth century it had been observed by a number of thoughtful people that the population of Europe seemed to be doubling at regular time intervals, a scaling phenomenon characteristic of exponential growth. Thomas Robert Malthus is most often quoted in this regard [28]; however, Thomas Jefferson and others had made similar observations earlier. It is curious that the most popular work on population growth was written by a cleric writing a discourse on moral philosophy, who had not made a single original contribution to the theory of how and why populations grow. The contribution of Malthus was an exploration of the consequences of the fact that a geometrically growing population always outstrips a linearly growing food supply, resulting in overcrowding and misery. Of course, there was no evidence for a linearly growing food supply, this was merely a convenient assumption that supported Malthus' belief that all social improvements only lead to an increase in the equilibrium population, which results in an increase in the sum total of human misery.

Consider the dynamical population variable Q_n with a discrete time index n denoting the iteration number or the discrete "time." The simplest mapping relates the dynamical variable in generation n to that in generation $n + 1$,

$$Q_{n+1} = \lambda Q_n. \tag{4.67}$$

Here the dynamical variable could be a population of organisms per unit area on a Petri dish, or the number of rabbits in successive generations of mating. It is the discrete version of the equation for the growth of human population postulated by the minister Malthus. The proportionality constant is given by the difference between the birth rate and the death rate and is therefore the net rate of change in the population, sometimes called the net birth rate. Suppose that the population has a level Q_0 in the initial generation. The linear recursion relation (4.67) yields the sequence of values

$$Q_1 = \lambda Q_0, \quad Q_2 = \lambda Q_1 = \lambda^2 Q_0, \quad \ldots, \tag{4.68}$$

so that in general we have

$$Q_n = \lambda^n Q_0, \tag{4.69}$$

just as we found in (2.4). This simple exponential solution exhibits a number of interesting properties in the context of population growth.

First of all, if the net birth rate λ is less than unity then the population decreases exponentially between successive generations. This is a consequence of the fact that, with $\lambda < 1$, the population of organisms fails to reproduce itself from generation to generation and therefore exponentially approaches extinction:

$$\lim_{n \to \infty} Q_n = 0 \text{ if } \lambda < 1. \tag{4.70}$$

On the other hand, if $\lambda > 1$, the population increases exponentially from generation to generation. This is a consequence of the fact that the population produces an excess at each generation, resulting in a population explosion. This is Malthus' exponential population growth:

$$\lim_{n \to \infty} Q_n = \infty \text{ if } \lambda > 1. \tag{4.71}$$

The only value of the parameter for which the population does not have these extreme tendencies asymptotically is $\lambda = 1$, for which, since the population reproduces itself exactly in each generation, we obtain the unstable situation

$$\lim_{n \to \infty} Q_n = Q_0 \text{ if } \lambda = 1. \tag{4.72}$$

In this way the initial population remains unchanged from one generation to the next, unless λ deviates slightly from unity, in which case the population diverges to zero or infinity.

Verhulst, himself a scientist, put forth a theory that mediated the pessimistic view of the cleric Malthus. He noted that the growth of real populations is not unbounded, since such factors as the availability of food, shelter and sanitary conditions all restrict, or at least influence, the growth of populations. In particular, he assumed the net birth rate to decrease with increasing population in a linear way and made the replacement

$$\lambda \to \lambda(1 - Q/\Theta), \tag{4.73}$$

where Θ is the population's saturation level. Thus, the linear recursion relation (4.67) is replaced with the nonlinear discrete logistic equation

$$Q_{n+1} = \lambda Q_n(1 - Q_n/\Theta). \tag{4.74}$$

It is clear that, when the population is substantially below the saturation level, that is $Q_n/\Theta \ll 1$, the growth is exponential since the nonlinear term is negligibly small compared with the linear term $\lambda Q_n \gg \lambda Q_n^2/\Theta$. However, at some point in time (iteration number) the ratio Q_n/Θ will be of order unity and the rate of population growth will be retarded. When $Q_n/\Theta = 1$ there are no more births and the population stops growing. Mathematically the regime $Q_n/\Theta > 1$ corresponds to a negative birth rate, but this does not make biological sense and so we restrict the region of interpretation of this particular model to $(1 - Q_n/\Theta) > 0$.

Finally, we reduce the number of parameters from two to one by introducing the scaled variable $Y_n = Q_n/\Theta$, the fraction of the saturation level achieved by the population. In terms of this ratio the logistic equation becomes

$$Y_{n+1} = \lambda Y_n(1 - Y_n), \tag{4.75}$$

where the maximum population fraction is unity. In Figure 4.8 the nearly continuous solution to the logistic map is given, showing a smooth growth to saturation. Note that the saturation level is achieved when $Y_{n+1} = Y_n$, so that the saturation level is $Y_{\text{sat}} = 1 - 1/\lambda$ rather than unity; or, in terms of the population variable, the saturation level is not Θ.

The form of the saturation curve in Figure 4.8 ought to be familiar. It appears in essentially all discussions of nonlinear phenomena that have a smooth switching behavior from one value of the dependent variable to another. In this parameter regime the mapping appears to be analogous to a discretization of the continuous logistic equation

$$\frac{dY}{dt} = kY(1 - Y). \tag{4.76}$$

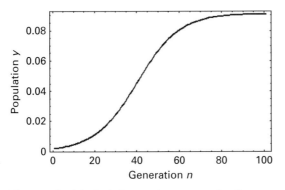

Figure 4.8. The growth of the logistic equation to saturation for a parameter value given by $\lambda = 1.1$ so that $Y_{\text{sat}} = 0.09$.

The solution to (4.76) is readily obtained through the transformation $Z = 1/Y$, which yields the linear equation

$$\frac{d}{dt}(Z - 1) = -k(Z - 1),$$ (4.77)

which immediately integrates to

$$\log\left[\frac{Z(t) - 1}{Z(0) - 1}\right] = -kt.$$ (4.78)

A little algebra gives the solution in terms of the original variable

$$Y(t) = \frac{Y(0)e^{kt}}{1 + (e^{kt} - 1)Y(0)},$$ (4.79)

where we see that asymptotically the saturation level is unity. This asymptotic value clearly shows that the logistic map is not simply a discretization of the continuous logistic equation; the relation between them is more subtle. Moreover, the early-time growth is exponential since, when $t \ll 1/k$, the second term in the denominator can be neglected to obtain

$$Y(t) \approx Y(0)e^{kt}$$ (4.80)

in accordance with the unbounded growth feared by Malthus and many of his contemporaries. Note, however, that this exponential behavior is indicative of the growth only at very early times.

A great deal has been written about how nonlinear maps become unstable in certain regimes of the control parameters, but we do not go into that literature here. Our intent is to show how an equation as benign looking as (4.75) can give rise to a wide variety of dynamical behaviors. In Figure 4.9 the iterations of this equation are graphed for four values of the control parameter λ. In Figure 4.9(a) with $\lambda = 2.9$ the population grows, oscillates about its asymptotic level and eventually is attracted to the saturation level 0.655. In Figure 4.9(b) the control parameter is $\lambda = 3.2$. This value gives rise to a bifurcation in the dynamics and the solution oscillates between two values, asymptotically producing a 2-cycle. Note that a bifurcation is a qualitative change in the dynamics.

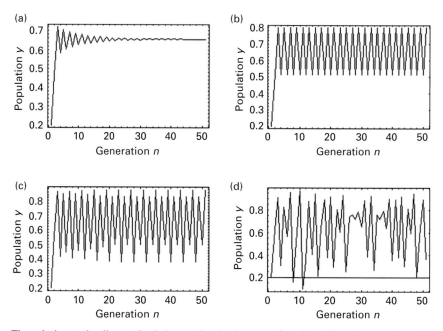

Figure 4.9.   The solution to the discrete logistic equation is given as a function of iteration number n for values of the control parameter: (a) $\lambda = 2.9$ in the region of a single asymptotic state; (b) $\lambda = 3.2$ yields a 2-cycle solution; (c) $\lambda = 3.5$ yields a region with 4-cycle solutions; and (d) $\lambda = 3.9$ is a region of bounded chaos.

In Figure 4.9(c) the value of the control parameter is $\lambda = 3.5$. This value of the control parameter gives rise to two bifurcations and a solution that oscillates among four values, asymptotically a 4-cycle. Finally, in Figure 4.9(d) the control parameter $\lambda = 3.9$ is selected. In this last case the dynamics are chaotic, which is to say that there is apparently a complete loss of regularity in the solution. This would appear to be complex dynamical behavior.

Thus, using a traditional definition of complexity, it would appear that chaos implies the generation of complexity from simplicity. It is also worth noting that chaos is a generic property of nonlinear dynamical webs, that is, it is ubiquitous; all networks change over time, and, because all naturally occurring dynamical networks are nonlinear, all such webs manifest chaotic behavior to a greater or lesser extent.

One way to interpret what is being represented by maps, such as the logistic equation, is that the time points in the map are equally spaced values of a continuous process. Consider a dance floor that is lit by a stroboscopic lamp, otherwise the dancers are in complete darkness. The motion of the dancers seems to occur in abrupt moves, but you can deduce where the dancers were during the periods of darkness because human motion is continuous. In the same way the dynamical unfolding of a continuous process can be recorded at a set of equally spaced intervals, say every second, or at unequally spaced time intervals, say by recording the amplitude each time the derivative of the time series vanishes, or using any of a number of other criteria to discretize continuous time series. The data points are related because they are each generated by the same dynamical web even though we have apparently thrown away a great deal of information, that is, the values of the time series between the measurements, when we are not watching. However, this is what is typically done in taking measurements, whether to determine the regularity in the beating of the heart, the uniformity of stride intervals or the patterns in breathing. Therefore, under some circumstances one can construct a map to generate the same set of data as the continuous generator and consequently the map is a discrete equivalent to the continuous dynamics, and, since they both give rise to the same data, they are indistinguishable.

In Figure 4.10 two examples of dynamical processes are depicted. The two time series look similar in their degree of irregularity. Bassingthwaighte *et al.* [6] point out that the two time series were constructed to have the same statistical properties, the same average value and approximately the same variance. Both time series look random, but we now know that not everything that fluctuates erratically in time is random in time. The time series on the left was generated using a computer random-number generator and would therefore traditionally be called noise. The time series on the right, on the other hand, is not random at all, in this sense, since it was generated by the logistic equation with a parameter value of $\mu = 3.95$. So how do we tell the randomness in the first time series from the chaos in the second time series?

A phenomenological technique for revealing hidden patterns in the time series is to use the data points as the coordinate axis for a phase space. In Figure 4.10 the phase-space coordinates are the data points separated by one time unit; that is, Y_{n+1} is plotted

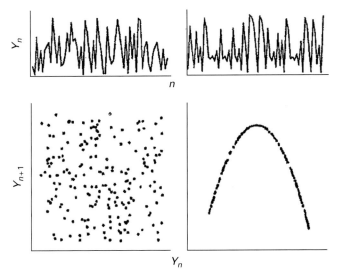

Figure 4.10.   Top: two time series that look random. The one on the left is random; each new value was chosen by a random-number generator. But the one on the right was generated by the logistic equation. Bottom: the true identity of each time series is revealed in the phase space where Y_{n+1} is plotted versus Y_n. The series on the left fills this two-dimensional phase space, indicating that it is random. The chaotic time series on the right is a one-dimensional line (map), indicating that it can be generated by a deterministic relationship having one dependent variable [6]. Reproduced with permission.

versus Y_n. On the left the splatter of points seems to randomly fill the two-dimensional space, strongly suggesting that the successive time points in the time series are truly independent of one another. However, on the right we see that successive data points fall on the curve that defines the map with which they were generated. The randomness in the upper-right time series has to do with the order in which the values are generated by the map; it is the order of appearance of the values of Y which looks random. All the values arise from the curve on the lower right, but the values on this curve do not appear in an ordered way but "randomly" jump from one value to the next. Chaos is the result of the sensitivity to initial conditions, so, no matter how small a change one makes in the initial condition, eventually the generated time series look totally different.

A coarse-grained version of this map, one in which values of Y in the interval $1/2 \leq Y \leq 1$ are assigned one value, say +1, and those in the interval $0 \leq Y \leq 1/2$ are assigned a second value, say 0, generates a Bernoulli sequence. Recall that the elements of a Bernoulli sequence are statistically independent of one another. Thus, the logistic map may be used as a random-number generator. The assumption that the sequence of zeros and ones generated in this way form a Bernoulli sequence can be tested.

Another use of the logistic map is to identify Y_n with the following quantities: in genetics, the change in the gene frequency between successive generations; in epidemiology, the fraction of the population infected at a given time; in psychology, the number of bits of information that can be remembered up to a generation n in certain learning theories; in sociology, the number of people having heard a rumor at time n and how the rumor is propagated in societies of various structures; see May [31] for a more extended discussion. Consequently, the potential application of these maps is restricted only by our imaginations. Conrad [16] suggested five functional categories to identify the role that chaos might play in biological networks: (1) search, (2) defense, (3) maintenance, (4) cross-level effects and (5) dissipation of disturbances. But it must also be acknowledged that using chaotic time series to model such phenomena has not been universally embraced.

One of the measures of chaos is the attractor-reconstruction technique (ART) in which the axes chosen for the phase space are empirical. These axes are intended to provide an indication of the number of independent variables necessary to describe the dynamical network. The idea is that a time series for a nonlinear dynamical process contains information about all the variables necessary to describe the network. If the number of variables is three then in the phase space the dynamics unfold on a three-dimensional manifold. If the time series is fractal the three-dimensional Euclidean space might contain a fractal manifold with fractal dimension in the interval $2 < D < 3$. Even in two dimensions the trajectory would be confined to a restricted region on which the higher-dimensional dynamics would be projected.

An example of such an erratic time series is the electroencephalogram (EEG) of the activity of the human brain. If $\{X_n\}$ is a discretely sampled EEG record then a two-dimensional phase space could be constructed from the data pairs $\{X_n, X_{n+\tau}\}$, where τ is an empirically determined integer. In Figure 4.11 EEG time series recorded under a variety of conditions are given on the left. On the right is the corresponding two-dimensional phase space obtained using ART. While the phase-space structures lack

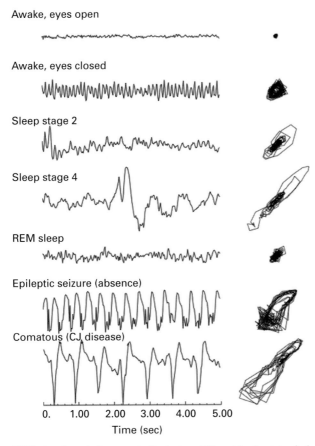

Awake, eyes open

Awake, eyes closed

Sleep stage 2

Sleep stage 4

REM sleep

Epileptic seizure (absence)

Comatous (CJ disease)

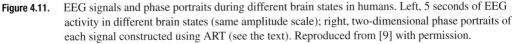

0.   1.00   2.00   3.00   4.00   5.00
Time (sec)

Figure 4.11. EEG signals and phase portraits during different brain states in humans. Left, 5 seconds of EEG activity in different brain states (same amplitude scale); right, two-dimensional phase portraits of each signal constructed using ART (see the text). Reproduced from [9] with permission.

the clarity of the mapping in Figure 4.10, they do not fill the phase space as would a purely random time series.

Complex dynamical webs have provided a new kind of insight into the evolution of such networks. As stated by Joe Ford [21],

... Determinism means that Newtonian orbits exist and are unique, but since existence-uniqueness theorems are generally not constructive, they assert nothing about the character of the Newtonian orbits they define. Specifically, they do not preclude a Newtonian orbit passing every computable test for randomness or being humanly indistinguishable from a realization of a truly random process. Thus, popular opinion notwithstanding, there is absolutely no contradiction in the term "deterministically random." Indeed, it is quite reasonable to suggest that the most general definition of chaos should read: chaos means deterministically random ...

Thus, if nonlinearities are ubiquitous then so too must be chaos. This led Ford to speculate on the existence of a generalized uncertainty principle based on the notion that the fundamental measures of physics are actually chaotic. The perfect clocks and meter sticks of Newton are replaced with "weakly interacting chaotic substitutes" so that the act of measurement itself introduces a small and uncontrollable error into the quantity being measured. Unlike the law of errors conceived by Gauss, which is based on linearity and the principle of superposition of independent events, the postulated errors arising from nonlinearities cannot be reduced by increasing the accuracy of one's measurements. The new kind of error is generated by the intrinsic chaos associated with physical being. Therefore, unlike the bell-shaped distribution resulting from the linear additive error concepts of Gauss, which have inappropriately been applied to many social, medical and psychological webs, we have inverse power-law distributions that require the underlying process to be contingent rather than independent, multiplicative rather than additive and nonlinear rather than linear.

4.3.2   Strange attractors

Most phenomena of interest cannot be described by a single variable, so even when we suspect that the dynamics being examined are chaotic, those dynamics are probably not described by the one-humped maps just discussed. It is often the case that we do not know how many variables are necessary to describe the web of interest. For example, in something as complex as the human brain we are restricted by what we can measure, not by what we know to be important. Thus, the EEG signal is often used to characterize the operation of the brain even though we do not know how this voltage trace recorded in the various EEG channels is related to brain function. This is a familiar situation to the experimenter who is attempting to understand the behavior of a complex web. A few variables are measured and from these data sets (usually time series) the fundamental properties of the web must be induced. Such induction is often made irrespective of whether or not the web is modeled by a set of deterministic dynamical equations. The difficulty is compounded when the time series is erratic because then one must determine whether the random variations are due to noise in the traditional sense of the web interacting with the environment, or whether the fluctuations are due to chaos, which is an implicit dynamical property of the web.

   Now we bring together the ideas of a chaotic map and the continuous dynamical equations of mechanics. Envision the one-dimensional mapping as being the intersection of a continuous trajectory with a line. If the map is chaotic, this implies that the original continuous trajectory is also chaotic. To generalize this remark to higher-order networks, consider the intersection of a higher-order dynamical process with a two-dimensional plane. A trajectory in three dimensions is sketched in Figure 4.12 and its intersection with a cross-hatched plane is indicated. For a Hamiltonian web the trajectory is a solution to the equations of motion and lies on the surface of a torus. However, we are interested only in the discrete times when this trajectory intersects the plane, the so-called Poincaré surface of section. The points $\{X_n, Y_n\}$ denote the intersections of the trajectory with the orthogonal axes x and y on the plane. These points can also be generated by the two-dimensional map

$$X_{n+1} = g_1(X_n, Y_n),$$
$$Y_{n+1} = g_2(X_n, Y_n). \qquad (4.81)$$

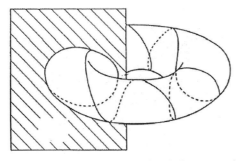

Figure 4.12. A trajectory in a three-dimensional phase space intersects a plane at a number of points that are labeled by the plane's coordinates. These points contain the measured information about the trajectory. Here an invariant torus is defined by the actions J_1 and J_2 and angles $\theta_1(t) = \Omega_1 t + \theta_1(0)$ and $\theta_2(t) = \Omega_2 t + \theta_2(0)$. The trajectory is periodic if the ratio of the frequencies Ω_1/Ω_2 is rational and the orbit intersects the plane at a finite number of points. The trajectory is quasi-periodic if the ratio of the frequencies is irrational and the orbit intersects the plane at a continuum of points that eventually sculpts out the torus.

If the quantities X_n and Y_n can be uniquely determined in terms of their values in the next iteration, X_{n+1} and Y_{n+1}, meaning a functional relation of the form

$$X_n = G_1(X_{n+1}, Y_{n+1}),$$
$$Y_n = G_2(X_{n+1}, Y_{n+1}),$$
(4.82)

then the two-dimensional map is invertible. If n is the time index of the map then invertibility is equivalent to time reversibility, so these maps are reversible in time, whereas the one-humped maps are not. The maps discussed here are analogous to Newtonian dynamics.

An example of a simple two-dimensional map is given by

$$X_{n+1} = f(X_n) + Y_n,$$
$$Y_{n+1} = \beta X_n,$$
(4.83)

where, if the mapping function $f(\cdot)$ is not invertible and $\beta = 0$, the dynamic equation reduces to the one-dimensional map of the last section. For any non-zero β, however, the map (4.83) is invertible since the inverse function can be defined as

$$X_n = Y_{n+1}/\beta,$$
$$Y_n = X_{n+1} - f(Y_{n+1}/\beta).$$
(4.84)

In this way a non-invertible map can be transformed into an invertible one, or, said differently, extending the phase space from one to two dimensions lifts the ambiguity present in the one-dimensional map.

Another property of interest is that the volume of phase space occupied by a dynamical web remains unchanged (conserved) during evolution. For example, the cream in your morning coffee can be stirred until it spreads from a concentrated blob of white in a black background into a uniformly tan mixture. The phase-space volume occupied by the cream does not change during this process of mixing. In the same way a volume V_n remains unchanged during iterations if the Jacobian of the iteration

$$dV_{n+1} = \begin{vmatrix} \dfrac{\partial X_{n+1}}{\partial X_n} & \dfrac{\partial Y_{n+1}}{\partial X_n} \\ \dfrac{\partial X_{n+1}}{\partial Y_n} & \dfrac{\partial Y_{n+1}}{\partial Y_n} \end{vmatrix} dV_n \qquad (4.85)$$

is unity. This unit value of the Jacobian is a property of conservative mechanical networks implying that the phase-space volume is conserved during the web's evolution. This conservation law is known in mechanics as Liouville's theorem. In the more general case this volume is not conserved and the phase-space volume contracts due to the occurrence of dissipation. For the two-dimensional map (4.84) the magnitude of the Jacobian is β, which determines the magnitude of the volume at consecutive iterations.

After n iterations of the map the volume changes by n factors of β so that if $|\beta| < 1$ the volume contracts by this factor with each application of the mapping. This volume contraction does not imply that the solution goes to a point in phase space, however, but only that the solution is attracted to a bounded region having a dimension lower than two. If the dimension of the attractor is non-integer then the attractor is said to be fractal. However, a fractal attractor need not be a *strange attractor*, or at least not strange in the mathematical sense of the term. Following common usage, we call an attractor strange if all the points in the initial volume converge to a single attractor and the initial points, which were arbitrarily close together, separate from one another exponentially with iteration number. This property of nearby trajectories exponentially separating is called *sensitive dependence on initial conditions* and gives rise to aperiodic behavior of orbits traversing the strange attractor.

Let us extend the logistic equation to two dimensions to test some of these ideas. First of all we introduce a new parameter defined by $\alpha = (\lambda/2 - 1)\lambda/2$ and, since $2 < \lambda \le 4$, the new parameter is in the range $0 < \alpha \le 2$. The transformation of variables is

$$2Z_n = (\lambda/2 - 1)Y_n + 1, \qquad (4.86)$$

so that the normalized logistic map, which was confined to the interval $[0, 1]$ now maps onto the interval $[-1, 1]$. This transformation allows us to construct the two-dimensional map first studied by Hénon [23]:

$$\begin{aligned} Z_{n+1} &= 1 - \alpha Z_n^2 + R_n, \\ R_{n+1} &= \beta Z_n. \end{aligned} \qquad (4.87)$$

In Figure 4.13 we indicate the loci of points for the Hénon map in which 10^4 successive points form the mapping with the parameter values $\alpha = 1.4$ and $\beta = 0.3$ initiated from a variety of choices of initial conditions (Z_0, R_0). Ott [35] points out that, as the map is iterated, points come closer and closer to the attractor, eventually becoming indistinguishable from it. This, however, is an illusion of scale.

If the boxed region in Figure 4.13 is magnified one obtains Figure 4.14(a), from which a great deal of the structure of the attractor can be discerned. Namely, what had appeared to be three unequally spaced lines in Figure 4.13 appear in Figure 4.14(a) as three distinct parallel intervals containing structure. Notice that the region from the box in Figure 4.14(a) magnified and shown in Figure 4.14(b) appears the same as the three parallel structures in Figure 4.14(a). This replication of the geometric structure

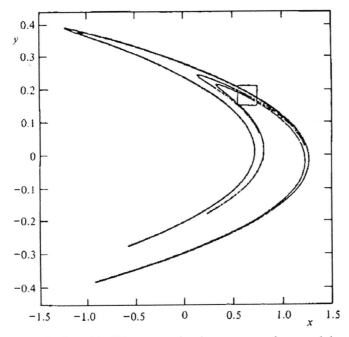

Figure 4.13. Iterated points of the Hénon map using the parameter values $\alpha = 1.4$ and $\beta = 0.3$. The boxed region is magnified to observe the fine structure of the attractor in Figure 4.14. Reproduced from [26] with permission.

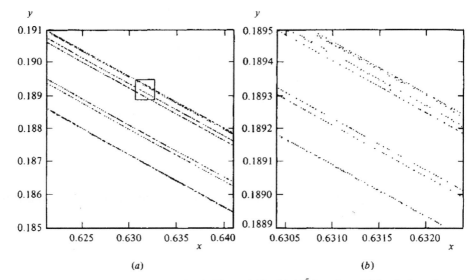

(a)   (b)

Figure 4.14. Left, magnification of the boxed region in Figure 4.13 with 10^5 iterations. Right, the boxed region of the left figure is magnified with over 10^6 iterations in the right figure. The self-similarity of the attractor structure is apparent from the repetition of the banded structure from one level of magnification to the next. Reproduced from [26] with permission.

of the attractor is repeated on successively smaller scales. Thus the Hénon attractor displays scale-invariant Cantor-set-like structure transverse to the local linear structure of the attractor. Ott concludes that because of this self-similar structure the attractor is probably strange. In fact it has been verified by direct calculation that initially nearby points exponentially separate with iteration number, thereby coinciding with at least one definition of a strange attractor.

Rössler's dynamical intuition of chaos involves the geometric operation of stretching and folding, much like a baker kneading dough for baking bread [40]. The conceptual baker in this mathematical example takes some dough and rolls it out on a floured bread board. When the dough has been rolled out thin enough he folds it back and rolls it out again. This operation is repeated over and over again. In this analogy the dough corresponds to a set of initial conditions that are taken sufficiently near to one another that they appear to form a continuous distribution of states. The stretching and folding operation represents the dynamical process undergone by each initial state. In this way the baker's transformation represents a mapping of the distribution of initial states onto a distribution of final states.

Two initially nearby orbits cannot separate forever on an attractor of finite size, therefore the attractor must fold over onto itself. Once folded, the attractor is again stretched and folded again. This process is repeated over and over, yielding an attractor structure with an infinite number of layers to be traversed by the various trajectories. The infinite richness of the attractor structure affords ample opportunity for trajectories to diverge and follow increasingly different paths. The finite size of the attractor insures that these diverging trajectories eventually pass close to one another again, but they do so on different layers of the attractor. Crutchfield *et al.* [17] visualized these orbits on a chaotic attractor as being shuffled by this process, much as a deck of cards is shuffled by a dealer. The randomness of the chaotic process is therefore a consequence of this shuffling; this stretching and folding creates folds within folds *ad infinitum*, resulting in a fractal structure in phase space, just as Hénon found for his attractor. The essential fractal feature of interest is that the greater the magnification of a region of the attractor, the greater the degree of detail that is uncovered. The geometric structure of the attractor itself is self-similar and its fractal character is manifest in the generated time series.

We have discussed the properties of mappings as mathematical objects. It is also of interest to see how these maps arise in classical mechanics, that branch of physics describing the motion of material objects. It is through the dynamics of physical phenomena that we can learn how chaos may influence our expectations regarding other complex situations.

4.4   Integrable and non-integrable Hamiltonians

We know that a Hamiltonian web has constant energy and that this conserved quantity determines the web's dynamics. Other symmetries play a similar role so that other constants of the motion simplify the dynamics because the phase-space trajectory $\{\mathbf{P}(t), \mathbf{Q}(t)\}$ would be forced to follow the hypersurface $F(\mathbf{p}, \mathbf{q}) =$ constant. Usually, these constants of motion are linked to obvious symmetries of the web: for example, translational symmetries lead to conservation of momentum and rotational symmetries result in angular-momentum conservation. In other cases the constants of motion are less evident and to find them or to prove their existence is a non-trivial problem. There does exist a simple and general way to determine whether a phase-space function $F(\mathbf{p}, \mathbf{q})$ is a constant of motion. In fact, taking its time derivative yields

$$\frac{dF}{dt} = \sum_j \left[\frac{\partial F}{\partial p_j} \dot{p}_j + \frac{\partial F}{\partial q_j} \dot{q}_j \right] \tag{4.88}$$

and using Hamilton's equations gives

$$\frac{dF}{dt} = \sum_j \left[-\frac{\partial F}{\partial p_j} \frac{\partial H}{\partial q_l} + \frac{\partial F}{\partial q_j} \frac{\partial H}{\partial p_l} \right] \equiv \{H, F\}, \tag{4.89}$$

where $\{A, B\}$ are Poisson brackets of the indicated functions. Therefore, a function $F(\mathbf{p}, \mathbf{q})$ is a constant of motion if its Poisson bracket with the Hamiltonian $H(\mathbf{p}, \mathbf{q})$ is identically zero:

$$\frac{dF}{dt} = \{H, F\} \equiv 0. \tag{4.90}$$

A system with N degrees of freedom is said to be integrable if there exist N constants of motion,

$$F_j(\mathbf{p}, \mathbf{q}) = c_j \quad (j = 1, 2, \ldots, N), \tag{4.91}$$

where the c_j are all constant. Each constant of motion reduces the web's number of degrees of freedom by one, so, if the web is integrable, the phase-space trajectory is confined to the surface of an N-dimensional manifold in a $2N$-dimensional phase space: this particular surface can be proven to have the topology of a torus. Figure 4.12 shows a simple two-dimensional torus that may restrict the motion of, for example, a web with two degrees of freedom and two constants of the motion. Two such constants are the total energy and the total angular momentum which are conserved in a rotationally symmetric potential. However, typical tori can be bent and twisted in very complicated ways in the phase space without violating the conservation rules.

A great deal of the time spent in attempting to understand a scientific problem is dedicated to determining its most appropriate representation: choosing the variables with which to model the web. Part of the consideration has to do with the symmetries and constraints on the network, both of which restrict the possible choices of representation. In this section we consider the best variables that can be used to describe dynamical webs that after some period of time return to a previously occupied state. Let us emphasize that such periodic motion does not imply that the web motion is harmonic. Think of a roller coaster: the car in which you sit periodically returns to its starting point, but there can be some exciting dynamics along the way. It is a general result of analytic dynamics that, if the web returns to a state once, then it returns to that state infinitely often. This principle is used to construct a very general representation of the dynamics, namely the action-angle variables.

The reason why action-angle variables are such a good representation of the dynamics of conservative Hamiltonian webs has to do with how complexity has traditionally been introduced into the dynamics of mechanical webs. Historically a complex physical network is divided into a simple part plus a perturbation. The simple part undergoes oscillatory motion and the deviation from this periodic behavior can be described by perturbation theory. If this is a correct picture of what occurs then action-angle variables are the best description, as shown below. A Hamiltonian can be written in terms of action-angle variables as $H_0(\mathbf{J})$ and the equations of motion take the simple form

$$\frac{dJ_j}{dt} = -\frac{\partial H_0}{\partial \theta_j} = 0, \tag{4.92}$$

$$\frac{d\theta_j}{dt} = \frac{\partial H_0}{\partial J_j} = \omega_j(\mathbf{J}) = \text{constant}. \tag{4.93}$$

We note in passing that the original set of canonically conjugate variables $(\mathbf{p}, \mathbf{q})$ is replaced by the new set of canonically conjugate variables $(\mathbf{J}, \theta)$, so that Hamilton's equations are valid in both representations. Equation (4.93) results from the fact that $\mathbf{J}$ is a constant as determined by (4.92), so the angle increases linearly with time. The solutions to Hamilton's equations in the action-angle representation are given by

$$J_j = \text{constant},$$
$$\theta_j(t) = \omega_j(\mathbf{J})t + \theta_j(0) \quad (j = 1, 2, \ldots, N). \tag{4.94}$$

In the simple torus example shown in Figure 4.12 it is easy to realize that, if the two angular frequencies ω_1 and ω_2 are commensurable, that is, their ratio is rational, then the trajectory eventually returns to its starting point and the motion is periodic. On the other hand, if the ratio of the two angular frequencies is irrational the trajectory never returns to its starting point and the motion is aperiodic. Complex webs can be separated into a piece described by action-angle variables, which is the simple part, and the deviation from this simple periodic behavior, which can be described using perturbation theory. The unperturbed Hamiltonian in the action-angle representation is given by

$$H_0(\mathbf{J}) = \sum_{j=1}^{N} J_j \omega_j \tag{4.95}$$

for an integrable web with N degrees of freedom.

If the web is integrable then the phase space is filled with invariant tori and a given trajectory remains on the particular torus selected by the initial conditions and its motion can be periodic or aperiodic. Complex webs deviate from such behavior and are described in analytic dynamics by perturbation theory. Perturbation theory arises from the notion that nonlinear effects can be treated as small deviations away from exactly solvable problems. For example, in the real world one can solve the two-body Kepler problem; that is, the universe consists of two bodies interacting under mutual gravitation only. This exactly solvable situation cannot be extended to a universe with three interacting bodies. The evolution of three bodies interacting through their mutual gravitational fields still eludes us. Thus, systematic approximation techniques for solving real-world problems are very important. In dynamics a strategy was devised for solving non-integrable Hamiltonian webs by defining integrable Hamiltonians that differed only slightly from non-integrable ones. The *real* problem is then said to be a perturbation of the *idealized* problem that can be rigorously solved. The difference between the two Hamiltonians is called the perturbation and this approach relies on the assumption that the perturbation is in some sense small.

4.4.1 The surface of section

In real complex physical webs it is often the case that the web is characterized by a large number of variables and the dimensionality of the phase space can exceed three. In such situations how can the trajectory be visualized? Here we have recourse to a technique briefly discussed previously that was originally suggested by Poincaré, called the Poincaré surface of section. Consider a web described by two spatial variables and their canonical momenta, so that the phase space is four-dimensional with axes x, y, p_x and p_y. If energy is conserved the motion takes place on the surface of a three-dimensional surface of constant energy:

$$E = H = \frac{p_x^2 + p_y^2}{2m} + V(x, y).$$
(4.96)

One variable, say p_y, can be expressed in terms of the constant energy and therefore eliminated from the dynamical description:

$$p_y = \pm \sqrt{2m \left[E - \frac{p_x^2}{2m} - V(x, y) \right]}.$$
(4.97)

Following motion on a three-dimensional surface can often be difficult in practice, so one needs a way to accurately characterize that motion on a two-dimensional plane such as a sheet of paper. This is where the surface of section becomes useful.

The trajectories determined by (4.96) are recorded on a surface of section along the x coordinate, denoted by S_x, given by the intersection of the three-dimensional manifold on which the dynamics unfold with the $y = 0$ plane and have the coordinates (x, p_x). This is a "slice" through the trajectories in phase space. If we specify the coordinates of the web on this slice S_x we have completely specified the state of the web. Apart from an overall sign, since x, p_x and $y = 0$ are specified on S_x, the momentum value p_y is given by (4.97). It is customary to take the positive sign and to define the y surface of section S_y in a similar way.

If the potential $V(x, y)$ supports bounded motion in the phase space, the trajectories will repeatedly pass through S_x with every traversal of the orbit. If the locations at which the orbit pierces S_x are recorded with each traversal we can build up a succession of intercept points as shown in Figure 4.15. The coordinates (x, p_x) of one point of interception can be related to those of the next interception by means of a map. Thus, an initial point $X_0 = (x_0, p_{x0})$ on S_x will generate a sequence of points $X_1, X_2, \ldots, X_N$. It is this sequence that constitutes the motion on the surface of section.

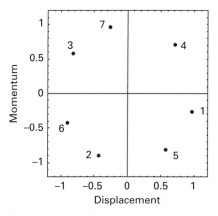

Figure 4.15. The succession of intercept points of the Poincaré surface of section. It is evident from the numbering that the intercept points are not sequential but occur more or less at random along the curve of the torus intersecting the plane.

If we denote the map by T then the sequence is generated by successive applications of the map to the initial point as follows:

$$X_1 = T(X_0),$$
$$X_2 = T(X_1) = T(T(X_0)) = T^2(X_0),$$
$$\vdots$$
$$X_N = T(X_{N-1}) = T^N(X_0).$$

(4.98)

It is clear that each of the points is generated from the initial point by applying the mapping a given number of times.

If the initial point is chosen to correspond to an orbit lying on a torus, the sequence lies on some smooth curve resulting from the intersection of that torus with the surface of section as indicated in Figure 4.16. If the chosen torus is an irrational number, that is, the frequency ratio ω_1/ω_2 is incommensurate, then a single orbit covers the torus densely. In other words, the orbit is ergodic on the torus. This ergodic behavior reveals itself on the surface of section by gradually filling in the curve with iterates until the curve appears to be smooth. Note, however, that the ordering of the intersection points is not sequential, but skips around the "curve" filling in at random locations. This twisting of the intersection point is determined by the ratio of frequencies, which is chosen here to be the square root of 13. If, on the other hand, the torus is rational, the orbit is closed and there will be only a finite number of intersections before the iterates repeat. Supposing that at intersection N we repeat the initial point $X_0 = X_N$, the number of iterates N is determined by ω_1/ω_2.

The fundamental property of Hamiltonian webs is that they correspond to area-preserving mappings. The utility of surfaces of section and the subsequent mapping of the web dynamics on them is that the iterates reveal whether the web is integrable or not. If the web is integrable we know that the orbit does not completely explore all of phase space but is confined to a surface with the topology of a torus. This torus intersects

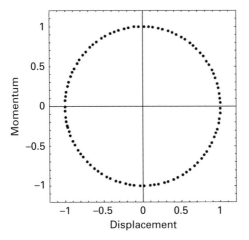

Figure 4.16. The initial point lies on the intersection of a torus with the surface of section and the mapping eventually traces out a continuous curve for the cross section of the torus.

the Poincaré surface in a smooth closed curve C. All the iterates of X_0 must lie on C and, as more iterates are generated, the form of the curve becomes apparent and the smooth curve is filled in. If the motion is periodic, which is to say that the orbit is closed due to the rational nature of the frequency ratio ω_1/ω_2, then $X_0 = X_N$. Therefore X_0 is a fixed point of the mapping T^N. If the web is integrable then such fixed points are all that exists on the torus. No matter from where one starts, eventually the initial orbit will close on itself. Therefore the entire curve C is made up of fixed points. Most curves C are invariant curves of the mapping because T maps each curve into itself,

$$T(C) = C. \tag{4.99}$$

The situation is quite different for non-integrable motion, since tori do not exist for non-integrable webs. Therefore the dynamics cover the energy-conserving surface in phase space. However, the crossing of any Poincaré surface of section will not be confined to a torus as it was in Figure 4.16. Subsequently we show that this lack of confinement is a consequence of the tori having been destroyed according to KAM theory. Trajectories in what had been regions of tori are now free to wander over larger regions of the phase space and this is realized on S_x as a random-looking "splatter" of iterates through which no smooth curve can be drawn; see, for example, Figure 4.17. Visual inspection of these figures does not suffice to determine whether the points are truly distributed randomly in the plane, but, with a sufficient number of intersections, it is clear that these points cannot be confused with the closed curves of integrable webs. Eventually one may find small areas of the surface of section being filled up so that a probability density function may be defined. We explore this subsequently using numerical examples.

The surface of section is an enormously valuable tool and, when computed for a large number of initial conditions on the same surface, it is able to give a picture of the complicated phase-space structure of both integrable and non-integrable Hamiltonian webs. But we should summarize, as did Brillouin [10], the weaknesses in the above discussion.

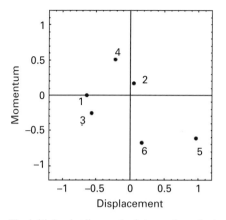

Figure 4.17. The initial point lies on the intersection of a torus with the surface of section but the mapping does not eventually trace out a continuous curve for the cross section of the torus. Instead one obtains a random display of points.

    A. The notion of an *irrational* number is non-physical. It is impossible to prove that a physical quantity is rational or irrational.

    B. The notion of commensurability is non-physical, when the quantities to be compared are known only to within some experimental errors.

    C. It is impossible to study the properties of a single (mathematical) trajectory. The physicist knows only *bundles of trajectories*, corresponding to slightly different initial conditions ... any information obtained from an experiment must be paid for by a corresponding increase of entropy in the measuring device: infinite accuracy could cost an infinite amount of entropy increase and require infinite energy! This is absolutely unthinkable.

4.4.2   Chaos and linear response theory

Let us introduce a class of nonlinear dynamical microscopic webs that reproduce Brownian motion at the macroscopic level. It is impossible to study all of the possible kinds of nonlinear networks which belong to this class, so we study a prototype of these nonlinear models. An important feature of our prototype network is the presence of a variable of interest w coupled to a thermal bath (in what follows this thermal bath is called the booster). The dynamical bath is here characterized by very few degrees of freedom that are nonlinearly coupled to one another and are therefore described by chaotic dynamics, and have no memory of the bath's initial conditions. The prototype model we study can be cast in the form [7, 8]

$$\frac{dw}{dt} = \xi,$$

$$\frac{d\xi}{dt} = R_\xi(\xi, \Pi, -\gamma w), \qquad (4.100)$$

$$\frac{d\pi_j}{dt} = \Phi_j(\xi, \Pi, -\gamma w); \qquad j = 1, 2, \ldots$$

The variable w is driven by the "doorway variable" ξ; that is, the web is coupled to the bath dynamics by means of this variable. The web variable is eventually identified with the "free" velocity of the Brownian particle, at which time the doorway variable is identified with the random force. The symbol Π denotes the set of additional phase-space variables necessary to complete the description of the bath dynamics and the "booster" refers to the dynamical network described by the equations

$$\frac{d\xi}{dt} = R_\xi(\xi, \Pi, \Theta),$$

$$\frac{d\pi_j}{dt} = \Phi_j(\xi, \Pi, \Theta); \quad j = 1, 2, \ldots, \tag{4.101}$$

where, for the moment, the function Θ representing the coupling between the web and the bath is a constant. If $\Theta = 0$ the booster is isolated and unperturbed; however, if $\Theta \neq 0$ then the booster is perturbed. In (4.100) the perturbation $\Theta = -\gamma w$ is the back-reaction of the booster to the web of interest.

A necessary condition for this prototype model is that it has no memory of initial conditions. This lack of memory implies that its dynamics are chaotic. Of course, no memory does not mean that the booster has an infinite number of degrees of freedom. A putative candidate could be a finite set of nonlinear oscillators in the chaotic regime. The response of the oscillators to an external perturbation is "naturally" different from zero. Broadly speaking, we expect that a chaotic network, if perturbed by an external force, rearranges itself to accommodate the perturbation.

Let us suppose that the value of the coupling between the web and bath is weak; the "macroscopic" time scale of the evolution of the variable w is then of order $[\gamma\tau_c]^{-1}$, and we expect that a time-scale separation holds for the dynamics of the web of interest and for the variables of the booster. From (4.100), the velocity $w(t)$ is the integral over time of the pseudo-stochastic variable $\xi(t)$ and, on applying the central limit theorem, we immediately conclude that $w(t)$ has Gaussian statistics, with a width that grows linearly in time. The properties of the web variable should not be too surprising: it is well known that chaotic processes can result in long-time diffusional processes, and successful methods to predict the corresponding diffusion coefficients have been developed [14]. However, although this research essentially solves the problem of a deterministic approach to Brownian motion, it refers only to the unperturbed case, that is, $\gamma = 0$, which corresponds to the case of no coupling, or pure diffusion. The coefficient of diffusion obtained in the unperturbed case reads

$$D = \left\langle \xi^2 \right\rangle_0 \tau_c, \tag{4.102}$$

where the characteristic time scale τ_c is defined as the integral over the autocorrelation function

$$\tau_c \equiv \int_0^\infty \Phi_\xi(t)dt. \tag{4.103}$$

Note that we have defined the microscopic time scale as the time-integral over the doorway variable's unperturbed autocorrelation function, which is assumed to be stationary.

We return to this discussion after having introduced the formal notion of LRT and justified some of the phenomenological terms used in the discussion above. For the moment we emphasize that if a variable w is weakly coupled to a nonlinear network, of any dimension, provided that this web is chaotic and ergodic, the resulting deterministic motion of the variable w conforms to that of a standard fluctuation–dissipation process found in most texts on statistical physics. The irregularities of the deterministic statistics are washed out by the large-time-scale separation between the variables in the web of interest and those in the chaotic booster. In other words, we find that the influence exerted by the chaotic and ergodic booster on the variable w is indistinguishable from that produced by a standard heat bath with infinitely many degrees of freedom.

Thus, we might expect complex social webs to exhibit Brownian motion even when the traditional heat-bath argument is not appropriate. This could well apply to the DDM model discussed earlier (Section 3.3.2). The uncertainty in the information on which decisions are based would not be due to an infinite number of variables in the environment, but would be a consequence of the nonlinear interactions among the finite number of variables describing society.

For the present approach to the standard fluctuation–dissipation process to apply to the variable of interest, the booster has to satisfy the basic condition of responding linearly to a constant external perturbation. If we adopt a mathematical mapping as a booster using the Poincaré surface of section from the last section, then we need to understand the problem of the linear response of a mapping to an external perturbation, which we take up subsequently. For the present motivational purposes we merely indicate some computer calculations that support the predictions of the theoretical approach discussed here, using a two-dimensional booster, in which the description of a chaotic web with a well-defined phase space is given by a microcanonical distribution. As a result of an external perturbation the phase-space manifold on which the dynamics unfolds is modified and the distribution is accordingly deformed, thereby making it possible to derive an analytic expression for the corresponding stationary web response. The details of this argument are given later.

We derive the standard fluctuation–dissipation-relation process from the following three-dimensional mapping:

$$w_{n+1} = w_n + \xi_{n+1},$$
$$\xi_{n+1} = \xi_n + f\left(\xi_n, \pi_n, -\gamma w_n\right),$$
$$\pi_{n+1} = \pi_n + g\left(\xi_n, \pi_n, -\gamma w_n\right).$$
$$(4.104)$$

The values w_n are to be regarded as being the values taken at discrete times by a continuous variable w, the same symbol as that used in (4.100). This choice is made because, as we shall see, (4.104) leads to dynamics for w_n that is virtually indistinguishable from the conventional picture of (3.52). Thus, we refer to w as the velocity of the Brownian particle. The functions f and g of this mapping are obtained as follows. A microscopic network with two degrees of freedom is assumed to have an unperturbed Hamiltonian, for example,

$$H = \frac{1}{2}m_1\pi^2 + \frac{1}{2}m_2v^2 + U(\xi, \zeta, \Theta), \tag{4.105}$$

where the two particles have masses m_1 and m_2. The Hamiltonian represents an oscillator with displacement ξ and velocity π interacting via a nonlinear interaction with another oscillator, with displacement ζ and velocity v. The equations of motion are given by

$$
\begin{aligned}
\frac{d\xi}{dt} &= \pi, \\
\frac{d\pi}{dt} &= -\frac{1}{m_1}\frac{\partial U(\xi, \zeta, \Theta)}{\partial \xi}, \\
\frac{d\zeta}{dt} &= v, \\
\frac{dv}{dt} &= -\frac{1}{m_2}\frac{\partial U(\xi, \zeta, \Theta)}{\partial \zeta}
\end{aligned}
\tag{4.106}
$$

and the masses are coupled to one another by means of the interaction potential

$$U(\xi, \zeta, \Theta) \equiv \frac{\xi^2}{2} + \frac{\zeta^2}{2} + \xi\zeta^2 - \frac{\zeta\xi^3}{3} + \xi^4\zeta^4 + \Theta, \tag{4.107}$$

where $\Theta = 0$ gives the unperturbed potential. The first oscillator is perturbed by a constant field Θ, acting as the contribution to the potential $-\gamma w$. Notice that the intersection of the solution trajectories with the plane $\zeta = 0$ results in a harmonic potential, thereby simplifying the subsequent calculation. We make the choice $m_1 = 1$ and $m_2 = 0.54$ and refer to (4.106) as the microscopic booster for our numerical example.

The connection between the continuous equation (4.106) and the mapping (4.104) is obtained by following the trajectory determined by the potential (4.107) from the pair (ξ_n, π_n) on the $\zeta = 0$ plane to the next intersection of that plane at (ξ_{n+1}, π_{n+1}). Throughout this motion the value of the perturbation field Θ is kept fixed at $\Theta = -\gamma w_n$. The subsequent crossing determines the new pair (ξ_{n+1}, π_{n+1}) and, through the first equation of (4.106), the new velocity w_{n+1}. In conclusion, the values (ξ_n, π_n) are obtained from the Poincaré surface of section of the continuous system (4.106) at $\zeta = 0$, with Θ undergoing an abrupt change at any and every crossing of this surface. The interval of time between the two consecutive crossings is set equal to unity for convenience.

We draw your attention to the fact that the time evolution of the variable w, as driven by this three-dimensional map, looks impressively similar to a Brownian-motion process. With the numerical calculation of the mapping we evaluate the equilibrium autocorrelation function of the velocity,

$$\Psi(t) = \frac{\langle w(0)w(t)\rangle_{eq}}{\langle w^2\rangle_{eq}} \tag{4.108}$$

and show that, as in Brownian motion, the autocorrelation function is an exponential with damping parameter λ. Then let us consider the equilibrium autocorrelation function of the second-order velocity variable

$$W(t) = w(t)^2 - \left\langle w^2 \right\rangle_{eq},$$ (4.109)

which has the autocorrelation function

$$\Xi\,(t) = \frac{\langle W(0)W(t)\rangle_{eq}}{\langle W^2\rangle_{eq}}.$$ (4.110)

If the quadratic velocity variable $W(t)$ has Gaussian statistics then the average in (4.110) factors and becomes the product of two autocorrelation functions; that is,

$$\Xi\,(t) = \Psi\,(t)^2\,.$$ (4.111)

Figure 4.18 shows that the condition (4.111) is satisfied.

Of course, to ensure the Gaussian character of the random velocity $w(t)$ one should study all the higher-order correlation functions; however, the computational calculation of these quantities becomes increasingly difficult with increasing order of the correlations considered. Thus, since we have not calculated these higher-order terms we cannot rigorously state that the web variable $w(t)$ does have Gaussian statistics on the basis of the computational fit to (4.111) alone. However, we believe that it is plausible that w does have Gaussian statistics. Since the decay of the autocorrelation function is exponential, we would be led to conclude on the basis of Doob's theorem that w is also Markov. On the other hand, a Markov and Gaussian process is known to evolve by means of a FPE. The damping of the FPE is determined by the damping of the autocorrelation function. The diffusion coefficient D is determined by the width of the equilibrium Gaussian distribution. We see numerically that the mapping (4.104) leads any initial condition to the same equilibrium distribution, and precisely that given by

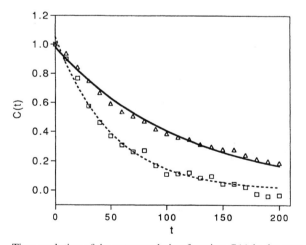

Figure 4.18. Time evolution of the autocorrelation function $C(t)$ in the case $\gamma = 0.01$. The triangles refer to $C(t) = \Psi(t)$, namely the equilibrium autocorrelation function of the velocity variable $w(t)$. The squares refer to $C(t) = \Xi(t)$, namely the equilibrium autocorrelation function of the squared velocity $W(t) = w(t)^2 - \langle w^2 \rangle_{eq}$. The full line is a fitting exponential with the damping coefficient 0.0111. The dashed line is an exponential with damping of twice that value. The theoretical prediction yields a damping coefficient of $\lambda = 0.01$ [7]. Reproduced with permission.

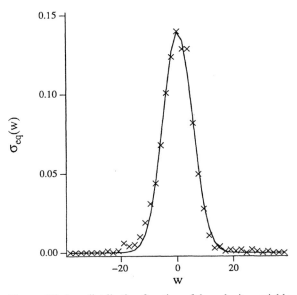

Figure 4.19. The equilibrium distribution function of the velocity variable w for network–bath coupling strength $\gamma = 0.01$. The crosses denote the results of the computer calculations of the mapping (4.104). The full line is a Gaussian function with the width 38.5, which has to be compared with the theoretical prediction 32.0 [7]. Reproduced with permission.

the Gaussian function of Figure 4.19. The width of this "experimental" function turns out to be $(\langle w^2 \rangle_{eq})_{fit} = 38.5$, compared with the theoretical value $\langle w^2 \rangle_{eq} = 32$ obtained later.

Thus, on the basis of our numerical calculations, and within the specific limits set by them on the determination of the higher-order moments, we can consider the system (4.104) to be equivalent to the Langevin equation (3.52).

One of the purposes of this book is to establish the theoretical reasons for the equivalence found between the two ways of determining the width of the Gauss distribution for a fluctuation–dissipation process. Another way to say this is that we know from the fluctuation–dissipation relation that the width of the Gauss distribution is proportional to the temperature of the environment for a Brownian particle. So why should the strength of the fluctuations produced by the booster give rise to a physical temperature? We address this question later.

The prototype booster

Let us again consider the network described by the equations of motion (4.100), that is, the network with the Hamiltonian (4.105), which we rewrite here:

$$H = \frac{1}{2}m_1\pi^2 + \frac{1}{2}m_2v^2 + U(\xi, \zeta, \Theta). \tag{4.112}$$

The behavior of this network at a fixed value of the mechanical energy E is investigated in the domain where the parameter values give rise to chaotic dynamics. Since the energy E is a constant of the motion, the four variables ξ, ζ, π and v are not independent of one another. The velocity variable v can be considered as a function of the other variables and of the energy E in the following way:

$$v = \sqrt{\frac{2}{m_2}\left[E - \frac{1}{2}m_1\pi^2 - U(\xi, \zeta, \Theta)\right]}. \tag{4.113}$$

In the three-dimensional phase space ξ, ζ, π the trajectories of the network lie inside a domain $\Omega(E)$ defined by the condition

$$v^2 \geq 0 \tag{4.114}$$

imposed by energy conservation. This domain is enclosed by the surface $S(E)$ defined by the region of the phase space where $v = 0$, namely

$$E - \frac{1}{2}m_1\pi^2 - U(\xi, \zeta, \Theta) = 0. \tag{4.115}$$

The surface $S(E)$ is shown schematically in Figure 4.20.

In the present situation the heat bath for our Brownian particle with velocity w is a booster obtained as a Poincaré map of the generating network (4.104). Consider the Poincaré map corresponding to the intersection of the trajectories inside the domain $\Omega(E)$ with the plane defined by $\zeta = \zeta^*$, where ζ^* denotes a generic fixed value of the variable ζ. The points of the Poincaré map lie within a manifold Σ given by the intersection of the domain $\Omega(E)$ with the plane $\zeta = \zeta^*$. The manifold Σ is formally defined by the condition

$$E - \frac{1}{2}m_1\pi^2 - U(\xi, \zeta^*, \Theta) \geq 0. \tag{4.116}$$

The manifold Σ is illustrated in Figure 4.21 for the case $\Theta = 0$. The Poincaré map is area-preserving so that the invariant measure is the flat (or Lebesgue) measure $d\sigma = d\xi\, d\pi$, where ξ and π are the coordinates on the plane $\zeta = \zeta^*$; that is, they are the independent variables describing the dynamics of the booster.

Owing to the chaotic and ergodic dynamics of the generating network given by (4.100), the dynamics of this Poincaré map is mixing, in which case μ is the unique

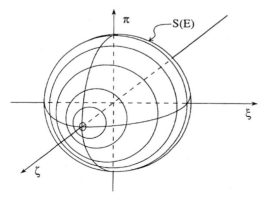

Figure 4.20. A schematic illustration of the domain $\Omega(E)$ enclosed by the surface $S(E)$ [7]. Reproduced with permission.

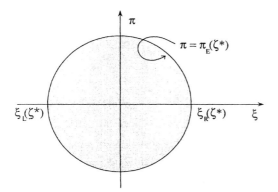

Figure 4.21. A schematic illustration of the intersection between the domain of $\Omega(E)$ and the plane $\zeta = \zeta^*$ (the Poincaré plane, which is the Σ domain) [7]. Reproduced with permission.

invariant measure on the manifold Σ and almost any distribution function relaxes to the flat distribution. Thus, the mean value of the booster doorway variable ξ is given by

$$\langle \xi \rangle = \frac{\displaystyle\iint_{\Sigma} \xi \, d\xi \, dv}{\displaystyle\iint_{\Sigma} d\xi \, dv} \tag{4.117}$$

and the mean-square value is

$$\langle \xi^2 \rangle = \frac{\displaystyle\iint_{\Sigma} \xi^2 \, d\xi \, dv}{\displaystyle\iint_{\Sigma} d\xi \, dv}, \tag{4.118}$$

both of which are relatively easy to obtain numerically.

Response to a weak external perturbation
Let us now investigate in some detail the effect of the perturbation on the booster. Consider the unperturbed Hamiltonian H_0 given by

$$H_0 = \frac{1}{2} m_1 \pi^2 + \frac{1}{2} m_2 v^2 + U_0(\xi, \zeta), \tag{4.119}$$

where the unperturbed potential is defined by

$$U_0(\xi, \zeta) \equiv U(\xi, \zeta, \Theta = 0) = \frac{\xi^2}{2} + \frac{\zeta^2}{2} + \xi \zeta^2 - \frac{\zeta \xi^3}{3} + \xi^4 \zeta^4, \tag{4.120}$$

so the web with $\Theta = 0$ has the zero subscripts. Assuming that the perturbation $H_1 = \varepsilon \xi$ is switched on at $t = 0$ and using the unit step function $\theta(t) = 0$ for $t < 0$ and $\theta(t) = 1$ for $t \geq 0$, the total Hamiltonian can be written as

$$H = H_0 + H_1(t)$$

$$= \frac{1}{2}m_1\pi^2 + \frac{1}{2}m_2v^2 + U_0(\xi, \zeta) + \theta(t)\,\varepsilon\xi, \tag{4.121}$$

where the unperturbed potential has the form given for the prototype (4.120).

Suppose that we are dealing with an ensemble of networks. Imagine that an individual unperturbed web of the ensemble of networks has coordinates $\xi(0)$, $\zeta(0)$, $\pi(0)$ and energy $E(0^-) = E$. Immediately after the abrupt application of the perturbation, the web behaves as if it has "felt" the influence of the additional potential $\varepsilon\xi$. If the external perturbation is turned on at $t = 0$ the individual webs in the ensemble will have the new energy

$$E(0^+) = \frac{1}{2}m_1\pi_0^2 + \frac{1}{2}m_2v_0^2 + U_0(\xi_0, \zeta_0, \Theta_0)$$

$$= E + \varepsilon\xi(0), \tag{4.122}$$

where we have tagged the phase-space variable with the initial time label. In the subsequent time evolution the webs will have an energy $E(0^+)$ fixed, and obviously for the invariant energy

$$E(0^+) = H_0(\xi, \zeta, \pi, v) + \varepsilon\xi. \tag{4.123}$$

From (4.122) and (4.123) it follows that the constant energy is given in terms of the perturbed Hamiltonian

$$E = H_0(\xi, \zeta, \pi, v) + \varepsilon[\xi - \xi(0)] \equiv H_0 + \tilde{H}_1. \tag{4.124}$$

For the remainder of the section the energy E will be a constant quantity in time, while the resulting interaction term changes in time with the motion of the doorway variable ξ, according to the definition (4.124) and to the specific value $\xi(0)$ that the doorway variable has at the time when the external perturbation is abruptly switched on.

It is important to note that if the interaction term $\tilde{H}_1$ is small enough the domain $\Omega(E)$ is only slightly modified from the unperturbed situation. By contrast, the individual trajectories are sensitively dependent on the perturbation, whatever its intensity, due to the assumption that the unperturbed network is in the chaotic regime.

Thus, from the geometric deformation of this domain we can derive the corresponding change in the mean value of the doorway variable of the Poincaré map, thereby determining the susceptibility χ of the booster. The additional potential $\varepsilon[\xi - \xi(0)]$ moves the extrema $\xi_R(\zeta^*)$ and $\xi_L(\zeta^*)$ of the Poincaré map domain depicted in Figure 4.21 towards the right (with $\varepsilon < 0$) or the left (with $\varepsilon > 0$) in a continuous way. For sufficiently small values of ε the average response to the perturbation can be written

$$\langle \xi \rangle_\varepsilon = \varepsilon\chi \tag{4.125}$$

and, using the formal expression for the mean value of the doorway variable (4.117), we obtain

$$\chi = \left.\frac{\partial \langle \xi \rangle_\varepsilon}{\partial \varepsilon}\right|_{\varepsilon=0} = \left.\frac{\partial}{\partial \varepsilon}\left(\frac{\displaystyle\iint_\Sigma \xi \, d\xi \, dv}{\displaystyle\iint_\Sigma d\xi \, dv}\right)\right|_{\varepsilon=0}. \tag{4.126}$$

With the choice of surface $\zeta^* = 0$ the condition (4.116) with the unperturbed potential given by (4.120) yields the following expression for the manifold Σ:

$$E - \frac{1}{2} m_1 \pi^2 - \frac{1}{2} \xi^2 - \varepsilon [\xi - \xi(0)] \geq 0, \tag{4.127}$$

which can be rewritten as the points interior to an ellipse

$$\frac{1}{2} m_1 \pi^2 + \frac{1}{2} (\xi - \varepsilon)^2 \leq 2E + 2\varepsilon \xi(0) + \varepsilon^2 \tag{4.128}$$

centered at $(\pi_c = 0, \xi_c = \varepsilon)$. The calculation of the Poincaré map for the unperturbed energy of the booster is set at $E = 0.8$. With this choice of the energy the booster is in a very chaotic regime and the intersections of the chaotic trajectories with the plane $\zeta^* = 0$ are randomly distributed over the accessible Poincaré surface of section as shown in Figure 4.22. Furthermore, the numerical calculation shows that the modulus of the greatest Lyapunov exponent is larger than 4. The network is proved to a very good approximation to be ergodic since the numerical calculation shows that the distribution of the points on the manifold is a flat function.

We are now in a position to determine the consistency of the calculations of the transport coefficients, these being the diffusion coefficient and the friction coefficient, with the theoretical predictions. According to the theory presented above, the derivation of the diffusion coefficient D rests on the properties of the Poincaré map of the unperturbed booster. More precisely, D is obtained by multiplying the correlation time τ_c defined by (3.45) by the unperturbed second moment of the doorway variable defined by (4.118). Insofar as the friction γ is concerned, we use the susceptibility χ of the booster, so that in the present calculation $\chi = 1$ and consequently, from (3.47), $\gamma = \Delta^2$. Thus, under the physical conditions for which Figure 4.21 was calculated, namely $\gamma = 0.01$ and

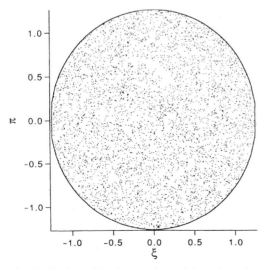

Figure 4.22. The distribution of the intersections of the trajectories of (4.117) with the plane $\zeta^* = 0$ in the chaotic case corresponding to the energy value $E = 0.8$ [7]. Reproduced with permission.

$\langle\xi^2\rangle_0 = D/\gamma = 32$, the theoretical predictions turn out to be in remarkably good agreement with the numerical result. Of course, the theory we have been using implies a time-scale separation between the web and bath. This means low values of the coupling strength Δ^2. How low the values must be for LRT to hold can be illustrated by making a comparison between the numerical and theoretical predictions of the mean-square web variable $\langle w^2\rangle_{eq}$ for various values of the coupling strength Δ^2. This quantity is the ratio of the diffusion coefficient D to the friction γ. Thus, we have the theoretical prediction

$$\langle w^2\rangle_{eq} = \frac{\langle\xi^2\rangle_0 \tau_c}{\Delta^2\chi}, \tag{4.129}$$

which is denoted by the straight-line segment in Figure 4.23. It is apparent from Figure 4.23 that the accuracy of the theoretical prediction is fairly good in the region of extremely weak coupling strengths. As expected, in the region of higher coupling strengths the departure from the FPE prediction becomes increasingly large.

It is remarkable that there is a finite region of parameter space in which LRT is valid for chaotic webs. This not only contributes to a fundamental controversy in statistical physics, but also suggests that it is not necessary to have the Hamiltonian formalism so necessary for physics to describe complex webs. It may be possible to have a valid Langevin representation for a complex web that is based on the interaction of one or a few web variables with the chaotic dynamics of the environment. This is of fundamental importance in constructing models in the social and life sciences.

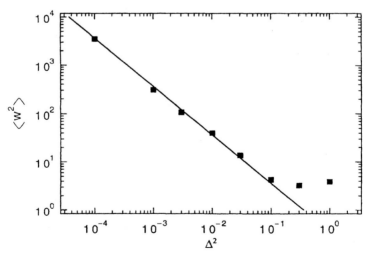

Figure 4.23.   The equilibrium mean-square web velocity is graphed as a function of the strength of coupling to the environment Δ^2. The black squares denote the result of the numerical solution of the mapping (4.104). The solid line refers to the theoretical prediction of (4.129). Note that $\langle\xi^2\rangle_0 = 0.53$, both from numerical calculations on the unperturbed booster and from analytic calculations, $\tau_c = 0.8$ as a result of numerical calculations on the unperturbed booster and $\chi = 1$ according to the analytic prediction (4.129) [7]. Reproduced with permission.

4.5 Précis

It is useful to observe that the central limit theorem and the normal distribution were the backbone of social and medical theory for a couple of hundred years, from sociophysics to homeostasis in the nineteenth century and well into the twentieth century. In the world according to Gauss, phenomena are dominated by averages, as we observed earlier. It is important to point out the fallacy of this view, but carping on this is not sufficient for our purposes because the normal perspective is *insidious*. The acceptance of the average as representative of a complex web is almost impossible to dislodge once it has been adopted. The difficulty is due, in part, to the impossibility of remembering how the phenomenon was understood before the average was taken as being representative.

The traditional random walk allows us to incrementally increase the complexity of a web, while retaining the simplicity of Gaussian statistics. Incorporating spatial non-locality and temporal anisotropy requires going beyond the historical random walks and incorporating long-time memory and spatial heterogeneity into the stepping process, which is done through the inclusion of fractal behavior in both space and time. The fractal random walk is now a few decades old, but its utility in the interpretation of complex webs is much closer to the present. The discrete form of the fractional calculus has been shown to be a consistent formalism for taking these effects into account and is one way to generate inverse power-law spectral densities for the web response to a random environment. Of course, phenomenological random walks, even those that are fractal, are not the only way for inverse power laws to enter into the dynamics of complex webs. Chaos resulting from nonlinear interactions in the dynamic equations is another.

One can find five functional categories to identify the role of randomness or chaos in complex webs: (1) search, (2) defense, (3) maintenance, (4) cross-level effects and (5) dissipation of disturbances, just as Conrad [16] suggested for chaos. The search function might include such phenomena as mutation, recombination and related genetic operations where randomness plays the role of generator of diversity. In addition to diversity, search processes also contribute to behavior, say for non-conscious micro-organisms. The function here is to enhance exploratory activity, whether dealing with macromolecules, membrane excitability, the dynamics of the cytoskeleton or, on a con-scious level, the human brain searching for creative solutions to difficult problems. The random dynamics becomes part of a strategy for exploring the web's phase space.

The function of defense distinguishes the diversity of behavior used to avoid preda-tors from that used to explore. An organism that moves about in an unpredictable way, making unexpected changes in speed and direction, is certainly more difficult to ensnare than one that moves in a straight line. This is apparent to any nature lover who has watched on public television the graceful but erratic running of a gazelle to avoid being taken down by a lion.

A third possible function for randomness is the prevention of entrainment in the peri-odic activity of physiologic processes such as the beating of the heart. The "normal sinus rhythm" taught in medical school is no more real than is normalcy; human heart rates are not strictly periodic but contain a broad spectrum of time intervals between beats. In another example, that of neurons, either absolutely regular pacemaker activity or highly explosive global neuronal firing patterns would develop in the absence of neural time-series fluctuations. The erratic component of the neural signal acts to maintain a functional independence of different parts of the nervous system. A complex web whose individual elements act more or less independently is more adaptable than one in which the separate elements are phase-locked. This same randomness-induced decoupling effect can be found in the immune network and in population dynamics.

One might also find cross-level effects in the context of population dynamics and in particular the significance of such effects for genetic structure and evolution. The basis for this argument is computer models of ecosystem dynamics in which organisms feed and reproduce in the modeling context, with certain phenotypic traits arising from genes surviving through interactions among the individual organisms by Darwinian variation and selection.

Finally, the disturbance of a biological web may relax back to regular behavior by means of a fluctuating trajectory absorbing the perturbation. In a formal context both fluctuations and dissipation in a network arise from the same source. The many degrees of freedom traditionally thought necessary to produce fluctuations and dissipation in physical networks may be effectively replaced by low-dimensional dynamical webs that are chaotic. If the web dynamics can be described by a *strange attractor* then all trajectories on such an attractor are equivalent and any disturbance of a trajectory is quickly absorbed through the *sensitive dependence of the trajectory on initial conditions* and a perturbed trajectory becomes indistinguishable from an unperturbed trajectory with different initial conditions. In this way chaos provides an effective mechanism for dissipating disturbances. The disturbance is truly lost in the noise but without changing the functionality of the strange attractor [16].

4.6   Problems

4.1 The Gaussian distribution
Use the probability density given by (4.8) to calculate the average location of the walker as well as her mean-square displacement. Use Stirling's approximation $n! \approx n \ln n$ $n! \approx n \ln n$ to the factorial to derive the Gaussian distribution from this equation as well.

4.2 Asymptotic properties of gamma functions
The asymptotic form of the expansion coefficient in the solution to the fractional random walk can be determined using Stirling's approximation for the factorial representation of the gamma functions. Prove that (4.55) follows from (4.54) in the asymptotic limit.

4.3 Anomalous diffusion
Physical webs whose second moment increases as a power law in time, for example,

$$\left\langle Q^2; t \right\rangle \propto t^{2H}$$

with $H \neq 1/2$ are said to exhibit anomalous diffusion. This is the case of the long-memory random walk just discussed. Show by direct calculation that (4.64) is true and that

$$\left\langle X_t^2 \right\rangle \propto t^{2H+2}.$$

4.4 Logistic maps

What is surprising about the chaotic behavior of the solutions to the logistic equation is that their successive bifurcation from single-valued saturation to chaos with increasing values of the control parameter is generic and independent of the particular one-parameter family of maps. Discuss the general dynamical properties of the solutions to (4.75) in terms of iterated maps

$$y_{n+1} = f(y_n),$$

where $f(y)$ is a nonlinear function, for example, the logistic map on the unit interval is

$$f(y) = \lambda y(1 - y).$$

One reference you might consider consulting is West [44], but there are many others. Please do not copy what is written, but demonstrate your understanding by putting the formalism in your own words. In particular, use three successive values of the logistic map in the chaotic region to empirically determine the value of the control parameter.

4.5 The nonlinear oscillator

Consider a quartic potential

$$U(q) = \frac{1}{4}\alpha q^4$$

whose potential energy looks similar to that shown in Figure 3.4 except that the sides of the potential are steeper. Use the analysis of the dynamics of the harmonic oscillator from Section 3.1 and construct the Hamiltonian in terms of the action-angle variables. Also find the action dependence of the frequency.

References

[1] R. J. Abraham and C. D. Shaw, *Dynamics – The Geometry of Behavior, Parts 1–4*, Santa Cruz, CA: Aerial Press (1982–1988).

[2] L. A. Adamic and B. A. Huberman, "The nature of markets in the World Wide Web," *Quart. J. Electron Commerce* **1**, 512 (2000).

[3] V. I. Arnol'd, *Russian Math. Surveys* **18**, 85–191 (1963); V. I. Arnol'd and A. Avez, *Ergodic Problems of Classical Mechanics*, New York: Benjamin (1968).

[4] L. Bachelier, *Annales scientifiques de l'Ecole Normale Supérieure, Suppl. (3) No. 1017* (1900): English translation by A. J. Boness, in *The Random Character of the Stock Market*, Cambridge, MA: P. Cootners, MIT Press (1954).

[5] A.-L. Barabási, *Linked*, New York: Plume (2003).

[6] J. B. Bassingthwaighte, L. S. Liebovitch and B. J. West, *Fractal Physiology*, Oxford: Oxford University Press (1994).

[7] M. Bianucci, R. Mannella, X. Fan, P. Grigolini and B. J. West, "Standard fluctuation–dissipation process from a deterministic mapping," *Phys. Rev. E* **47**, 1510–1519 (1993).

[8] M. Bianucci, R. Mannella, B. J. West and P. Grigolini, "From dynamics to thermodynamics: linear response and statistical mechanics," *Phys. Rev. E* **51**, 3002 (1995).

[9] S. El Boustani and A. Destexhe, "Brain dynamics at multiple scales: can one reconcile the apparent low-dimensional chaos of macroscopic variables with the seemingly stochastic behavior of single neurons?," *Int. J. Bifurcation and Chaos* **20**, 1687–1702 (2010).

[10] L. Brillouin, *Scientific Uncertainty, and Information*, New York: Academic Press (1964).

[11] A. Boder, R. Kumar, F. Maghoul *et al.*, "Power-law distribution and graph theoretic methods on the web," *Comp. Net.* **33**, 309–320 (2000).

[12] M. Buchanan, *Nexus*, New York: W. W. Norton & Co. (2002).

[13] S. Chandresakar, "Stochastic problems in physics and astronomy," *Rev. Mod. Phys.* **15**, 1–89 (1943).

[14] B. V. Chirikov, "A universal instability of many-dimensional oscillator systems," *Phys. Rep.* **52**, 265 (1979).

[15] A. Clauset, C. R. Shalizi and M. E. J. Newman, "Power-law distributions in empirical data," *SIAM Rev.* **51**, 661–703 (2009).

[16] M. Conrad, "What is the use of chaos?," in *Chaos*, ed. A. V. Holden, Princeton, NJ: Princeton University Press (1986), pp. 3–14.

[17] J. P. Crutchfield, D. Farmer, N. H. Packard and R. S. Shaw, "Chaos," *Scientific American* **46**, 46–57 (1987).

[18] A. de Moivre, *The Doctrine of Chances: A Method of Calculating the Probability of the Events of Play*, London (1718); 3rd. edn. (1756); reprinted by Chelsea Press (1967).

[19] I. Ekeland, *Mathematics and the Unexpected*, Chicago, IL: University of Chicago Press (1988).

[20] P. Erdös and A. Rényi, "Random graphs," *Publ. Math. Inst. Hungarian Acad. Sci.* **5**, 17 (1960).

[21] J. Ford, "Directions in classical chaos," in *Directions in Chaos*, Vol. 1, ed. H. Bai-lin, Singapore: World Scientific (1987), pp. 1–16.

[22] J. Gillis and G. H. Weiss, "Expected number of distinct sites visited by a random walk with an infinite variance," *J. Math. Phys.* **11**, 1308 (1970).

[23] M. Hénon, "A two-dimensional mapping with a strange attractor," *Comm. Math. Phys.* **50**, 69 (1976).

[24] J. T. M. Hosking, "Fractional differencing," *Biometrika* **68**, 165–176 (1981).

[25] B. D. Hughes, E. W. Montroll and M. F. Shlesinger, "Fractal random walks," *J. Statist. Phys.* **28**, 111–126 (1982).

[26] E. Atlee Jackson, *Perspectives of Nonlinear Dynamics*, Vol. 2, New York: Cambridge University Press (1990).

[27] A. N. Kolmogorov, "A new invariant for transitive dynamical systems," *Dokl. Akad. Nauk* **119**, 861 (1958).

[28] R. T. Malthus, *Population: The First Essay* (1798), Ann Arbor, MI: University of Michigan Press (1959).

[29] B. B. Mandelbrot, *Fractals and Scaling in Finance*, New York: Springer (1997).

[30] R. N. Mantegna and H. Eugene Stanley, *An Introduction to Econophysics*, Cambridge: Cambridge University Press (2000).

[31] R. M. May, "Simple mathematical model with very complicated dynamics," *Nature* **261**, 459–467 (1976).

[32] E. W. Montroll and W. W. Badger, *Introduction to the Quantitative Aspects of Social Phenomena*, New York: Gordon and Breach (1974).

[33] E. W. Montroll and B. J. West, "On an enriched collection of stochastic processes," in *Fluctuations Phenomena*, 2nd edn., eds. E. W. Montroll and J. L. Lebowitz, Amsterdam: North-Holland (1987), pp. 61–206; first published in 1979.

[34] J. Moser, *Stable and Random Motions in Dynamical Systems*, Princeton, NJ: Princeton University Press (1973).

[35] E. Ott, *Chaos in Dynamical Systems*, New York: Cambridge University Press (1993).

[36] C. Pesee, "Long-range dependence of financial time series data," *Proc. World Acad. Sci., Eng. and Technol.* **34**, 163 (2008).

[37] E. E. Peters, *Chaos and Order in the Capital Markets*, New York: Wiley (1996).

[38] H. Poincaré, *The Foundations of Science*, 1st edn. (1913), translated by G. B. Halsted, New York: The Science Press (1929).

[39] L. F. Richardson, "Statistics of deadly quarrels," in *The World of Mathematics*, Vol. 2, ed. J. B. Newman, New York: Simon and Schuster (1956), pp. 1254–1263.

[40] O. E. Rössler, "An equation for continuous change," *Phys. Lett. A* **57**, 397 (1976).

[41] I. Stewart, *Does God Play Dice?*, Cambridge, MA: Basil Blackwell (1989).

[42] D. J. Watts, *Small Worlds*, Princeton, NJ: Princeton University Press (1999).

[43] D. J. Watts, *Six Degrees*, New York: W. W. Norton (2003).

[44] B. J. West, *Physiology, Promiscuity and Prophecy at the Millennium: A Tale of Tails*, Singapore: World Scientific (1999).

[45] B. J. West, *Where Medicine Went Wrong*, Singapore: World Scientific (2006).

[46] Q. Wright, *A Study of War*, Chicago, IL: Chicago University Press (1942).

5 Non-analytic dynamics

In this chapter we investigate one procedure for describing the dynamics of complex webs when the differential equations of ordinary dynamics are no longer adequate, that is, the webs are fractal. We described some of the essential features of fractal functions earlier, starting from the simple dynamical processes described by functions that are fractal, such as the Weierstrass function, which are continuous everywhere but are nowhere differentiable. This idea of non-differentiability suggests introducing elementary definitions of fractional integrals and fractional derivatives starting from the limits of appropriately defined sums. The relation between fractal functions and the fractional calculus is a deep one. For example, the fractional derivative of a regular function yields a fractal function of dimension determined by the order of the fractional derivative. Thus, changes in time of phenomena that are described by fractal functions are probably best described by fractional equations of motion. In any event, this perspective is the one we developed elsewhere [31] and we find it useful here for discussing some properties of complex webs.

The separation of time scales in complex physical phenomena allows smoothing over the microscopic fluctuations and the construction of differentiable representations of the dynamics on large space scales and long time scales. However, such smoothing is not always possible. Examples of physical phenomena that resist this approach include turbulent fluid flow [23]; the stress relaxation of viscoelastic materials such as plastics, rubber and taffy [21]; and finally phase transitions [17, 27] as well as the discontinuous physical phenomena discussed in the first chapter. Recall that the physical phenomena included the sizes of earthquakes, the number of solar flares and sunspots in a given period, and the magnitude of volcanic eruptions, but also intermittent biological phenomena, including the variability of blood flow to the brain, yearly births to teenagers and the variability of step duration during walking.

Metaphorically, complex phenomena, whose dynamics cannot be described by ordinary differential equations of motion, leap and jump in unexpected ways to obtain food, unpredictably twist and turn to avoid capture, and suddenly change strategy to anticipate changes in an enemy that can learn. To understand these and other analogous processes in the social and life sciences, we find that we must adopt a new type of modeling, one that is *not* in terms of ordinary or partial differential equations of motion. It is clear that the fundamental elements of complex physical phenomena, such as phase transitions, the deformation of plastics and the stress relaxation of polymers, satisfy Newton's laws. In these phenomena the evolution of individual particles is described by ordinary differential equations that control the dynamics of individual particle trajectories. It is equally clear that the connection between the fundamental laws of motion controlling the individual particle dynamics and the observed large-scale dynamics cannot be made in any straightforward way.

In a previous section we investigated the scaling properties of processes described by certain stochastic differential equations. The scaling in the web response was a consequence of the inverse power-law correlation. We now consider a fractional Langevin equation in which the fractional derivatives model the long-time memory in the web dynamics. It is determined that the solutions to such equations describe multifractal statistics and we subsequently apply this model to a number of complex life phenomena, including cerebral blood flow and migraines.

5.1 Transition to fractional dynamics

The differentiable nature of macroscopic physical phenomena, in a sense, is a natural consequence of the microscopic randomness and of the related non-differentiability as well, due to the key role of the central limit theorem (CLT). Recall that in the CLT the quantities being added together are statistically independent, or at most weakly dependent on one another, in order for the theorem to be applicable and normal statistics to emerge. Once a condition of time-scale separation is established, in the long-time limit the memory of the non-differentiable character of the microscopic dynamics is lost and Gaussian statistics result. This also means that ordinary differential equations can again be used on the macroscopic scale, even if the microscopic dynamics are incompatible with the adoption of ordinary calculus methods.

On the other hand, if there is no time-scale separation between macroscopic and microscopic levels of description, the non-differentiable nature of the microscopic dynamics is transmitted to the macroscopic level. This can, of course, result from a non-integrable Hamiltonian at the microscopic level so that the microscopic dynamics are chaotic and the time scale associated with their dynamics diverges. Here we express the generalized Langevin equation (3.46) as

$$\frac{dV(t)}{dt} = -\Delta^2 \int_0^t \Phi_\xi(t - t')V(t')dt' + \xi(t), \tag{5.1}$$

where Δ is the coupling coefficient between the Brownian particle velocity and the booster, $\xi(t)$ is the doorway variable and $\Phi_\xi(t)$ is the autocorrelation function for the doorway variable. We choose the autocorrelation function to be a hyperbolic distribution of the form

$$\Phi_\xi(t) = \left(\frac{T}{T + t}\right)^{\mu - 1} \tag{5.2}$$

and the microscopic time scale is given by

$$\tau = \int_0^\infty \Phi_\xi(t)dt = \int_0^\infty \left(\frac{T}{T + t}\right)^{\mu - 1} dt = \frac{T}{\mu - 2} \tag{5.3}$$

for $\mu > 2$. However, with the power-law index in the interval $1 < \mu < 2$ the microscopic time scale diverges. If we multiply (5.1) by the velocity, average over the fluctuations and assume that the velocity is statistically independent of the doorway variable $\langle V(0)\xi(t)\rangle_\xi = 0$ we obtain

$$\frac{d\Phi_0}{dt} = -\Delta^2 \int_0^t \Phi_\xi(t-t')\Phi_0(t')dt', \tag{5.4}$$

where the velocity autocorrelation function is given by

$$\Phi_0(t) \equiv \frac{\langle V(0)V(t)\rangle_\xi}{\langle V^2\rangle_\xi}. \tag{5.5}$$

Inserting the hyperbolic autocorrelation function into (5.4) yields

$$\frac{d\Phi_0(t)}{dt} = -\Delta^2 \int_0^t \left(\frac{T}{T+t-t'}\right)^{\mu-1}\Phi_0(t')dt', \tag{5.6}$$

which can be approximated by using $t - t' \gg T$,

$$\frac{d\Phi_0(t)}{dt} \approx -T^{\mu-1}\Delta^2 \int_0^t \frac{1}{(t-t')^{\mu-1}}\Phi_0(t')dt'. \tag{5.7}$$

The evolution equation for the velocity autocorrelation function is simplified by introducing a limiting procedure such that as T becomes infinitesimally small and Δ becomes infinitely large the indicated product remains finite:

$$\Upsilon = \lim_{T\to 0, \Delta\to\infty} T^{\mu-1}\Delta^2. \tag{5.8}$$

This limiting procedure is a generalization of a limit to an integral equation taken by Van Hove and applied in this context by Grigolini *et al.* [6]. Adopting this ansatz reduces (5.7) to the integro-differential equation

$$\frac{d\Phi_0(t)}{dt} = -\Upsilon \int_0^t \frac{1}{(t-t')^{\mu-1}}\Phi_0(t')dt'. \tag{5.9}$$

Notice that in general the velocity autocorrelation function is related to the waiting-time distribution density $\psi(t)$ of the process under study as

$$\Phi_0(t) = 1 - \int_0^t \psi(t')dt' \tag{5.10}$$

and therefore, on taking the time derivative of (5.10), we obtain from (5.9)

$$\psi(t) = \Upsilon \int_0^t \frac{1}{(t-t')^{\mu-1}}\Phi_0(t')dt'. \tag{5.11}$$

In subsequent sections we show that one form of the fractional integral of interest to us is given by

$$D_t^{-\alpha}[f(t)] = \frac{1}{\Gamma(\alpha)} \int_0^t \frac{f(t')}{(t-t')^{1-\alpha}}dt' \tag{5.12}$$

and consequently, identifying $1 - \alpha = \mu - 1$, we can rewrite (5.9) as

$$\frac{d}{dt}\Phi_0(t) = -\Gamma(3 - \mu)\Upsilon D_t^{\mu-2}[\Phi_0(t)]. \tag{5.13}$$

Equation (5.13) can also be expressed as

$$D_t^{3-\mu}[\Phi_0(t)] = -\lambda^{3-\mu}\Phi_0(t), \tag{5.14}$$

where we have introduced the dissipation constant

$$\lambda = [\Gamma(3 - \mu)\Upsilon]^{1/(3-\mu)}. \tag{5.15}$$

However, we note that $3 - \mu > 1$ for $\mu < 2$ and therefore the fractional derivative is greater than unity and we write

$$3 - \mu = 1 + \varepsilon. \tag{5.16}$$

The solution to (5.14) can be obtained using Laplace transforms, as we explain later. The Laplace transform of the fractional derivative can be written, see page 160 of West et al. [31],

$$\mathcal{LT}\left[D_t^{1+\varepsilon}[\Phi_0(t)]; u\right] = u^\varepsilon\left[u\widehat{\Phi}_0(u) - \Phi_0(0)\right]. \tag{5.17}$$

We use throughout the notation $\hat{f}(u)$ to denote the Laplace transform of the function $f(t)$,

$$\hat{f}(u) = \int_0^\infty dt\, f(t)e^{-ut}, \tag{5.18}$$

and take the Laplace transform of the fractional dynamical equation (5.14) so that, using (5.17), we obtain

$$u^{1+\varepsilon}\widehat{\Phi}_0(u) - u^\varepsilon\Phi_0(0) = -\lambda^{1+\varepsilon}\widehat{\Phi}_0(u),$$

which, on rearranging terms, gives

$$\widehat{\Phi}_0(u) = \frac{1}{u}\frac{1}{(u/\lambda)^{1+\varepsilon} + 1}\left(\frac{u}{\lambda}\right)^{1+\varepsilon}\Phi_0(0). \tag{5.19}$$

The asymptotic-in-time form of the solution to the fractional differential equation can be obtained from the small u form of (5.19),

$$\lim_{u \to 0}\widehat{\Phi}_0(u) = \frac{1}{u}\left(\frac{u}{\lambda}\right)^{1+\varepsilon}\Phi_0(0), \tag{5.20}$$

whose inverse Laplace transform yields

$$\Phi_0(t) \approx \frac{\Phi_0(0)}{(\lambda t)^{1+\varepsilon}}. \tag{5.21}$$

The velocity autocorrelation function therefore decays asymptotically as an inverse power law in time with an index given by $3 - \mu$.

Consequently we see that the influence of the environment on the evolution of the Brownian particle, that is, the influence of the chaotic booster, is to asymptotically replace the exponential decay of the velocity correlation that appears in traditional statistical physics with an inverse power-law decorrelation. This replacement is due to the divergence of the microscopic time scale resulting in the loss of time-scale separation between microscopic and macroscopic dynamics. Therefore the chaos on the microscopic scale is transmitted as randomness to the macroscopic level and the derivatives and integrals of the ordinary calculus are replaced with fractional derivatives and integrals on the macroscopic level. We now briefly discuss some of the formal and useful properties of the fractional calculus.

5.1.1   Fractional calculus

It is useful to have in mind the formalism of the fractional calculus before interpreting models using this formalism to explain the dynamics of complex webs. The fractional calculus dates back to a question l'Hôpital asked Leibniz, the co-inventor with Newton of the ordinary calculus, in a 1695 letter. L'Hôpital wanted to know how to interpret a differential operator when the index of the operator is not an integer. Specifically, he wanted to know how to take the 1/2-power derivative of a function. Rather than quoting the response of Leibniz, let us consider the ordinary derivative of a monomial, say the nth derivative of t^m for $m > n$:

$$D_t^n[t^m] = m(m-1)\ldots(m-n+1)t^{m-n}$$

$$= \frac{m!}{(m-n)!}t^{m-n}, \tag{5.22}$$

where the operator D_t is the ordinary derivative with respect to t and we interpret D_t^n to mean taking this derivative n times. Formally the derivative can be generalized by replacing the ratio of factorials with gamma functions,

$$D_t^n[t^m] = \frac{\Gamma(m+1)}{\Gamma(m+1-n)}t^{m-n}, \tag{5.23}$$

and, proceeding by analogy, replacing the integers n and m with the non-integers α and β:

$$D_t^\alpha[t^\beta] = \frac{\Gamma(\beta+1)}{\Gamma(\beta+1-\alpha)}t^{\beta-\alpha}. \tag{5.24}$$

In a rigorous mathematical demonstration this blind replacement of gamma functions would not be tolerated since the definition of the gamma function needs to be generalized to non-integer arguments. However, we never pretended to be giving a rigorous mathematical demonstration, so we take solace in the fact that this generalization can be done and the details of how to do that need not concern us here. But if you are interested there are texts that address the mathematical subtleties [24], excellent books on the engineering applications [14] and monographs that focus on the physical interpretation of such non-integer operators [31].

Using this new definition of derivatives given by (5.24), we can solve the 1/2-derivative problem posed to Leibniz. Consider the definition (5.24) for $\alpha = -\beta = 1/2$ so that

$$D_t^{1/2}[t^{-1/2}] = \frac{\Gamma(-1/2 + 1)}{\Gamma(-1/2 + 1 - 1/2)} t^{-1/2 - 1/2}$$
$$= \frac{\Gamma(1/2)}{\Gamma(0)} t^{-1} = 0, \tag{5.25}$$

where the last equality arises from the gamma function $\Gamma(0) = \infty$. Thus, a particular function is effectively a constant with regard to a certain fractional derivative; that is, the fractional derivative of a monomial with a matching index vanishes, just like the ordinary derivative of a constant.

Consider a second example. This time let us examine the fractional derivative of a constant. Take the index of the monomial $\beta = 0$ so that the 1/2-derivative of a constant is given by (5.24) to be

$$D_t^{1/2}[1] = \frac{\Gamma(0 + 1)}{\Gamma(0 + 1 - 1/2)} t^{-1/2}$$
$$= \frac{1}{\sqrt{\pi t}}. \tag{5.26}$$

Thus, we see that the 1/2-derivative of a constant does not vanish, but is a function of the independent variable. This result implies the broader conclusion that what is considered a constant in the ordinary calculus is not necessarily a constant in the fractional calculus.

Now let us consider the response of Leibniz to the original question and determine the 1/2-derivative of the monomial t, $\beta = 1$:

$$D_t^{1/2}[t] = \frac{\Gamma(1 + 1)}{\Gamma(1 + 1 - 1/2)} t^{1-1/2}$$
$$= \sqrt{\frac{t}{\pi}}, \tag{5.27}$$

the result obtained by Leibniz. Mandelbrot quotes Leibniz' letter[1] from which we extract the piece:

... John Bernoulli seems to have told you of my having mentioned to him a marvelous analogy which makes it possible to say in a way the successive differentials are in geometric progression. One can ask what would be a differential having as its exponent a fraction. You see that the result can be expressed by an infinite series. Although this seems removed from Geometry, which does not yet know of such fractional exponents, it appears that one day these paradoxes will yield useful consequences, since there is hardly a paradox without utility.

After 310 years of sporadic development, the fractional calculus is now becoming sufficiently well developed and well known that books and articles are being devoted to its "utility" in the physical sciences [9, 26, 31]. Of most interest to us is what the

---

[1] This letter was translated by B. Mandelbrot and is contained in the "Historical sketches" of his second book [16].

fractional calculus can contribute to an understanding of complex webs in the physical, social and life sciences.

Another way to introduce fractional operators without the torture of rigorous mathematics is through Cauchy's formula for n-fold integration of a function $f(t)$ over a fixed time interval (a, t):

$$\int_a^t d\tau_1 \int_a^{\tau_1} d\tau_2 \int_a^{\tau_2} d\tau_3 \cdots \int_a^{\tau_{n-1}} d\tau_n \, f(\tau_n) = \frac{1}{(n-1)!} \int_a^t d\tau (t-\tau)^{n-1} f(\tau).$$

(5.28)

The n-fold integral on the left is denoted by the integral or anti-derivative operator

$$_aD_t^{-n}[f(t)] \equiv \frac{1}{\Gamma(n)} \int_a^t d\tau (t-\tau)^{n-1} f(\tau),$$

(5.29)

where for convenience we have again introduced the gamma function and extended the notation to include the lower limit of the integral as an index on the lower left side of the operator. The fractional integral analogue to this equation is given, just as in the case of the derivative, by the same operator expression with a non-integer index,

$$_aD_t^{-\alpha}[f(t)] \equiv \frac{1}{\Gamma(\alpha)} \int_a^t d\tau (t-\tau)^{\alpha-1} f(\tau),$$

(5.30)

with the restriction $a \leq t$. Note that the gamma function remains well defined for non-integer and non-positive values of its argument by analytic continuation into the complex domain. The corresponding fractional derivative is defined by

$$_aD_t^{\alpha}[f(t)] \equiv D_t^n \left[_aD_t^{\alpha-n}[f(t)]\right]$$

(5.31)

with the restrictions on the fractional index $\alpha - n < 0$ and $\alpha - n + 1 > 0$. Consequently, for $0 < \alpha < 1$ we have $n = 1$. Equation (5.30) is the Riemann–Liouville (RL) formula for the fractional operator; it is the integral operator when interpreted as (5.30) and it is the differential operator interpreted as (5.31) when $\alpha > 0$.

Fractal function evolution

Recall the fractal function introduced by Weierstrass and generalized by Mandelbrot and others. Let us rewrite the generalized Weierstrass function (GWF) here,

$$W(t) = \sum_{n=-\infty}^{\infty} \frac{1}{a^n} [1 - \cos(b^n t)],$$

(5.32)

and recall further that the parameters a and b are related by

$$a^\mu = b$$

(5.33)

and that in terms of the fractal dimension $D = 2 - \mu$ the GWF can be written as

$$W(t) = \sum_{n=-\infty}^{\infty} \frac{1}{b^{(2-D)n}} [1 - \cos(b^n t)].$$

(5.34)

The RL fractional integral of the GWF is given by [22]

$$W^{(-\alpha)}(t) \equiv {}_{-\infty}D_t^{-\alpha}[W(t)] = \frac{1}{\Gamma(\alpha)} \int_{-\infty}^{t} \frac{W(\tau)d\tau}{(t-\tau)^{1-\alpha}} \qquad (5.35)$$

for $0 < \alpha < 1$, which, after some not-so-straightforward analysis, yields

$$W^{(-\alpha)}(t) = \sum_{n=-\infty}^{\infty} \frac{1}{b^{(2-D+\alpha)n}}[1 - \cos(b^n t)]. \qquad (5.36)$$

It is clear that the fractional integration has shifted the fractal dimension of the GWF from D to $D - \alpha$, thereby reducing the fractal dimension and making the function smoother; this is consistent with our intuition about integration.

Similarly the RL fractional derivative of the GWF is given by [22]

$$W^{(\alpha)}(t) \equiv {}_{-\infty}D_t^{\alpha}[W(t)] = \frac{1}{\Gamma(\alpha)} \frac{d}{dt} \int_{-\infty}^{t} \frac{W(\tau)d\tau}{(t-\tau)^{\alpha}} \qquad (5.37)$$

for $0 < \alpha < 1$, which, again after some not-so-straightforward analysis, integrates to

$$W^{(\alpha)}(t) = \sum_{n=-\infty}^{\infty} \frac{1}{b^{(2-D-\alpha)n}}[1 - \cos(b^n t)]. \qquad (5.38)$$

Consequently, we see that the fractional derivative has shifted the fractal dimension of the GWF from D to $D + \alpha$, thereby increasing the fractal dimension and making the function more erratic; this is consistent with our intuition about derivatives.

Note that we have shown that a fractal function, whose integer derivative diverges, has a finite fractional derivative. It is therefore reasonable to conjecture that a dynamical process described by such a fractal function might have fractional differential equations of motion. This assumption is explored below.

5.1.2   Deterministic fractional dynamics

Of course, the fractional calculus does not in itself constitute a theory for complex webs in the physical, social or life sciences; however, such theories are necessary in order to interpret the fractional derivatives and integrals in the appropriate contexts. We therefore begin by examining a simple relaxation process described by the rate equation

$$D_t[Q(t)] + \lambda Q(t) = 0, \qquad (5.39)$$

where $t > 0$ and the relaxation rate λ determines how quickly the process returns to its equilibrium state. The solution to this equation is given in terms of its initial value by

$$Q(t) = Q(0)e^{-\lambda t}, \qquad (5.40)$$

which is unique. Consequently, everything we can know about this simple network starting from an initial condition is contained in the relaxation rate. Equation (5.40) can also describe a stochastic process if the initial condition is interpreted as the rate $Q(0) = \lambda$, the rate of occurrence of a random event, so that $Q(t)$ is the exponential probability density and the process being described is Poisson.

An alternative representation of the relaxation equation can be expressed in terms of the anti-derivative operator by

$$Q(t) - Q(0) = -\lambda D_t^{-1}[Q(t)]. \tag{5.41}$$

This equation suggests replacing the anti-derivative operator to describe a more complex network with the RL fractional integral operator

$$Q(t) - Q(0) = {_0}D_t^{-\alpha}[-\lambda^\alpha Q(t)] \tag{5.42}$$

and, by inverting the fractional integral operator, replacing the relaxation equation with the fractional relaxation equation

$$_0D_t^\alpha[Q(t)] + \lambda^\alpha Q(t) = \frac{t^{-\alpha}}{\Gamma(1 - \alpha)} Q(0). \tag{5.43}$$

Here the initial value is a time-dependent inhomogeneous term and the relaxation time is raised to a power to match the index of the fractional derivative in order to maintain a consistent system of units. This rather intimidating-looking equation has a remarkable solution with application to a wide variety of complex phenomena. But it should be emphasized that we are incorporating the history of the dynamical web into its solution; what happens at a given time depends on what happened at earlier times, not just on the initial value.

Equations of the form (5.43) are mathematically well defined, and strategies for solving them have been developed by a number of investigators, particularly by Miller and Ross [18], whose book is devoted almost exclusively to solving such equations when the fractional index α is rational. Here we allow the index to be irrational and posit the solution to the dynamical equation through Laplace transforms. Consider the Laplace transform of (5.43),

$$\widehat{Q}(u) = \frac{Q(0)}{u} \frac{u^\alpha}{\lambda^\alpha + u^\alpha}, \tag{5.44}$$

whose inverse is the solution to the fractional relaxation equation. Inverting Laplace transforms with irrational powers, such as (5.44), can be non-trivial and special techniques for this purpose have been developed, but we do not review those techniques here. The inverse Laplace transform of (5.44) yields the solution to the fractional relaxation equation in terms of the Mittag-Leffler function (MLF)

$$E_\alpha(-(\lambda t)^\alpha) = \sum_{k=0}^\infty \frac{(-1)^k}{\Gamma(1 + k\alpha)} (\lambda t)^{k\alpha} \tag{5.45}$$

given by

$$Q(t) = Q(0) E_\alpha(-(\lambda t)^\alpha). \tag{5.46}$$

To verify that the series (5.46) and (5.44) are Laplace-transform pairs, consider the Laplace transform of the MLF

$$\int_0^\infty e^{-ut} E_\alpha(-(\lambda t)^\alpha) dt = \sum_{k=0}^\infty \frac{(-1)^k \lambda^{k\alpha}}{\Gamma(1 + k\alpha)} \int_0^\infty e^{-ut} t^{k\alpha} dt,$$

so that evaluating the integral on the rhs yields

$$\int_0^\infty e^{-ut} E_\alpha(-(\lambda t)^\alpha)dt = \sum_{k=0}^\infty \frac{(-1)^k \lambda^{k\alpha}}{\Gamma(1+k\alpha)} \frac{\Gamma(1+k\alpha)}{u^{k\alpha+1}}$$

$$= \frac{1}{u}\sum_{k=0}^\infty (-1)^k \left(\frac{\lambda}{u}\right)^{k\alpha} = \frac{u^{\alpha-1}}{\lambda^\alpha + u^\alpha}.$$

Note that when $\alpha = 1$ this expression is just the Laplace transform for the exponential and, indeed,

$$E_\alpha(-(\lambda t)^\alpha)\big|_{\alpha=1} = e^{-\lambda t} \tag{5.47}$$

and consequently the MLF is considered a generalization of the exponential function. This limit of the MLF is consistent with the fractional relaxation equation reducing to the ordinary relaxation equation in this limit since the inhomogeneous term in (5.43) vanishes at $\alpha = 1$.

The MLF has been used to model the stress relaxation of materials for over a century and has a number of interesting properties in both the short-time and the long-time limits. In the short-time limit it yields the Kohlrausch–Williams–Watts law for stress relaxation in rheology given by

$$\lim_{t\to 0} E_\alpha(-(\lambda t)^\alpha) = e^{-(\lambda t)^\alpha}, \tag{5.48}$$

which is also known as the stretched exponential. This name is a consequence of the fact that the fractional index is less than unity, so the exponential relaxation is stretched out in time, with the relaxation being slower than ordinary exponential relaxation for $\alpha < 1$. In a probability context the MLF is known as the Weibull distribution and is used to model crack growth within materials, as mentioned previously.

In the long-time limit the MLF yields an inverse power law, known as the Nutting law,

$$\lim_{t\to\infty} E_\alpha(-(\lambda t)^\alpha) = \frac{1}{(\lambda t)^\alpha}. \tag{5.49}$$

Consequently, the initial relaxation for the MLF is slower than the ordinary exponential, but asymptotically in time the relaxation slows down considerably, becoming an inverse power law. Here again, the long-time limiting form of the MLF is the well-known probability distribution observed for multiple phenomena in the first few chapters. We discuss the probability implications of this solution later.

Figure 5.1 displays the general MLF as well as its two asymptotes; the dashed curve is the stretched exponential and the dotted curve depicts the inverse power law. It is apparent from the discussion that there is a long-time memory associated with the fractional relaxation process characterized by the inverse power law. This asymptotic behavior replaces the short-time memory of the exponential relaxation in ordinary relaxation. Moreover, it is apparent that the MLF smoothly joins the two empirically determined asymptotic distributions of Kohlrausch–Williams–Watts and Nutting.

Magin [14] gives an excellent discussion of the properties of the MLF in both the time and the Laplace-transform domains. Moreover, he discusses the solutions to the

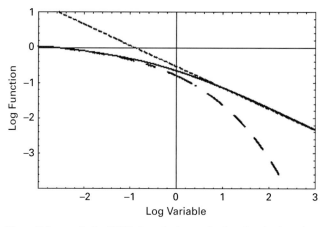

Figure 5.1.   The solid curve is the MLF, the solution to the fractional relaxation equation. The dashed curve is the stretched exponential (Kohlrausch–Williams–Watts law) and the dotted curve is the inverse power law (Nutting law).

fractional equations of motion that arise in a variety of contexts, emphasizing that the inverse power law is more than curve fitting and consequently the fractional calculus which gives rise to inverse power-law solutions may well describe the dynamics of certain complex webs.

5.2   Some applications

5.2.1   The fractional Poisson distribution

In Section 3.3 we considered the occurrence of events as a dynamical counting process and discussed the Poisson process as being renewal. Here we extend that discussion and consider the fractional dynamics introduced in the previous section as describing a new kind of renewal process. Mainardi *et al.* [15] refer to this as a Mittag-Leffler renewal process because the MLF replaces the exponential in characterizing the time interval between events. This is a reasonable extension since the MLF

$$\Psi(t) \equiv E_\beta\left(-(\lambda t)^\beta\right) = \sum_{n=0}^{\infty} \frac{(-1)^n}{\Gamma(1+n\beta)} (\lambda t)^{n\beta} \tag{5.50}$$

reduces to the exponential for $\beta = 1$. The function $\Psi(t)$ given by (5.50) can therefore be interpreted as the probability that an event has not occurred in the time interval $(0, t)$. The corresponding pdf is given by

$$\psi(t) \equiv -\frac{d\Psi(t)}{dt}. \tag{5.51}$$

Another quantity involving the counting process $N(t)$ is the probability that k events occur in the closed time interval $[0, t]$,

$$P(k; t) = \Pr[N(t) = k] = P(t_k \le t, t_{k+1} > t). \tag{5.52}$$

From this definition it is evident that

$$\Psi(t) = P(k = 0; t). \tag{5.53}$$

It is interesting to note that when the time denotes the lifetime of a manufactured object then $\Psi(t)$ is called the survival probability and correspondingly $1 - \Psi(t)$ is the probability of failure. This terminology has been extended beyond renewal theory and the predicted lifetimes of manufactured objects to other, more inclusive contexts, such as the failure of complex webs, and was taken up in our discussion of random walks.

The expected number of events $N(t)$ is called the renewal function $m(t)$ and is defined by

$$m(t) = \langle N(t) \rangle = \sum_{k=1}^{\infty} P(t_k \leq t), \tag{5.54}$$

where t_k is the time of occurrence of the kth event. The renewal function can also be expressed as the convolution

$$m(t) = \int_0^t \psi(t - t')[1 - m(t')]dt', \tag{5.55}$$

whose Laplace transform is

$$\widehat{m}(u) = \frac{\widehat{\psi}(u)}{u\left[1 - \widehat{\psi}(u)\right]}. \tag{5.56}$$

The Laplace transform of the waiting-time distribution is given by

$$\widehat{\psi}(u) = 1 - u\widehat{\Psi}(u) \tag{5.57}$$

and, using the Laplace transform of the MLF,

$$\mathcal{LT}\left[t^{k\beta} E_\beta^{(k)}\left(-(\lambda t)^\beta\right); u\right] = \frac{k! u^{\beta-1}}{\left(u^\beta + \lambda^\beta\right)^{k+1}}; \quad u > \lambda, \tag{5.58}$$

we obtain for the waiting-time distribution function with $k = 0$

$$\widehat{\psi}(u) = 1 - \frac{u^\beta}{u^\beta + \lambda^\beta}. \tag{5.59}$$

Consequently, the Laplace transform of the renewal function is

$$\widehat{m}(u) = \frac{\lambda^\beta}{u^{\beta+1}} \tag{5.60}$$

and in the time domain

$$m(t) = \frac{(\lambda t)^\beta}{\Gamma(1 + \beta)}. \tag{5.61}$$

Thus the renewal function increases with time more slowly than a Poisson process ($\beta = 1$).

The probability that $N(t) = k$ is determined by the convolution equation

$$P(N(t) = k) = \int_0^t \Psi(t - t')\psi_k(t')dt', \qquad (5.62)$$

which is the product of the probability that k events occur in the interval $(0, t')$ and no events occur in the time interval (t', t). The Laplace transform of $P(N(t) = k)$ is given by

$$\widehat{P}_k(u) = \widehat{\psi}(u)^k \widehat{\Psi}(u) = \left[1 - \frac{u^\beta}{u^\beta + \lambda^\beta}\right]^k \frac{u^{\beta-1}}{u^\beta + \lambda^\beta}$$

$$= \frac{\lambda^{k\beta} u^{\beta-1}}{\left(u^\beta + \lambda^\beta\right)^{k+1}} \qquad (5.63)$$

and the inverse Laplace transform is

$$P(N(t) = k) = \frac{(\lambda t)^{k\beta}}{k!} E_\beta\left(-(\lambda t)^\beta\right), \qquad (5.64)$$

which is a clear generalization of the Poisson distribution with parameter λt and is called the *β-fractional Poisson distribution* by Mainardi *et al.* [15]. Of course (5.64) becomes an ordinary Poisson distribution when $\beta = 1$.

5.2.2   The Weibull distribution

In an earlier section we asserted without proof that the MLF was a stretched exponential at early times and an inverse power law at late times. It is rather easy to prove these asymptotic forms. Consider first the Laplace transform of the MLF,

$$\widehat{E}_\beta(u) = \frac{u^{\beta-1}}{u^\beta + \lambda^\beta}. \qquad (5.65)$$

The short-time limit $t \to 0$ of the MLF is obtained from the $u \to \infty$ limit of (5.65) by means of a Taylor expansion,

$$\lim_{u \to \infty} \widehat{E}_\beta(u) = \frac{1}{u} \sum_{k=0}^{\infty} (-1)^k \left(\frac{\lambda}{u}\right)^{k\beta}, \qquad (5.66)$$

whose inverse Laplace transform is

$$\Psi_W(t) = \sum_{k=0}^{\infty} (-1)^k \frac{(\lambda t)^{k\beta}}{\Gamma(k)} = e^{-(\lambda t)^\beta}; \quad \frac{1}{\lambda} \gg t. \qquad (5.67)$$

The subscript W on the survival probability is used because the pdf

$$\psi_W(t) = -\frac{d\Psi_W(t)}{dt} = \beta \lambda^\beta t^{\beta-1} e^{-(\lambda t)^\beta} \qquad (5.68)$$

is the Weibull distribution (WD).

The WD was empirically developed by Weibull in 1939 [28] to determine the distribution of crack sizes in materials and consequently in this context the independent variable is not the time but the size of the crack. The probability that a crack in a material occurs in the interval $(t, t + dt)$ is given by $\psi_W(t)dt$.

The WD was subsequently developed as the pdf for the minimum extreme values among measurements modeled by an initial distribution bounded by a smallest value. In material science, crack initiation dominates the phenomenon of fracture in brittle solids, which initiated the development of models in which defects, point sources of stress concentration, are statistically distributed in a unit volume of a solid. As Bury [3] points out, each of these defects determines the local strength of the material and consequently the overall strength of the material is equal to the minimum local strength. A chain is no stronger than its weakest link. The asymptotic model of specimen strength is the WD and is given by [3]

$$F_W(s) = 1 - \Psi_W(s) = 1 - \exp\left[-\left(\frac{s}{\sigma}\right)^{\beta}\right], \tag{5.69}$$

where s is the stress, β is a dimensionless shape parameter and σ is a scale parameter with the dimensions of stress.

The modern statistical modeling of the strength of materials is based on the WD, but it had not previously been linked to the fractional calculus and the asymptotic form of the MLF.

5.3 Fractional stochastic equations

5.3.1 The fractional Langevin equation

We now generalize the fractional differential equation to include a random force $\xi(t)$ and in this way obtain a fractional Langevin equation

$$_0D_t^{\alpha}[Q(t)] + \lambda^{\alpha} Q(t) = \frac{t^{-\alpha}}{\Gamma(1 - \alpha)} Q(0) + \xi(t). \tag{5.70}$$

Here we have introduced two distinct kinds of complexity into the dynamical web. The fractional relaxation takes into account the complicated webbing of the interactions within the process of interest and the random force intended to represent how this complex web interacts with its environment. Note that on the face of it this approach to modeling complexity stacks one kind of complexity on another without attempting to find a common cause. It remains to be seen whether such a common cause exists.

The solution to the fractional Langevin equation is obtained in the same way as the simpler deterministic fractional dynamical equation, by using Laplace transforms. The transformed equation is

$$\widehat{Q}(s) = \frac{Q(0)}{u} \frac{u^{\alpha}}{\lambda^{\alpha} + u^{\alpha}} + \frac{\widehat{\xi}(u)}{\lambda^{\alpha} + u^{\alpha}}, \tag{5.71}$$

where we need to know the statistical properties of the random force in order to properly interpret the inverted Laplace transform. Note the difference between the u-dependences of the two coefficients of the rhs of (5.71). The inverse Laplace transform of the first term yields the Mittag-Leffler function as found in the homogeneous case. The inverse Laplace transform of the second term is the convolution of the random force and a stationary kernel.

The explicit inverse of (5.71) yields the formal solution [31]

$$Q(t) = Q(0)E_\alpha(-(\lambda t)^\alpha) + \int_0^t d\tau \frac{E_{\alpha,\alpha}(-(\lambda(t-\tau))^\alpha)\xi(\tau)}{(t-\tau)^{1-\alpha}}, \tag{5.72}$$

where we see that the fluctuations from the random force are smoothed or filtered by the generalized MLF, as well as by the inverse power law. The kernel in this convolution is given by the series

$$E_{\alpha,\beta}(z) \equiv \sum_{k=0}^{\infty} \frac{z^k}{\Gamma(\beta + k\alpha)}; \quad \alpha, \beta > 0, \tag{5.73}$$

which is the generalized MLF and for $\beta = 1$ this series reduces to the standard MLF. Consequently both the homogeneous and the inhomogeneous terms in the solution to the fractional Langevin equation can be expressed in terms of these MLF series. The average of the solution (5.72) over an ensemble of realizations of a zero-centered random force is given by

$$\langle Q(t)\rangle_\xi = Q(0)E_\alpha(-(\lambda t)^\alpha), \tag{5.74}$$

which we see coincides with the deterministic solution to the fractional diffusion equation (5.46).

In the case $\alpha = 1$, the MLF reduces to the exponential, so that the solution to the fractional Langevin equation becomes that for an Ornstein–Uhlenbeck (OU) process,

$$Q(t) = Q(0)e^{-\lambda t} + \int_0^t d\tau\, e^{-\lambda(t-\tau)}\xi(\tau), \tag{5.75}$$

as it should. Equation (5.75) is simple to interpret since the dissipation has the typical form of a filter and it smoothes the random force in a familiar way.

The properties of the solutions to such stochastic equations can be determined only after the statistics of the random force have been specified. The usual assumption made is that the random force is zero-centered, delta-correlated in time with Gaussian statistics. From these assumed statistics and the fact that the equation of motion is linear, even if fractional, the statistics of the response must also be Gaussian, and consequently its properties are completely determined by its first two moments. It is interesting to note that the variance for the dissipation-free ($\lambda = 0$) fractional Langevin equation can be calculated to yield

$$\left\langle \left[Q(t) - \langle Q(t)\rangle_\xi\right]^2 \right\rangle_\xi \propto t^{2\alpha - 1}, \tag{5.76}$$

where the proportionality constant is given in terms of the strength of the random force. This time dependence of the variance agrees with that obtained for anomalous diffusion if we make the identification

$$2H = 2\alpha - 1, \tag{5.77}$$

where, given that the fractional operator index is in the interval $1 > \alpha > 1/2$,

$$1/2 \geq H > 0. \tag{5.78}$$

Consequently, the process described by the dissipation-free fractional Langevin equation is anti-persistent.

This anti-persistent behavior of the time series was observed by Peng *et al.* [20] for the differences in time intervals between heartbeats. They interpreted this result, as did a number of subsequent investigators, in terms of random walks with $H < 1/2$. However, we can see from (5.76) that the fractional Langevin equation without dissipation is an equally good, or one might say better, description of the underlying dynamics. The scaling behavior alone cannot distinguish between these two models; what is needed is the complete statistical distribution of the empirical data rather than just the time-dependence of one or two moments.

5.3.2 Multiplicative random force

So far we have learned that the dynamics of a conservative dynamical web can be modeled using Hamilton's equations of motion. A dynamical web with many degrees of freedom can be used to construct a Langevin equation from a Hamiltonian model for a simple dynamical network coupled to the environment. The equations of motion for the coupled network are manipulated so as to eliminate the degrees of freedom of the environment from the dynamical description of the web of interest. Only the initial state of the environment remains in the Langevin description, where the random nature of the driving force is inserted through the choice of distribution of the initial states of the environment. The random driver is typically assumed to be a Wiener process – that is, to have Gaussian statistics and no memory.

When the web dynamics depends on what occurred earlier, that is, the environment has memory, the simple Langevin equation is modified. The generalized Langevin equation takes memory into account through an integral term whose memory kernel is connected to the autocorrelation of the random force. Both the simple and the generalized Langevin equations are monofractal if the fluctuations are monofractal, so the web response is a fractal random process if the random force is a fractal random process and the dynamics are linear. But the web response can be a fractal random process even when the random force is a Wiener process, if the web dynamics are fractional.

A different way of modeling the influence of the environment by a random force does not require the many degrees of freedom used in the traditional approach. Instead the nonlinear dynamics of the bath become a booster due to chaos, and the back-reaction of the booster to the web of interest provides the mechanism for dissipation. This picture of the source of the Langevin equation sidesteps a number of the more subtle statistical physics issues and suggests a more direct way to model fluctuations in complex social and biological webs.

While the properties of monofractals are determined by the global Hurst exponent, there exists a more general class of heterogeneous signals known as multifractals which are made up of many interwoven subsets with different local Hurst exponents h. The statistical properties of these subsets may be characterized by the distribution of fractal dimensions $f(h)$. In order to describe the scaling properties of multifractal signals it is necessary to use many local h-exponents.

The incremental increases in web complexity have progressed under our eyes from linear to nonlinear, then back to linear, but with fractional dynamics, and from ordinary Langevin to generalized Langevin to fractional Langevin equations when the random influence of the environment is taken into account. However, even these techniques are not adequate for describing multifractal statistical processes. Multifractals have not a single fractal dimension, but multiple fractal dimensions, which could, for example, change with the development of the network in time. One strategy for modeling a phenomenon whose fractal dimension changes over time, a multifractal process, is to make the operator's fractional index a random variable. To make these ideas concrete, consider the dissipation-free fractional Langevin equation

$$_0D_t^\eta[Q(t)] - \frac{t^{-\eta}}{\Gamma(1-\eta)}Q(0) = \xi(t). \tag{5.79}$$

Equation (5.79) has been shown to be derivable from the construction of a fractional Langevin equation for a free particle coupled to a heat bath when the inertial term is negligible [13]. The formal solution to this fractional Langevin equation is

$$\Delta Q(t) \equiv Q(t) - Q(0) = \int_0^t K_\alpha(t-\tau)\xi(\tau)d\tau; \quad \alpha = 1 - \eta, \tag{5.80}$$

where the integral kernel is defined as

$$K_\alpha(t-\tau) = \frac{1}{\Gamma(\alpha)}\frac{1}{(t-\tau)^\alpha}. \tag{5.81}$$

The form of (5.80) for multiplicative stochastic processes and its association with multifractals have been noted in the phenomenon of turbulent fluid flow through a space rather than a time integration kernel [1].

The random-force term on the rhs of (5.80) is selected to be a zero-centered, Gaussian random variable and therefore to scale as

$$\xi(\lambda t) = \lambda^H \xi(t), \tag{5.82}$$

where the Hurst exponent is in the range $0 < H \leq 1$. In a similar way the kernel scales as follows:

$$K_\alpha(\lambda t) = \frac{1}{\Gamma(\alpha)}\frac{1}{(\lambda t)^\alpha} = \lambda^{-\alpha}K_\alpha(t), \tag{5.83}$$

so that the solution to the fractional multiplicative Langevin equation scales as

$$\Delta Q(\lambda t) = \lambda^{H+1-\alpha}\Delta Q(t). \tag{5.84}$$

In order for this solution to be multifractal we assume that the parameter α is a random variable. To construct the traditional measures of multifractal stochastic processes we calculate the qth moment of the solution by averaging over both the random force and the random parameter to obtain

$$\left\langle |\Delta Q(\lambda t)|^q \right\rangle = \lambda^{(H+1)q} \left\langle \lambda^{-q\alpha} \right\rangle_\alpha \left\langle |\Delta Q(t)|^q \right\rangle_\xi$$
$$= \left\langle |\Delta Q(t)|^q \right\rangle_\xi \lambda^{\zeta(q)}, \tag{5.85}$$

yielding the qth-order correlation coefficient $\zeta(q)$. We assume the statistics of the fractional index α to be Gaussian such that

$$\left\langle \lambda^{-q\alpha} \right\rangle_\alpha = e^{-\sigma q^2 \ln \lambda}. \tag{5.86}$$

Thus, we obtain the correlation coefficient

$$\zeta(q) = (q+1)H - \sigma q^2. \tag{5.87}$$

The solution to the fractional Langevin equation is monofractal when the correlation coefficient is linear in q, that is, when $\sigma = 0$; otherwise the process is multifractal.

Properties of multifractals

Partition functions [4] have been the dominant measure used to determine the multifractal behavior of time series covered by a uniform mesh of cell size δ. The typical scaling behavior of the partition function $Z_q(\delta)$ in the limit of vanishing grid scale [4, 5] $\delta \to 0$,

$$Z_q(\delta) \approx \delta^{\tau(q)}, \tag{5.88}$$

determines the mass exponent $\tau(q)$. However, there is no unique way to determine the partition function and consequently a variety of data-processing methods has been used for its calculation. Wavelet transforms have been used in studying the properties of cerebral blood flow [32], as well as in the investigation of stride-interval time-series data [25], but the latter time series have also been analyzed using a random-walk approach to determining the partition function. However, reviewing these data-processing techniques would take us too far from our path, so we restrict our discussion to the results.

The mass exponent is related to the generalized dimension $D(q)$ by the relation

$$\tau(q) = (1-q)D(q), \tag{5.89}$$

where $D(0)$ is the fractal or box-counting dimension, $D(1)$ is the information dimension obtained by applying l'Hôpital's rule to the ratio

$$D(1) = \lim_{q \to 1} \frac{\tau(q)}{(1-q)} = -\tau'(q) \tag{5.90}$$

and $D(2)$ is the correlation dimension [5]. The moment q therefore accentuates different aspects of the underlying dynamical process. For $q > 0$, the partition function $Z_q(\delta)$ emphasizes large fluctuations and strong singularities through the generalized dimensions, whereas for $q < 0$, the partition function stresses the small fluctuations and the weak singularities. This property of the partition function deserves a cautionary note because the negative moments can easily become unstable, introducing artifacts into the calculation, and it was to handle various aspects of the instability that the different ways of calculating the partition function were devised.

A monofractal time series can be characterized by a single fractal dimension. In general, time series have a local fractal exponent h that varies over the course of the trajectory. The function $f(h)$, called the multifractal or singularity spectrum, describes how the local fractal exponents contribute to such time series. Here the scaling index h and its spectral distribution $f(h)$ are independent variables, as are the moments q and the mass exponent $\tau(q)$. The general formalism of Legrendre-transform pairs interrelates these two sets of variables by the relation, using the sign convention in Feder [5],

$$f(h) = qh + \tau(q). \tag{5.91}$$

The local Hölder exponent h varies with the q-dependent mass exponent through the equality

$$h(q) = -\frac{d\tau(q)}{dq} = -\tau'(q), \tag{5.92}$$

so the singularity spectrum can be written as

$$f(h(q)) = -q\tau'(q) + \tau(q). \tag{5.93}$$

Below we explore these relations using data sets to calculate $\tau(q)$ and its derivative $\tau'(q)$.

5.3.3 Some physiologic multifractals

Migraine and cerebral blood flow

Migraine headaches have been the bane of humanity for centuries, afflicting such notables as Caesar, Pascal, Kant, Beethoven, Chopin and Napoleon. However, their etiology and pathomechanism have not to date been satisfactorily explained. West *et al.* [32] demonstrate that the characteristics of the time series associated with cerebral blood flow (CBF) significantly differ between normal healthy individuals and migraineurs, concluding that migraine is a functional neurological disorder that affects CBF autoregulation. A healthy human brain is perfused by the laminar flow of blood through the cerebral vessels providing brain tissue with substrates, such as oxygen and glucose. The CBF is relatively stable, with typical values in a narrow range even for relatively large changes in systemic pressure. This phenomenon is known as cerebral autoregulation. Transcranial Doppler ultrasonography enables high-resolution measurement of middle-cerebral-artery flow velocity (MCAfv) and, even though this technique does not allow direct determination of CBF values, it does help to elucidate the nature and role of vascular abnormalities associated with migraine.

Here we examine the signature of migraine pathology in the dynamics of cerebral autoregulation through the multifractal properties of the human MCAfv time series. As we mentioned, the properties of monofractals are determined by the global Hurst exponent, but, as mentioned, there exists a more general class of heterogeneous signals, multifractals, made up of many interwoven subsets with different local Hurst exponents. The statistical properties of these subsets may be characterized by the distribution of fractal dimensions $f(h)$, which is determined by the mass exponent $\tau(q)$. We apply the partition-function measure and numerically evaluate the mass exponent

$$\tau(q) = \frac{\ln Z_q(\delta)}{\ln \delta} \tag{5.94}$$

for a measured MCAfv time series.

West *et al.* [32] measured MCAfv using an ultrasonograph. The 2-MHz Doppler probes were placed over the temporal windows and fixed at a constant angle and position. The measurements were taken continuously for approximately two hours with the subjects at supine rest. The study sample comprised eight migraineurs and five healthy individuals. It was checked that those in the migraine group had no illnesses that would affect cerebral autoregulation. Figure 5.2(a) shows the averaged exponent and singularity spectrum of the aggregated MCAfv time series for ten measurements of five

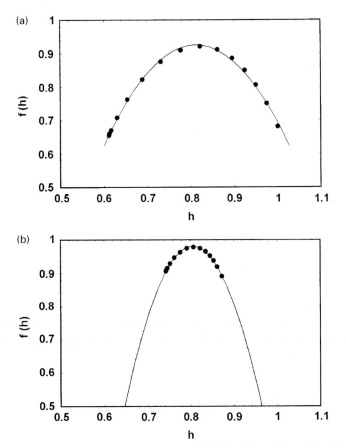

Figure 5.2. (a) The average singularity spectrum $f(h)$ of healthy subjects (filled circles) for middle-cerebral-artery blood flow. The spectrum is the average of ten measurements of five subjects. The solid curve is the best least-square fit of the parameters in (5.98) to the data. (b) The average singularity spectrum $f(h)$ of migraineurs (filled circles). The spectrum is the average of fourteen measurements of eight subjects. The solid curve is the best least-square fit of the parameters in (5.98) to the data [32]. Reproduced with permission.

healthy individuals, whereas Figure 5.2(b) depicts the average exponent and singularity spectrum for fourteen measurements of eight migraineurs. The convex form of the singularity spectrum indicates that the CBF is multifractal. However, the degree of variability, as measured by the width of the singularity distribution, is vastly restricted for the migraineurs compared with healthy individuals. Note that the partition function was calculated by West *et al.* [32] using wavelet transforms.

The fractional multiplicative Langevin equation from the previous subsection provides one possible explanation of this effect. Consider the relation between the two scaling exponents

$$\zeta(q) = 2 - \tau(q), \tag{5.95}$$

so that, since from (5.87) $\zeta(0) = H$, we have

$$\tau(0) = 2 - H, \tag{5.96}$$

resulting in the well-known relation between the fractal dimension and the global Hurst exponent

$$D(0) = 2 - H. \tag{5.97}$$

When the multiplicative stochastic parameter in the fractional Langevin equation has Gaussian statistics we obtain, by inserting $\zeta(q)$ into (5.93), the singularity spectrum

$$f(h) = f(H) - \frac{(h - H)^2}{2\sigma}, \tag{5.98}$$

for which the mode of the spectrum is located by

$$f(H) = 2 - H \tag{5.99}$$

and the quadratic term arises from $q = (h - H)/(2\sigma)$. The solid curves in Figure 5.2 are the best least-square fits of (5.98) to the empirical singularity spectrum.

A significant change in the multifractal properties of the MCAfv time series of the migraineurs is apparent. Namely, the interval for the singularity distribution $f(h)$ on the local Hurst exponent h is vastly constricted. This is reflected by the small value of the width of the singularity spectrum squared, $\sigma = 0.013$ for the migraineurs, which is almost three times smaller than the value for the control group $\sigma = 0.038$. In both cases the distribution is centered at $H = 0.81$. The parameters H and σ were determined via the least-square fit to (5.98). This collapse of the singularity spectrum suggests that the underlying process has lost its flexibility. The biological advantage of multifractal processes is that they are highly adaptive, so that in this case the brain of a healthy individual adapts to the multifractality of the interbeat interval time series. Here again we see that disease, in this case migraine, may be associated with the loss of complexity and consequently the loss of adaptability, thereby suppressing the normal multifractality of blood-flow time series.

Stride-interval multifractality

Walking consists of a sequence of steps and the corresponding time series consists of the time intervals for these steps, say, the time from one sequential striking of the right heel to the ground to the next. In normal relaxed walking it is often assumed that the stride interval is constant. However, it has been known for over a century that there is variation in the stride interval of approximately 3%–4%. This random variability is so small that the biomedical community has historically considered these fluctuations to be an uncorrelated random process. In practice this means that the fluctuations in gait were thought to convey no useful information about the underlying motor-control web. On the other hand, Hausdorff *et al.* [8] demonstrated that stride-interval time series exhibit long-time correlations and suggested that walking is a self-similar, fractal activity. In their analysis of stride-interval data Scafetta *et al.* [25] concluded that walking is in fact multifractal.

We apply the partition-function measure to a stride-interval time series [7] and numerically evaluate the mass exponent using (5.89). In this analysis the partition function was evaluated by constructing a random-walk sequence at a given level of resolution [32]. The results of this analysis are depicted in Figure 5.3(a). Rigorously, the expression for the mass exponent requires $\delta \to 0$, but we cannot do that with the data, so there is some error in the results depicted. In Figure 5.3(a) we show the mass exponent averaged over ten subjects, since their data individually do not look too different from the curve shown. It is clear from the figure that the mass exponent is not linear in the moment index q. In Table 5.1 we record the fitting coefficients for each of the ten time series using the quadratic polynomial in the interval $-3 \leq q \leq 3$,

$$\tau(q) = a_0 + a_1 q + a_2 q^2. \tag{5.100}$$

The fit to the data using (5.100) is indicated by the solid curve in Figure 5.3(a).

Table 5.1. The fitting parameters for the scaling exponent $\tau(q)$ are listed for the stride-interval time series. The column $-a_1$ is the fractal dimension for the time series. In each case these numbers agree with those obtained elsewhere using a different method [7].

Walker	a_0	$-a_1$	a_2
1	1.03	1.26	0.13
2	0.99	1.41	0.08
3	1.05	1.31	0.14
4	1.05	1.26	0.12
5	1.00	1.12	0.07
6	1.01	1.07	0.05
7	1.02	1.17	0.09
8	1.09	1.29	0.14
9	1.02	1.14	0.08
10	1.01	1.17	0.09
Average	1.03	1.19	0.10

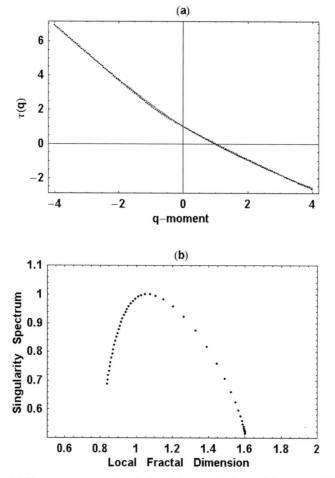

Figure 5.3.   (a) The mass exponent is calculated from stride-interval data using (5.94), indicated by the dots, and the solid curve is the fit to the data obtained using (5.100). The singularity spectrum given in (b) is determined from the data using (5.93) and is done independently of the fitting used in (a) [7]. Reproduced with permission.

The singularity spectrum can now be determined using the Legendre transformation by at least two different methods. One procedure is to use the fitting equation substituted into (5.93). We do not do this here, but we note in passing that, if (5.100) is inserted into (5.92), the fractal dimension is determined by the $q = 0$ moment to be

$$h(0) = -\tau'(0) = -a_1. \tag{5.101}$$

The values of the parameter a_1 listed in Table 5.1 agree with the fractal dimensions obtained elsewhere by using a scaling argument for the same data [7].

A second method for determining the singularity spectrum is to numerically determine both $\tau(q)$ and its derivative. In this way we calculate the multifractal spectrum directly from the data using (5.93). It is clear from Figure 5.3(b) that we obtain the canonical form of the spectrum; that is, $f(h)$ is a convex function of the scaling parameter h. The peak of the spectrum is determined to be the fractal dimension, as it should be. Here again we have an indication that the interstride-interval time series describes a multifractal process. But we stress that we are only using the qualitative properties of the spectrum for $q < 0$, due to the sensitivity of the numerical method to weak singularities.

5.4 Spatial heterogeneity

5.4.1 Random fields

Random fields play a central role in the modeling of complex geophysical webs because of the natural heterogeneity of oil deposits, hydrology, earthquakes, meteorology, agriculture and even climate change. In a geophysical context the smoothness of the random field can be characterized in terms of the fractal dimension obtained from the scaling behavior of the spatial autocorrelation. Lim and Teo [12] have considered the case in which the spatial correlation strength has a hyperbolic form for large time lags. We discuss their work and the complex phenomena it was designed to describe in this section.

Let us consider a two-dimensional plane $\mathbf{x} = (x, y)$ on which a random function $Y(\mathbf{x})$ is defined as the solution to the fractional stochastic differential equation

$$\left[-\Delta + \lambda^2\right]^{\alpha/2} Y(\mathbf{x}) = \xi(\mathbf{x}), \tag{5.102}$$

where $\alpha > 0$, the Laplace operator in two dimensions is

$$\Delta \equiv \frac{\partial^2}{\partial x^2} + \frac{\partial^2}{\partial y^2} \tag{5.103}$$

and the random field is zero-centered, delta-correlated and of unit strength with two-dimensional Gaussian statistics

$$\langle \xi(\mathbf{x}) \rangle = 0, \qquad \langle \xi(\mathbf{x})\xi(\mathbf{x}') \rangle = 2\delta(\mathbf{x} - \mathbf{x}'). \tag{5.104}$$

It probably bears stressing that what is discussed with regard to the solution to (5.102) is readily generalized to dimensions of arbitrary size, but the notation is somewhat more compact in two dimensions.

The solutions to the equations of motion in time were found using Laplace transforms. The equations in space are typically dealt with by implementing Fourier transforms, so we introduce the notation

$$\widetilde{F}(\mathbf{k}) = \mathcal{FT}\{F(\mathbf{x}); \mathbf{k}\} \equiv \int_{\mathcal{R}^2} e^{i\mathbf{k}\cdot\mathbf{x}} F(\mathbf{x}) d^2x, \tag{5.105}$$

with the inverse Fourier transform denoted by

$$F(\mathbf{x}) = \mathcal{FT}^{-1}\{\widetilde{F}(\mathbf{k}); \mathbf{x}\} \equiv \frac{1}{(2\pi)^2} \int_{\mathcal{R}^2} e^{-i\mathbf{k}\cdot\mathbf{x}} \widetilde{F}(\mathbf{k}) d^2k. \tag{5.106}$$

Consequently, by taking the Fourier transform of (5.102) and using the fact that

$$\mathcal{FT}\{-\Delta F(\mathbf{x}); \mathbf{k}\} = k^2 \widetilde{F}(\mathbf{k}),$$ (5.107)

we obtain the quantity

$$\mathcal{FT}\left\{\left[-\Delta + \lambda^2\right]^{\alpha/2} Y(\mathbf{x}); \mathbf{k}\right\} = \left(k^2 + \lambda^2\right)^{\alpha/2} \widetilde{Y}(\mathbf{k}) = \widetilde{\xi}(\mathbf{k}).$$ (5.108)

Therefore the Fourier transform of the solution of (5.102) is

$$\widetilde{Y}(\mathbf{k}) = \frac{\widetilde{\xi}(\mathbf{k})}{(k^2 + \lambda^2)^{\alpha/2}}.$$ (5.109)

The solution to (5.102) can be written, using the inverse Fourier transform, as

$$Y(\mathbf{x}) = \frac{1}{(2\pi)^2} \int_{R^2} \frac{\widetilde{\xi}(\mathbf{k}) e^{-i\mathbf{k}\cdot\mathbf{x}}}{(k^2 + \lambda^2)^{\alpha/2}} d^2k$$ (5.110)

or explicitly taking the inverse Fourier transform in (5.110) [12] allows us to write the spatial convolution integral

$$Y(\mathbf{x}) = \frac{\lambda^{1-\alpha/2}}{2^{\alpha/2} \pi \Gamma(\alpha/2)} \int_{R^2} \frac{K_{1-\alpha/2}(\lambda|\mathbf{x} - \mathbf{x}'|)}{|\mathbf{x} - \mathbf{x}'|^{1-\alpha/2}} \xi(\mathbf{x}') d^2x',$$ (5.111)

where $K_\nu(z)$ is the modified Bessel function of the second kind in the Euclidean norm $|\mathbf{x}| = \sqrt{x^2 + y^2}$, $\lambda > 0$ is a scale parameter controlling the spatial range of the kernel and $\nu = 1 - \alpha/2$ is the parameter determining the level of smoothness of the statistical fluctuations in the solution. Using the lowest-order approximation to the Bessel function,

$$K_\nu(z) \sim \sqrt{\frac{\pi}{2z}} e^{-z},$$ (5.112)

yields

$$Y(\mathbf{x}) = \frac{\lambda^{(1-\alpha)/2}}{2^{(1+\alpha)/2} \sqrt{\pi} \Gamma(\alpha/2)} \int_{R^2} \frac{e^{-\lambda|\mathbf{x}-\mathbf{x}'|}}{|\mathbf{x} - \mathbf{x}'|^{(3-\alpha)/2}} \xi(\mathbf{x}') d^2x',$$ (5.113)

indicating an exponentially truncated inverse power-law potential that attenuates the influence of remote random variations on a given location. It is clear that (5.113) can be expressed in terms of a fractional integral.

The power spectrum for the spatial process (5.113) is given by the Fourier transform of the autocorrelation function

$$S(\mathbf{k}) = \int_{R^2} d^2x \, e^{i\mathbf{k}\cdot\mathbf{x}} C(\mathbf{x}),$$ (5.114)

which, using the solution (5.111) and the statistics for the fluctuations (5.104), yields

$$S(\mathbf{k}) = \frac{1}{(2\pi)^2} \frac{1}{\left(k^2 + \lambda^2\right)^\alpha}.$$ (5.115)

Equation (5.115) is the power spectrum of a hyperbolic spatial response to the Gaussian random field and it is evidently an inverse power law for $k \gg \lambda$. This is the kind of spectrum that is observed in turbulent fluid flow where the spatial eddies give rise to an inverse power-law energy spectrum of the form $k^{-5/3}$.

It is possible to generalize stochastic differential equations even further by introducing a second fractional exponent into (5.102) to obtain

$$\left[(-\Delta)^{\gamma} + \lambda^2\right]^{\alpha/2} Y_{\alpha,\gamma}(\mathbf{x}) = \xi(\mathbf{x}), \tag{5.116}$$

where we now index the solution with subscripts for each of the exponents. Here the new index is restricted to the interval $0 < \gamma \leq 1$ and defines a Riesz fractional derivative in terms of its Fourier transform to be [31]

$$\mathcal{FT}[(-\Delta)^{\gamma} F(\mathbf{x}); \mathbf{k}] = |\mathbf{k}|^{2\gamma} \widetilde{F}(\mathbf{k}). \tag{5.117}$$

As Lim and Teo [12] point out, the operator in (5.116) can be regarded as a shifted Riesz fractional derivative, which has the formal series expansion

$$\left[(-\Delta)^{\gamma} + \lambda^2\right]^{\alpha/2} = \sum_{j=1}^{\infty} \binom{\alpha/2}{j} \lambda^{\alpha-2j} (-\Delta)^{\gamma j}, \tag{5.118}$$

so that its Fourier transform is

$$\mathcal{FT}\left[\left[(-\Delta)^{\gamma} + \lambda^2\right]^{\alpha/2} F(\mathbf{x}); \mathbf{k}\right] = \sum_{j=1}^{\infty} \binom{\alpha/2}{j} \lambda^{\alpha-2j} |\mathbf{k}|^{2\gamma j} \widetilde{F}(\mathbf{k}). \tag{5.119}$$

Consequently, the solution to (5.116) can be expressed by the integral

$$Y_{\alpha,\gamma}(\mathbf{x}) = \frac{1}{(2\pi)^2} \int_{\mathcal{R}^2} \frac{\widetilde{\xi}(\mathbf{k})e^{-i\mathbf{k}\cdot\mathbf{x}}}{\left(k^{2\gamma} + \lambda^2\right)^{\alpha/2}} d^2k \tag{5.120}$$

and this function is well defined as an ordinary random field only when $\alpha\gamma > 1$ since it is only in this case that the $\mathcal{R}^2$ norm remains finite over the entire domain. To see this restriction on the power-law indices, consider the covariance

$$C_{\alpha,\gamma}(\mathbf{x}, \mathbf{x}') = \langle Y_{\alpha,\gamma}(\mathbf{x}) Y_{\alpha,\gamma}(\mathbf{x}') \rangle, \tag{5.121}$$

so that, on substituting (5.120) into (5.121) and using the delta-correlated property of the Gaussian field,

$$C_{\alpha,\gamma}(\mathbf{x}, \mathbf{x}') = \frac{2}{(2\pi)^2} \int_{\mathcal{R}^2} \frac{e^{-i\mathbf{k}\cdot(\mathbf{x}-\mathbf{x}')}}{\left(k^{2\gamma} + \lambda^2\right)^{\alpha}} d^2k. \tag{5.122}$$

The covariance is therefore isotropic,

$$C_{\alpha,\gamma}(\mathbf{x}, \mathbf{x}') = C_{\alpha,\gamma}(\mathbf{x} - \mathbf{x}'), \tag{5.123}$$

and the variance is obtained when $\mathbf{x} = \mathbf{x}'$. However, from the integral we determine by direct integration that the variance diverges as $k \to \infty$ when $\alpha\gamma < 1$. Consequently, the solution $Y_{\alpha,\gamma}(\mathbf{x})$ is not just a random field when $\alpha\gamma < 1$ and the traditional methods for its analysis no longer apply. Consequently, the concept of a random field must be generalized to include those without central moments. But this is not the place to make such generalizations.

5.4.2   Continuous-time random walks (CTRWs)

One of the most useful generalizations of the random-walk model was made by Montroll and Weiss [19] in 1965, when they included random intervals between successive steps in the walking process to account for local structure in the environment. In this way the time interval between successive steps $\tau_n = t_{n+1} - t_n$ becomes a random variable. This generalization is referred to as the continuous-time random-walk (CTRW) model since the walker is allowed to step at a continuum of times and the time between successive steps is considered to be a second random variable, the vector value of the step being the first random variable. The CTRW has been used to model a number of complicated statistical phenomena, from the microscopic structure of individual lattice sites for heterogeneous media to the stickiness of stability islands in chaotic dynamical networks [33]. In our continuing discussion of random walks we remain focused on the dynamical variable.

The CTRW explicitly assumes that the sequence of time differences $\{\tau_n\}$ constitutes a set of independent identically distributed random variables. The probability that the random variable τ lies in an interval $\{t, t + dt\}$ is given by $\psi(t)dt$, where $\psi(t)$ is the now-familiar waiting-time distribution function. The name reflects the fact that the walker waits for a given time at a lattice site before stepping to the next site. The length of the sojourn is characteristic of the structure of the medium, as we discussed in the previous chapter on memory and fractional dynamical equations, where the waiting-time distribution was found to be a renewal process.

The waiting-time distribution is also used to define the probability that a step has not been taken in the time interval $(0, t)$. Thus, if the function $W(\mathbf{q}, t)$ is the probability that a walker is at the lattice site $\mathbf{q}$ at a time t immediately after a step has been taken, and $P(\mathbf{q}, t)$ is the probability of a walker being at $\mathbf{q}$ at time t, then in terms of the survival probability

$$P(\mathbf{q}, t) = \delta_{\mathbf{q},0}\Psi_0(t) + \int_0^t \Psi(t - t')W(\mathbf{q}, t')dt'. \tag{5.124}$$

If a walker arrives at $\mathbf{q}$ at time t' and remains there for a time $(t - t')$, or the walker has not moved from the origin of the lattice, $P(\mathbf{q}, t)$ is the average over all arrival times with $0 < t' < t$. Note that the waiting time at the origin might be different, so that $\Psi_0(t)$ can be determined from the transition probabilities. It should be pointed out that the function $P(\mathbf{q}, t)$ is a probability when $\mathbf{q}$ is a lattice site and a probability density when $\mathbf{q}$ is a point in the spatial continuum.

The probability function $W(\mathbf{q}, t)$ itself satisfies the recurrence relation

$$W(\mathbf{q}, t) = p_0(\mathbf{q}, t) + \sum_{\mathbf{q}'} \int_0^t p(\mathbf{q} - \mathbf{q}', t - t')W(\mathbf{q}', t')dt', \tag{5.125}$$

where $p(\mathbf{q}, t)dt$ is the probability that the time between two successive steps is between t and $t + dt$, and the step length extends across $\mathbf{q}$ lattice points. The transition probability satisfies the normalization condition

$$\sum_{\mathbf{q}} p(\mathbf{q}, t) = \psi(t), \tag{5.126}$$

where here, as before, $\psi(t)$ is the waiting-time distribution density and the integral over time yields the overall normalization

$$\int_0^\infty \sum_{\mathbf{q}} p(\mathbf{q}, t)dt = \int_0^\infty \psi(t)dt = 1. \tag{5.127}$$

Montroll and Weiss made the specific assumption that the length of a pause and the size of a step are mutually independent, an assumption we use temporarily, but avoid making later in the discussion. For the moment, assuming space-time independence in the jump-length probability yields the product of the waiting-time distribution density and the jump-length probability

$$p(\mathbf{q}, t) = p(\mathbf{q})\psi(t), \tag{5.128}$$

in which case the random walk is said to be factorable. The Fourier–Laplace transform of the transition probability yields

$$p^*(\mathbf{k}, u) \equiv \sum_{\mathbf{q}} e^{i\mathbf{k}\cdot\mathbf{q}} \int_0^\infty e^{-ut} p(\mathbf{q}, t)dt = \widetilde{p}(\mathbf{k})\widehat{\psi}(u), \tag{5.129}$$

where $\widetilde{p}(\mathbf{k})$ denotes both the discrete and the continuous Fourier transforms of the jump probability. We use W^* to denote the double Fourier–Laplace transform of W throughout. Consequently, using the convolution form of (5.125), we obtain

$$W^*(\mathbf{k}, u) = \frac{p_0^*(\mathbf{k}, u)}{1 - p^*(\mathbf{k}, u)} \tag{5.130}$$

and for the double transform of the probability from (5.124) we have, using (5.130),

$$P^*(\mathbf{k}, u) = \widehat{\Psi}_0(u) + \frac{\widehat{\Psi}(u) p_0^*(\mathbf{k}, u)}{1 - p^*(\mathbf{k}, u)}. \tag{5.131}$$

In the case in which the transition from the origin does not play a special role $p_0^*(\mathbf{k}, \mathbf{u}) = p^*(\mathbf{k}, \mathbf{u})$, $\widehat{\Psi}_0(u) = \widehat{\Psi}(u)$ and (5.131) simplifies to

$$P^*(\mathbf{k}, u) = \frac{\widehat{\Psi}(u)}{1 - p^*(\mathbf{k}, u)}, \tag{5.132}$$

which can be further simplified when the space-time transition probability factors. The relation between the survival probability $\Psi(t)$ and the waiting-time distribution density is given by (5.51) and the relation between the Laplace transform of the two quantities by (5.57). Consequently, the factorable network (5.132) simplifies to

$$P^*(\mathbf{k}, u) = \frac{1 - \widehat{\psi}(u)}{u\left[1 - \widetilde{p}(\mathbf{k})\widehat{\psi}(u)\right]}. \tag{5.133}$$

This is the Montroll–Weiss equation for the standard CTRW.

The further interpretation of (5.133) is accomplished by restricting the solution to the situation that the walker starts from the origin so that $P(\mathbf{q}, t = 0) = \delta(\mathbf{q})$, and consequently the probability density function is the propagator for the CTRW process. This interpretation is more apparent in the space-time form of the phase-space equation, so we introduce the Laplace transform of the memory kernel,

$$\widehat{\phi}(u) = \frac{u\widehat{\psi}(u)}{1 - \widehat{\psi}(u)}, \tag{5.134}$$

into (5.133), which, after some algebra, yields

$$P^*(\mathbf{k}, u) = \frac{1}{u + \widehat{\phi}(u)[1 - \widetilde{p}(\mathbf{k})]}. \tag{5.135}$$

Equation (5.135) can be put into the simpler form

$$u P^*(\mathbf{k}, u) - 1 = -\widehat{\phi}(u)[1 - \widetilde{p}(\mathbf{k})]P^*(\mathbf{k}, u), \tag{5.136}$$

whose inverse Fourier–Laplace transform yields the integro-differential equation for the propagator

$$\frac{\partial P(\mathbf{q}, t)}{\partial t} = \int_0^t dt' \, \phi(t - t') \left\{ -P(\mathbf{q}, t') + \int p(\mathbf{q} - \mathbf{q}') P(\mathbf{q}', t') d\mathbf{q}' \right\}. \tag{5.137}$$

Equation (5.137) is the Montroll–Kenkre–Shlesinger master equation [10] and it is clearly non-local in both time and space. In time the non-locality is determined by the memory kernel, which in turn is determined by the waiting-time distribution function, that is, the inverse Laplace transform of (5.134). The spatial non-locality is determined by the jump probability, which is the inverse Fourier transform of the structure function.

Equation (5.133) is the simplest situation of a factorable random walk, when the origin of time coincides with the beginning of the waiting time. The Fourier–Laplace-transformed probability can be used as the moment-generating function. Recall that the Fourier transform of the probability density is the characteristic function. For example, the first moment of the process, in one dimension, can be written as

$$\mathcal{LT}\{\langle q; t \rangle; u\} = -i \frac{\partial P^*(k, u)}{\partial k}\bigg|_{k=0}. \tag{5.138}$$

Inserting (5.133) into (5.138) and taking the appropriate derivatives yields the Laplace transform of the first moment,

$$\mathcal{LT}\{\langle q; t \rangle; u\} = \frac{\mu_1 \widehat{\psi}(u)}{u\left[1 - \widehat{\psi}(u)\right]}. \tag{5.139}$$

The first moment is obtained from the definition of the structure function

$$\mu_1 = -i \frac{\partial \widetilde{p}(k)}{\partial k}\bigg|_{k=0} = \sum_q q p(q). \tag{5.140}$$

We now expand the exponential in the definition of the Laplace transform of the waiting-time density to obtain

$$\widehat{\psi}(u) = \int_0^\infty dt \, \psi(t) e^{-ut} = 1 - \langle t \rangle u + \frac{1}{2} \langle t^2 \rangle u^2 + \cdots, \tag{5.141}$$

which, on its insertion into (5.139), keeping only the lowest-order terms in u, and inverse Laplace transforming, yields the average displacement

$$\langle q; t \rangle \approx \frac{\mu_1}{\langle t \rangle} t \tag{5.142}$$

and the constant average velocity is given by

$$\frac{d\langle q; t \rangle}{dt} \approx \frac{\mu_1}{\langle t \rangle}. \tag{5.143}$$

In the same way the second moment of the process, in one dimension, can be written

$$\mathcal{LT}\{\langle q^2; t \rangle; u\} = -\left. \frac{\partial^2 P^*(k, u)}{\partial k^2} \right|_{k=0} \tag{5.144}$$

and evaluating the second moment using the structure function

$$\mu_2 = \left. \frac{\partial^2 \widetilde{p}(k)}{\partial k^2} \right|_{k=0} = \sum_q q^2 p(q) \tag{5.145}$$

inserted into (5.144) and taking the appropriate derivative yields

$$\mathcal{LT}\{\langle q^2; t \rangle; u\} = \frac{\mu_2 \widehat{\psi}(u)}{u\left[1 - \widehat{\psi}(u)\right]} + \frac{1}{u}\left[\frac{\mu_1 \widehat{\psi}(u)}{1 - \widehat{\psi}(u)}\right]^2. \tag{5.146}$$

We again make use of the Taylor expansion of the exponential function in the Laplace transform of the waiting-time distribution in (5.141), which we insert into (5.146), keeping only the lowest-order term in u. The inverse Laplace transform of the resulting expression yields the variance

$$\langle q^2; t \rangle - \langle q; t \rangle^2 \approx \frac{\left(\mu_2 - \mu_1^2\right) \langle t \rangle^2 + \left(\langle t^2 \rangle - \langle t \rangle^2\right)}{\langle t \rangle^3} t. \tag{5.147}$$

Note that we are assuming that the second moment of the waiting-time distribution is finite. There are two contributions to the variance of the displacement of the random walker; one is due to the variance in the step size and the other is due to the variance in the waiting-time distribution. If either of these two variances diverges the resulting variance of the random walk also diverges. We have assumed that the second moment of the waiting-time distribution is finite and in so doing both the mean and the variance of the random walk are determined to increase linearly with time. Consequently, the dimensionless variable, the ratio of the standard deviation to the mean, goes to zero as the square root of the time,

$$\frac{\sqrt{\langle q^2; t \rangle - \langle q; t \rangle^2}}{\langle q; t \rangle} \approx \frac{1}{\sqrt{t}}, \tag{5.148}$$

a quantity known as the relative dispersion in the biomedical literature [2]. In the next section we determine the consequences of having a waiting-time distribution with a diverging average sojourn time.

5.4.3   Infinite moments

The finite moments of the random-walk process in the CTRW formalism have been shown to be a direct consequence of the waiting-time distribution having finite moments. This is made explicit in the Taylor expansion of the Laplace transform given in (5.141). However, there are examples in the physical, social and life sciences where the central moments diverge. This divergence of the moments in space and/or time is a consequence of non-locality in the corresponding distribution functions. In time this non-local effect is called memory and is the influence of past behavior on future activity. In particular, when the first moment of the distribution diverges the memory extends over a very long time compared with that of an exponential relaxation, which is to say that there is no characteristic relaxation time for the statistical process.

This observation is not particularly profound when we examine how things work in the real world. Pretty much everything that happens is dependent on history; whether being late for work is overlooked as an anomaly or is considered just one more incident in a train of irresponsible actions depends on your work record; similarly, whether your wife treats your forgetting an anniversary as the foible of an absent-minded but loving husband depends on the warmth of the day-to-day exchanges. But, as much as we recognize this dependence of our personal lives on history, it has proven to be too difficult to include it formally in our mathematical models. It is not the case that we have never been able to do it, but that when there is success it is considered extremely significant.

In an analogous way the non-local effect in space implies that what is occurring at one point in space is not merely dependent on what is happening in the immediate neighborhood of that point, but depends on what is happening in very distant spatial regions as well. Here we have the spatial analogue of memory, spatial anisotropy. In our private lives we know that the time it takes to get from point A to point B depends on the path taken. In society these paths are laid out for us. We do not drive the wrong way on a road or go "up the down staircase." We learn very young how to traverse the physical as well as the psychological paths to minimize resistance. Zipf thought that this was a universal property of human behavior resulting in the inverse power-law distribution. We cannot change our direction of motion when we wish, only when we are allowed to do so. The constraints on us depend not only on where we are, but also on how where we are is connected to where we want to go. This connectedness imposes non-local constraints on our movements.

These structural constraints are included in modeling the dynamics through the waiting-time and step-length distributions. When the random walker is moving on a desert plain these functions have a slow decay. However, when the walker is in an urban environment or in other contexts having rich structure these functions may decay much more rapidly. Therefore we extend the preceding analysis to cases where the waiting-time distribution is hyperbolic, the transition probability density is hyperbolic or both.

Consider the inverse power-law waiting-time distribution function with no finite moments and the asymptotic form

$$\psi(t) \sim \frac{1}{t^{\beta+1}} \tag{5.149}$$

with $0 < \beta < 1$. When the lowest moment of a distribution diverges, as in the $\beta < 1$ case, it is not possible to simply Taylor expand the Laplace transform of the waiting-time distribution function. To circumvent this formal difficulty the Laplace transform of the waiting-time distribution can be written

$$\widehat{\psi}(u) = 1 - \left[1 - \widehat{\psi}(u)\right] = 1 - \int_0^\infty \left(1 - e^{-ut}\right)\psi(t)dt. \tag{5.150}$$

On introducing the scaling variable $y = ut$ we obtain for the integral

$$\int_0^\infty \left(1 - e^{-ut}\right)\psi(t)dt = \int_0^\infty \left(1 - e^{-y}\right)\psi\left(\frac{y}{u}\right)\frac{dy}{u} = u^\beta \int_0^\infty \left(1 - e^{-y}\right)\psi(y)dy, \tag{5.151}$$

where the right-most integral, obtained using (5.149), is finite for the specified interval of the scaling index. Therefore we obtain the expansion for the Laplace transform of the waiting-time distribution function

$$\widehat{\psi}(u) = 1 - Au^\beta, \tag{5.152}$$

so the scaling index is positive definite and $\widehat{\psi}(u) \to 1$ as $u \to 0$. The Laplace transform of the memory function associated with this Laplace transform waiting-time distribution, to lowest order in the Laplace variable, is given by

$$\lim_{u \to 0} \widehat{\phi}(u) \propto u^{1-\beta}. \tag{5.153}$$

Consequently, using a Tauberian theorem,

$$\frac{1}{u^n} \Leftrightarrow \frac{t^{n-1}}{\Gamma(n)}, \tag{5.154}$$

which it is simple to show that asymptotically in time the memory function is given by

$$\lim_{t \to \infty} \phi(t) \propto t^{\beta-2}, \tag{5.155}$$

which is an inverse power law with an index $1 < 2 - \beta < 2$.

This argument can be repeated for the jump probability when it too is no longer an analytic function, which is to say that the central moments diverge as in the case above. We consider the situation of a symmetric transition probability $p(\mathbf{q}) = p(-\mathbf{q})$, but for simplicity of presentation we restrict the argument to one spatial dimension. In the context of random walks this probability determines the length of the successive steps within the walk and here we take it to be given by the inverse power law in the step length,

$$p(q) = \frac{B}{|q|^{\alpha+1}}, \tag{5.156}$$

just as obtained earlier for the Weierstrass walk. A random-walk process in which each of the steps is instantaneous, but with (5.156) giving the probability of a step of length q occurring, gives rise to a limit distribution first discussed by Paul Lévy, that being the Lévy distribution [11], which we mentioned previously.

As Weiss [29] points out, knowledge of the asymptotic behavior of the transition probability in (5.156) is sufficient for us to calculate a single correction term to the normalization condition for the characteristic function ($\widetilde{p}(k = 0) = 1$) in the Taylor expansion of the characteristic function about $k = 0$. We proceed as with the waiting-time distribution and write

$$\widetilde{p}(k) = 1 - \left[1 - \widetilde{p}(k)\right] = 1 - \int_{-\infty}^{\infty} [1 - \cos(kq)]p(q)dq, \tag{5.157}$$

where we have used the symmetry properties of the probability distribution to replace the exponential with the cosine. We are most interested in the region around $k = 0$, but it is clear that we cannot expand the cosine since the second moment of the distribution diverges. We can use the same form of transformation of variables $y = kq$ as used earlier to write the integral term as

$$\int_{-\infty}^{\infty} [1 - \cos(kq)]p(q)dq = \int_{-\infty}^{\infty} [1 - \cos y]p\left(\frac{y}{k}\right)\frac{dy}{k}. \tag{5.158}$$

Here again the asymptotic form of the distribution dominates the small-k behavior, so we can write

$$\int_{-\infty}^{\infty} [1 - \cos(kq)]p(q)dq = B|k|^{\alpha} \int_{-\infty}^{\infty} \frac{1 - \cos y}{|y|^{\alpha+1}}dy. \tag{5.159}$$

The integral on the rhs of (5.159) converges for $0 < \alpha < 1$, so that the leading term in the expansion of the characteristic function for the steps in the random walk given by the inverse power law yields

$$\widetilde{p}(k) \approx 1 - C|k|^{\alpha}, \tag{5.160}$$

where the constant C is the product of B and the finite integral to the right in (5.159). The demonstration given here is for an inverse power-law index in the interval $0 < \alpha < 1$, but the first-order correction to the characteristic function (5.160) can be shown to be valid for the interval $0 < \alpha < 2$; see, for example, the discussion in Weiss [29].

Note that we now have first-order expansions for the Laplace transform of the waiting-time distribution function (5.152) and the Fourier transform of the jump probability (5.160) with non-integer exponents in the expansion variables. In general these are approximate expressions because we can no longer use the Taylor expansion due to the divergence of the central moments of the corresponding distributions. The corresponding probability density function is then given by the inverse Fourier–Laplace transform of the equation resulting from inserting (5.152) and (5.160) into (5.135) to obtain

$$P^*(\mathbf{k}, u) \approx \frac{u^{\beta-1}}{u^{\beta} + C'|\mathbf{k}|^{\alpha}}, \tag{5.161}$$

where the constant is given by $C' = C/A$. In general it is not an easy task to obtain the inverse Fourier–Laplace transform of (5.161) for arbitrary indices α and β. Recall the algebra of exponents given for the analysis of such equations by the fractional calculus [31], from which we can conclude that the double inverse transform of (5.161) yields a fractional diffusion equation in space and time. We refer the reader to Magin [14] and West *et al.* [31] for a general discussion of fractional diffusion equations.

5.4.4   Scaling solutions

One technique for solving (5.161) uses the properties of Fourier transforms and Laplace transforms to obtain the scaling solutions to this equation and may be implemented without performing either of the transforms explicitly. First note the scaling of the E-dimensional Fourier transform of an analytic spatial function,

$$
f(a\mathbf{q}) = \int_{-\infty}^{\infty} \frac{d^E k}{(2\pi)^E} e^{-i\mathbf{k}\cdot a\mathbf{q}} \, \widetilde{f}(\mathbf{k}) = \frac{1}{a^E} \int_{-\infty}^{\infty} \frac{d^E k'}{(2\pi)^E} e^{-i\mathbf{k}'\cdot\mathbf{q}} \, \widetilde{f}\left(\frac{\mathbf{k}'}{a}\right), \tag{5.162}
$$

and the Laplace transform of an analytic temporal function

$$
g(bt) = \int_0^{\infty} du \, \hat{g}(u) e^{-ubt} = \frac{1}{b} \int_0^{\infty} du' \, \hat{g}\left(\frac{u'}{b}\right) e^{-u't}. \tag{5.163}
$$

From the last two equations we conclude that the space-time probability density $P(a\mathbf{q}, bt)$ has the Fourier–Laplace transform $P^*(\mathbf{k}/a, \; u/b)/(ba^E)$, which, on comparison with (5.161), yields the relationship

$$
\frac{1}{ba^E} P^*\left(\frac{\mathbf{k}}{a}, \frac{u}{b}\right) = \frac{1}{a^E} P^*\left(\frac{b^{\beta/\alpha}\mathbf{k}}{a}, u\right), \tag{5.164}
$$

whose inverse Fourier–Laplace transform is

$$
P(a\mathbf{q}, bt) = \frac{1}{b^{E\beta/\alpha}} P\left(\frac{a\mathbf{q}}{b^{\beta/\alpha}}, t\right). \tag{5.165}
$$

The scaling relations (5.164) and (5.165) can be solved in E dimensions by assuming that the probability density function satisfies a scaling relation of the form

$$
P(a\mathbf{q}, bt) = \frac{1}{b^{E\mu} t^{\mu}} F_{\alpha,\beta}\left(\frac{a\mathbf{q}}{b^{\mu} t^{\mu}}\right) \tag{5.166}
$$

which when inserted into (5.165) yields

$$
\frac{1}{b^{E\mu} t^{\mu}} F_{\alpha,\beta}\left(\frac{a\mathbf{q}}{b^{\mu} t^{\mu}}\right) = \frac{1}{b^{E\beta/\alpha} t^{\mu}} F_{\alpha,\beta}\left(\frac{a\mathbf{q}}{b^{\beta/\alpha} t^{\mu}}\right). \tag{5.167}
$$

The solution to (5.167) is obtained by equating indices to obtain $\mu = \beta/\alpha$, resulting in the scaling solution in the CTRW formalism in E dimensions

$$
P(\mathbf{q}, t) = \frac{1}{t^{\beta/\alpha}} F_{\alpha,\beta}\left(\frac{\mathbf{q}}{t^{\beta/\alpha}}\right) \tag{5.168}
$$

in terms of the scaling or similarity variable $\mathbf{q}/t^{\beta/\alpha}$. We find this form of the solution to be extremely useful in subsequent discussions. For example, West and Nonnenmacher [30] have shown that a symmetric E-dimensional Lévy-stable distribution of the form

$$P_{\mathrm{L}}(\mathbf{q}, t) = \int \frac{d^E k}{(2\pi)^E} \exp[i\mathbf{k} \cdot \mathbf{q} - bt^\beta |\mathbf{k}|^\alpha] \tag{5.169}$$

satisfies the scaling relation (5.168). Consequently, the scaling solution (5.168) embraces a broad range of scaling statistics, and we obtain the incidental result that the Lévy statistical distribution $P_{\mathrm{L}}(\mathbf{q}, t)$ is a solution to the CTRW formalism.

5.5   Reiteration

This chapter encapsulates a number of approaches to describing the dynamics of complex webs when the pieces of the network burst like bubbles or the present configuration strongly depends on what happened in the distant past, or the measures of variability overwhelm our ability to make predictions. Non-analytic dynamics examines ways of studying the changes in web properties over time when the differential equations of motion are no longer suitable.

The fractional calculus is one systematic method for determining how a fractal function changes in space and time. The ordinary derivative of a fractal function diverges and consequently the process described by such a function cannot have ordinary differential equations of motion. On the other hand, the fractional derivative of a fractal function is another fractal function, suggesting that a fractal process described by such a function may well have fractional equations of motion. In the deterministic case the ordinary rate equation whose solution is an exponential is replaced by a fractional rate equation whose solution is a Mittag-Leffler function (MLF). The memory incorporated into the time dependence of the MLF depicts a process that has an asymptotic inverse power-law relaxation. This is the slow response of taffy to chewing and the slow but inexorable spreading of a fracture in a piece of material.

The fractional calculus opens the door to a wide variety of modeling strategies, from the generalization of a simple Poisson process to the dynamics of the distribution of cracks. Of course, there is also the extension of stochastic differential equations to the fractional form, where the history-dependent dynamics determines the response to present-day fluctuations. This means that the way a web responds to uncertainty today is dependent on the dynamical history of the web. The longer the history the more mitigated is the present uncertainty; in particular, such dependence can explain anomalous diffusion such as that observed in heart-rate variability.

The next step beyond fractal functions or processes is multifractals, that is, time series in which the fractal dimension changes over time. The distribution of fractal dimensions is necessary to capture the full dynamical complexity of some webs. One approach to modeling multifractals is through the fractional calculus in which the fractional index is itself a random quantity. The distribution of fractal dimensions can then be determined through the scaling behavior of the solution to the filtering of the random force in the fractional Langevin equation. This approach was useful in determining the statistical properties of migraine pathology through the regulation of cerebral blood flow, as well as the weak multifractality of walking.

The fractional calculus has also been shown to describe the heterogeneity of random fields and the propagation of probability densities in inhomogeneous anisotropic physical webs. The CTRW of Montroll and Weiss has experienced a renaissance in the modern explanation of the behavior of complex networks. In particular, the scaling of phenomena in space and time has been shown to be a consequence of fractional diffusion equations, one of the more popular solutions of which is the Lévy distribution.

5.6   Problems

5.1 Generalized Weierstrass Function
The fractional derivatives of the GWF can yield deep insight into the utility of the fractional calculus for understanding complex phenomena and consequently it is worthwhile going through the algebra of the analysis at least once. Carry out the integration indicated by (5.35) explicitly to derive (5.36), as well as the integration indicated by (5.37) to derive (5.38). The details of this analysis as well as additional discussion are given in [22].

5.2 Scaling of the second moment
Complete the discussion of the second moment of the solution to the fractional Langevin equation (5.72) by evaluating the integral for the variance (5.76) using the delta-correlated property of the Gaussian random force. If this proves to be too difficult, carry out the calculation for the $\lambda = 0$ case, but do it for the autocorrelation function rather than the second moment. If you really want to impress your professor, do both.

5.3 The fractal autocorrelation function
Of course, the solution (5.111) when measured is often too erratic to be useful without processing. The quantity typically constructed in a geophysical context is the autocorrelation function or covariance function

$$C(\mathbf{x} - \mathbf{x}') = \langle Y(\mathbf{x})Y(\mathbf{x}') \rangle_\xi.$$

Construct an analytic expression for this function on the basis of the above problem assuming the statistical properties (5.104). Note that there is a simple way to do this and one that is not so simple.

References

[1] E. Bacry, J. Delour and J. F. Muzy, "Multifractal random walk," *Phys. Rev. E* **64**, 026103-1 (2001).
[2] J. B. Bassingthwaighte, L. S. Liebovitch and B. J. West, *Fractal Physiology*, Oxford: Oxford University Press (1994).
[3] K. V. Bury, *Statistical Models in Applied Science*, New York: John Wiley (1975).

[4] K. Falconer, *Fractal Geometry*, New York: John Wiley (1990).

[5] J. Feder, *Fractals*, New York: Plenum Press (1988).

[6] P. Grigolini, A. Rocco and B. J. West, "Fractional calculus as a macroscopic manifestation of randomness," *Phys. Rev. E* **59**, 2603 (1999).

[7] L. Griffin, D. J. West and B. J. West, "Random stride intervals with memory," *J. Biol. Phys.* **26**, 185–202 (2000).

[8] J. M. Hausdorff, C.-K. Peng, Z. Ladin *et al.* "Is walking a random walk? Evidence for long-range correlations in stride interval of human gait," *J. Appl. Physiol.* **78** (1), 349–358 (1995).

[9] R. Hilfer, ed., *Applications of Fractional Calculus in Physics*, Singapore: World Scientific (2000).

[10] V. M. Kenkre, E. W. Montroll and M. F. Shlesinger, "Generalized master equations for continuous-time random walks," *J. Statist. Phys.* **9**, 45 (1973).

[11] P. Lévy, *Théorie de l'addition des variables aléatoires*, Paris: Gauthier-Villars (1937)

[12] S. C. Lim and L. P. Teo, "Generalized Whittle–Matérn random field with application to modeling," *J. Phys.: Math. Theor.* **42**, 105202 (2009).

[13] E. Lutz, "Fractional Langevin equation," *Phys. Rev. E* **64**, 051106 (2001).

[14] R. L. Magin, *Fractional Calculus in Bioengineering*, Connecticut: Begell House (2006).

[15] F. Mainardi, R. Gorenflo and A. Vivoli, "Renewal processes of Mittag-Leffler and Wright type," *Fractional Calculus and Applied Analysis* **8**, 7–38 (2005).

[16] B. B. Mandelbrot, *The Fractal Geometry of Nature*, San Francisco, CA: W. J. Freeman (1983).

[17] P. Meakin, *Fractals, Scaling and Growth Far from Equilibrium*, Cambridge: Cambridge University Press (1998).

[18] K. S. Miller and B. Ross, *An Introduction to the Fractional Calculus and Fractional Differential Equations*, New York: John Wiley & Sons (1993).

[19] E. W. Montroll and G. H. Weiss, "Random walks on lattice. II," *J. Math. Phys.* **6**, 167–181 (1965).

[20] C. K. Peng, J. Mietus, J. M. Hausdorff *et al.*, "Long-range autocorrelations and non-Gaussian behavior in the heartbeat," *Phys. Rev. Lett.* **70**, 1343 (1993).

[21] Yu. N. Rabotnov, *Elements of Hereditary Solid Mechanics*, Moscow: Mir (1980).

[22] A. Rocco and B. J. West, "Fractional calculus and the evolution of fractal phenomena," *Physica A* **265**, 535 (1999).

[23] D. Ruelle, *Chaotic Evolution and Strange Attractors*, Cambridge: Cambridge University Press (1989).

[24] S. G. Samko, A. A. Kilbas and O. I. Marichev, *Fractional Integrals and Derivatives*, Amsterdam: Gordon and Breach Science Publishers (1993).

[25] N. Scafetta, L. Griffin and B. J. West, "Hölder exponent spectra for human gait," *Physica A* **328**, 561–583 (2003).

[26] I. Sokolov, J. Klafter and A. Blumen, "Fractional kinetics," *Physics Today*, Nov., 48–54 (2002).

[27] H. E. Stanley, *Introduction to Phase Transitions and Critical Phenomena*, Oxford: Oxford University Press (1979).

[28] W. Weibull, "A statistical theory of strength of materials," *Ing. Vetenskaps Akad. Handl. No. 151* (1939).

[29] G. Weiss, *Aspects and Applications of the Random Walk*, Amsterdam: North-Holland (1994).

[30] B. J. West and T. Nonnenmacher, "An ant in a gurge," *Phys. Lett. A* **278**, 255–259 (2001).

[31] B. J. West, M. Bologna and P. Grigolini, *Physics of Fractal Operators*, New York: Springer (2003).

[32] B. J. West, M. Latka, M. Glaubic-Latka and D. Latka, "Multifractality of cerebral blood flow," *Physica A* **318**, 453–460 (2003).

[33] G. M. Zalslavsky, R. Z. Sagdeev, D. A. Usikov and A. A. Chernikov, *Weak Chaos and Quasi-Regular Patterns*, Cambridge: Cambridge University Press (1991).

6 A brief recent history of webs

Complex webs are not all complex in the same way; how the distribution in the number of connections on the Internet is formed by the conscious choices of individuals must be very different in detail from the new connections made by the neurons in the brain during learning, which in turn is very different mechanistically from biological evolution. Therefore different complex webs can be categorized differently, each representation emphasizing a different aspect of complexity. On the other hand, a specific complex web may be categorized in a variety of ways, again depending on what aspect of entanglement is being emphasized. For example, earthquakes are classified according to the magnitude of displacement they produce, in terms of the Richter scale, giving rise to an inverse power-law distribution in a continuous variable measuring the size of local displacement, the Gutenberg–Richter law. Another way to categorize earthquake data is in terms of whether quakes of a given magnitude occur within a given interval of time, the Omori law, giving rise to an inverse power-law distribution in time. The latter perspective, although also continuous, yields the statistics of the occurrence of individual events. Consequently, we have probability densities of continuous variables and those of discrete events, and which representation is selected depends on the purpose of the investigator. An insurance adjustor might be interested in whether an earthquake of magnitude 8.0 is likely to destroy a given building at a particular location while a policy is still in effect. Insurance underwriting was, in fact, one of the initial uses of probabilities and calculating the odds in gambling was another.

One of the first books on the theory of probability was Abraham de Moivre's *The Doctrine of Chances*, with the subtitle *A Method of Calculating the Probability of the Events of Play* [14]. De Moivre was a Huguenot refugee from France who took shelter in England and subsequently became a friend of Sir Isaac Newton. He was a brilliant mathematician who managed to survive by tutoring mathematics, calculating annuities and devising new games of chance for which he would calculate the odds of winning. A mathematician could make a living devising new gambling games for the aristocracy in the seventeenth century, if not by teaching them the secrets of the discipline [36]. Results from this 300-year-old treatise form the basis of much of what is popularly understood today about probability and we have already used a number of his results in the study of random walks. We mention these things to make you aware of the fact that the uncertainty of outcome which is so obvious in gambling actually permeates our lives no matter how we struggle to control the outcome of events. In this chapter we outline one more calculus for quantifying and manipulating the uncertainty in the decisions we make and in putting metrics on the natural variability in our elaborate predictions of the future behavior of the complex webs in which we are entangled.

In the literature of complex webs there is increasing focus on the search for the most useful web topology. A conviction shared by a significant portion of the scientists involved in research underlying the search for a network science is that there exists a close connection between web topology and web function. There is also increasing interest in web dynamics and how that dynamics is related to web topology and function. It is convenient to stress that the concept of web dynamics that is leading the research work of many groups is the time dependence of web topology that is based on the model put forward by Barabási and Albert (BA) introduced in earlier chapters [6]. Web dynamics is interpreted in this chapter to mean the time-dependent change in the properties of the units making up the nodes of the web. In the next chapter we examine a model in which the elements of the web dynamically interact with one another in ways that change their properties.

The condition of scale-free webs refers to the distribution density $\theta(k)$ of the number of links k connecting the nodes, which is typically an inverse power law

$$\theta(k) \propto \frac{1}{k^\alpha}. \tag{6.1}$$

The inverse power-law parameter α of the distribution of node links must not be confused with the parameter μ of the waiting-time distribution density $\psi(t)$ discussed in earlier chapters. However, it is interesting to ask whether a relation exists between the two distributions and their power-law parameters. This is one of the problems addressed in this and the next chapter. We limit our discussion in this chapter to pointing out that according to the model of *preferential attachment* $\alpha = 3$, as we discuss presently, but experimental and theoretical arguments indicate that the power-law index is generally in the interval $1 \le \alpha \le 3$. The highest-degree nodes, those nodes with the greatest number of connections, are often called "hubs" and are thought to serve specific purposes in their webs, although what that purpose is depends greatly on the domain of the index.

Figure 6.1 suggests that there are clusters with hubs, the hubs being the local leaders that may facilitate a cluster reaching a consensus that can then be transmitted to other clusters by means of long-range interactions. However, there is apparently no reason why local clusters may not be characterized by all-to-all coupling, which should make it easier to reach consensus locally. Is a local leader necessary in order to realize local

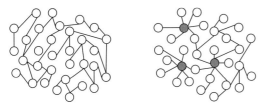

Figure 6.1.   In the literature of complex networks there is the widespread conviction that the real networks on the right are, unlike the random network on the left, characterized by scale-free distributions. This property implies the existence of hubs, shown as filled circles on the right, namely nodes with a number of links much higher than that of the typical nodes.

consensus? It is evident that all-to-all coupling is not realistic when applied to the entire web; it is too slow and consumes too many resources. It is more plausible if the coupling concerns small clusters, and when these are truncated at a given distance T_{trunc}. Is it possible to establish a condition that the best web topology is that which produces the largest value of the truncation distance? It is plausible that the best web topology that fits this condition does not coincide with the "best" topologies devised by other methods [5].

Finally, note that one of the most important properties of complex networks is the size of the web; that is, the number of jumps from one node to any other node of the web is surprisingly small. It is evident that, if there are all-to-all coupling clusters, the length l of a cluster is $l = 1$; that is, any node in the web can be reached from any other node in a single step. If the clusters are suitably connected by long-distance links, we can produce webs of relatively small access size, that is, of length $l \approx 6$. However, it is important to take into account that not all the links have the same importance, which is a fact of significance in sociological networks [19]. In this chapter we adopt for most of the discussion the simple assumption of uniform links; that is, each link has the same dynamical behavior and exerts the same influence on all the other links. This is not a realistic assumption and we eventually move beyond it to a more realistic supposition in the next chapter.

6.1   Web growth

The understanding of scale-free inverse power-law webs by physicists began with small-world theory, a theory of social interactions in which social ties can be separated into two primary kinds: strong and weak. Strong ties exist within a family and among the closest of friends, those you call in case of emergency and contact to tell about a promotion. These are the people who make up most of our lives. Then there are the weak ties, such as those we have with many of our colleagues at work with whom we chat, but never reveal anything of substance, friends of friends, business acquaintances and most of our teachers.

Figure 6.1 depicts schematically a random web and a scale-free web. It is evident that scale-free webs are characterized by clusters and that each cluster contains at least one hub, that being a node interacting with all, or nearly all, the other nodes of the same cluster. The power-law distribution strongly influences the web topology. It turns out that the major hubs are closely followed by smaller ones. These nodes, in turn, are followed by other nodes with even smaller degrees and so on. This hierarchy of connectedness allows fault-tolerant behavior. Since failures typically occur at random and the vast majority of nodes will be those with small degree, the likelihood that a hub would be affected by a random failure is almost negligible. Even if such events as the failure of a single hub occur, the web will not lose its connectedness, which is guaranteed by the remaining hubs. On the other hand, if we choose a few major hubs and take them out of the web, the network simply falls apart and the interconnected web is turned into a set of rather isolated graphs. Thus, hubs are the strength of scale-free networks as well as being their Achilles' heel.

Clusters form among those individuals with strong interactions, forming closely knit groups, clusters in which everyone knows everyone else. These clusters are formed from strong ties, but then the cliques are coupled to one another through weak social contacts. The weak ties provide contact from within a cluster to the outside world. It is the weak ties that are all-important for interacting with the world at large, say for getting a new job. A now classic paper by Granovetter, "The strength of weak ties" [19], explains how it is that the weak ties to near strangers are much more important in getting a new job than are the stronger ties to one's family and friends. In this "small world" there are short cuts that allow connections from one tightly clustered group to another tightly clustered group very far away. With relatively few of these long-range random connections it is possible to link any two randomly chosen individuals with a relatively short path. This has become known as the *six-degrees-of-separation phenomenon* [34]. Consequently, there are two basic elements necessary for the small-world model, clustering and random long-range connections.

6.1.1 Scale-free webs

Small-world theory is the precursor to scale-free networks. Recent research into the study of how networks are formed and grow over time reveals that even the smallest preference introduced into the selection process has remarkable effects on the final structure of the web. Here the selection procedure is the process by which a node entering the web chooses to connect with another node. Two mechanisms seem to be sufficient to obtain many of the inverse power-law distributions that are observed in the world. One of the mechanisms abandons the notion of independence between successive choices within a network. In fact, the mechanism is known in sociology as the Matthew effect, and was taken from the Book of Matthew in the Bible:

For unto every one that hath shall be given, and he shall have abundance: but from him that hath not shall be taken away even that which he hath.

Or, in more familiar terms, this is the principle that the rich get richer and the poor get poorer. In a computer-network context this principle implies that the node with the greater number of connections attracts new links more strongly than do nodes with fewer connections, thereby providing a mechanism by which a web can grow as new nodes are added. It is worth pointing out that this mechanism is not new and was identified as the *Gibrat principle* in 1935 by Simon [30] and as *cumulative advantage* in 1976 by Price [28] as well as being called the *Matthew effect* by Merton [23] in 1968 and, most recently, *preferential attachment* by Barabási and Albert [6] in 1999.

The thread of the argument presented by Barabási and Albert [6] to incorporate the growth of a network and the dependence of the links made by the new nodes to the previously existing nodes goes as follows. At time $t = 0$ there are m_0 nodes with no links to connect them. At later times $t_1, t_2, \ldots$ new vertices are established and with each new vertex another m new edges are formed. We assume $m \leq m_0$ so that each new node can accommodate all its edges with the nodes already belonging to the network. The choice of the first m elements from the preexisting m_0 elements to connect to the

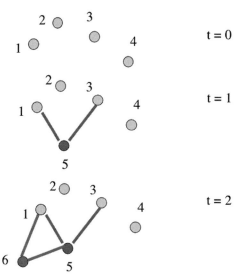

Figure 6.2.   This graph illustrates how the BA model works. At time $t = 0$ the network consists of four nodes with no link. At time $t = 1$ the new node 5 arrives with two links. The new node selects randomly two of the four nodes already present in the network, in this case nodes 1 and 3. At time $t = 2$ the new node 6 arrives and selects two of the five nodes already present in the environment. Node 5 has already two links. Thus the new node selects it, with one of its two links. The nodes 1 and 3 have one link, whereas nodes 2 and 4 do not have any. Thus, node 6 may select either node 1 or node 3 with equal probability. In this case it selects node 1.

new element is made at random. The choice of the next m nodes is not completely random, but is carried out by selecting with higher probability the elements that have the larger number of previously existing links; see, for example, a geometric sketch of this argument in Figure 6.2.

At a given integer time t we have N nodes, with m_0 playing the role of the initial state of the web,

$$N(t) = m_0 + t, \qquad (6.2)$$

with K edges,

$$K(t) = mt. \qquad (6.3)$$

Note that at any time step the number of edges per node increases by $2m$. In fact, any new edge arriving increases by one the number of links of an element already belonging to the network, and brings an additional link for the newly arriving node. Thus, the total number of connections at any time is given by

$$\sum_j k_j = 2mt, \qquad (6.4)$$

where k_j is the number of links to the jth element at time t.

The time rate of change in the number of links to the jth element at time t is given by

$$\frac{dk_j}{dt} = m\Theta(k_j), \qquad (6.5)$$

where $\Theta(k_j)$ is the probability that a new element makes a link to the jth node, and is given by the relative frequency form

$$\Theta(k_j) = \frac{k_j}{\sum_j k_j}. \tag{6.6}$$

Thus, $m\Theta(k_j)$ is the fraction of new nodes that make a link to the jth node; that is, they are attracted to the jth node at time t. We replace the sum in (6.6) by (6.4) to obtain in place of (6.5)

$$\frac{dk_j}{dt} = \frac{k_j}{2t}, \tag{6.7}$$

which has the general solution

$$k_j(t) = m\left(\frac{t}{t_j}\right)^\beta, \tag{6.8}$$

where t_j is the time at which the jth node is established and

$$\beta = 1/2. \tag{6.9}$$

From the form of the solution for the increase in the number of connections with time we see that the older elements have a shorter t_j and consequently a larger number of links than do the younger nodes. In the literature the number of links is called the degree of the node. Equation (6.8) quantifies the rich-get-richer mechanism referred to above.

To determine how the Pareto distribution follows from the solution (6.8) we introduce $P(k_j < k)$, the probability that the number of connections k for a generic node exceeds $k_j(t)$. An explicit expression for this quantity can be determined by inverting the general solution to obtain for the time of establishing the jth node

$$t_j = t\left(\frac{m}{k_j}\right)^{1/\beta} \tag{6.10}$$

and noting that if $k > k_j$ then

$$t_j > t\left(\frac{m}{k_j}\right)^{1/\beta}. \tag{6.11}$$

Thus, the probability for the connectivity can be obtained from the probability in terms of time,

$$P(k_j < k) = P\left(t_j > t\left(\frac{m}{k_j}\right)^{1/\beta}\right). \tag{6.12}$$

Consequently, the probability that, out of the total number of nodes at time t, k_j are connected to the jth node is

$$P\left(t_j > t\left(\frac{m}{k_j}\right)^{1/\beta}\right) = \int_{t(m/k_j)^{1/\beta}}^t \frac{dt_j}{N(t)} = \frac{t}{m_0 + t}\left[1 - \left(\frac{m}{k_j}\right)^{1/\beta}\right]. \tag{6.13}$$

The probability that $k > k_j$ can be expressed in terms of the probability density introduced earlier in a different context,

$$P(k_j < k) = \int_0^k \theta(k')dk',$$ (6.14)

so that

$$\theta(k) = \frac{dP(k_j < k)}{dk} = \frac{dP\left(t_j > t\left(\frac{m}{k_j}\right)^{1/\beta}\right)}{dk}$$

$$= \frac{t}{m_0 + t}\frac{1}{\beta}\frac{m^{1/\beta}}{k^{1+1/\beta}}.$$ (6.15)

In the asymptotic limit and recalling that $\beta = 1/2$ we obtain for the probability density

$$\theta(k) = \frac{2m^2}{k^3},$$ (6.16)

which establishes $\alpha = 3$ for what is called a scale-free network in the literature.

The scale-free nature of complex webs affords a single conceptual picture spanning scales from those in the World Wide Web to those within an organization. As more people are added to an organization, the number of connections between existing members depends on how many links already exist. In this way the oldest members, those who have had the most time to establish links, grow preferentially in connectedness. Thus, some members of the organization have substantially more connections than do the average, many more than predicted by any bell-shaped curve. These are the individuals out in the tail of the distribution, the gregarious individuals who seem to know everyone. Of course, $\alpha = 3$ is not characteristic of most empirical data, so we subsequently return to the model in a more general context as first discussed by Yule [38].

6.1.2   Communication webs

There has been a steady increase in the understanding of a variety of complex webs over the past decade. The present resurgence of interest in the field began with the pioneering paper of Watts and Strogatz [35], who established that real-world networks deviate significantly from the totally random webs considered half a century earlier [15]. They observed that real networks exhibit significant clustering that was not observed in the early theory and that these strongly interacting clusters were weakly coupled together, leading to the small-world theory. Barabási and Albert (BA) pointed out that small-world theory lacked web growth and introduced preferential attachment, resulting in a complex network having an inverse power-law connectivity distribution density with power-law index $\alpha = 3$ [6]. BA referred to such complex webs as scale-free because the distribution in the number of connections k has the form

$$p(\lambda k) = g(\lambda)p(k)$$

which we discussed in Chapter 2. This relation indicates that a change in scale $\lambda k \to k$ results in merely an overall change in amplitude A of the distribution $A \to g(\lambda)A$.

Consequently, the functional form of the distribution is the same at every scale in k and there is no single scale with which to characterize the web. We emphasize that this definition is a mathematical idealization and may require modification when dealing with empirical data, such as time series of finite length and with finite resolution. Subsequent investigations have focused on defining parameters with which to characterize scale-free webs, such as the clustering coefficient, the betweenness coefficient, the average size of the web and other such properties. Moreover, assumptions regarding how new links are formed that were intended to modify and replace preferential attachment have been made. But, before we introduce more formalisms, let us examine a concrete example of these ideas in the realm of communications.

The way people treat one another has always been a great mystery and understanding that mystery is the basis of great literature, art, drama and psychology. Consequently, when someone with the pedigree of a scientist, for example, Sigmund Freud, comes along, with his insight into the psyche and his gift for writing, society listens. However, given that psychoanalysis is a science, it can be tested, and in that regard many of Freud's ideas about the human psyche have not withstood the test of time. On the other hand, another area of psychology stemming at least in part from nineteenth-century psychophysics, although somewhat less glamorous than psychoanalysis, has been making steady progress. At about the same time as psychophysics was being introduced by Fechner [16], sociophysics appeared as well. These two disciplines, as their names imply, were attempts to apply the quantitative methods of the physical sciences to the human domain. One aspect of this interaction between people is communication and one question concerning communication is whether the dynamical element of humans has changed as society has changed.

The nineteenth century saw the transformation of society from the rural to the urban. In this period most private communication was done by letter writing, with items of correspondence sometimes taking months to reach their destination. In this slower-paced time there were many inconveniences in correspondence; one needed pen, paper, ink, sealing wax, literacy (in itself rare) in both oneself and the recipient of the letter, as well as access to a mail carrier. Consequently, it is interesting to consider whether people of that age are any different in any fundamental way from people of today who send and receive dozens of emails every day. This question was addressed by Oliveira and Barabási [27] using the lifelong correspondence of two scientists, Albert Einstein (1879–1955) and Charles Darwin (1809–1882). Both these men of science were prolific letter writers with an average number of letters sent approaching one for each day of life.

Oliveira and Barabási determined that the patterns of Einstein's and Darwin's letter writing follow the same scaling laws as the electronic transmission of today, namely an inverse power-law. On the other hand, given the difference in time scales between letter writing and emails the power-law indices of the two methods of communication are found to be distinctly different. They interpreted this consistency in the existence of an inverse power-law distribution for the two methods of communication as evidence for a new class of phenomenon in human dynamics even though the indices are different.

The statistical analysis of the Darwin and Einstein correspondence motivated Barabási [8] to look for a social rather than mathematical [1] explanation of these results. Barabási proposed a model based on executing L distinct tasks, such as writing a letter or sending an email. Each task has a random priority index x_i $(i = 1, \ldots, L)$ extracted from a probability density function $\rho(x)$ independently of each other. The dynamical rule is the following: with probability $0 \leq p \leq 1$ the most urgent task is selected, while with complementary probability $1 - p$ the selection of the task is random. The selected task is executed and removed from the list. At this point the completed task is replaced by a new task with random priority extracted again from $\rho(x)$. It is evident that when $p = 0$ the duration of the wait before the next letter in the list is answered depends only on the total number of letters L and the time τ;

$$\Psi(\tau) = \left(1 - \frac{1}{L}\right)^{\tau} \approx e^{-\tau/\tau_0}, \tag{6.17}$$

with the parameter in the exponential determined by the total number of letters

$$\tau_0 = \frac{1}{L}. \tag{6.18}$$

In the language of queuing theory, if there are L customers, each with the same probability of being served, the waiting-time distribution density is the exponential function

$$\psi(\tau) = \frac{1}{\tau_0} e^{-\tau/\tau_0}. \tag{6.19}$$

This waiting-time distribution is consistent with certain results of queuing theory. Abate and Whitt have shown [1] that the survival probability $\Psi(t)$ is asymptotically given by

$$\Psi(t) = \exp\left[\exp[-\eta(t - t_0)]\right] \approx 1 - \alpha e^{-\eta t}, \tag{6.20}$$

where the double exponential is the well-known Gumbel distribution for extreme events [20]. This distribution is obtained when the data in the underlying process consist of independent elements such as those given by a Poisson process. Another survival probability is given by

$$\Psi(t) \approx 1 - \frac{\alpha}{t^{3/2}} e^{-\eta t}, \tag{6.21}$$

where the parameters are different from those in the first case. This second distribution is of special interest because it yields the waiting-time distribution density

$$\psi(t) = \frac{\alpha \eta}{t^{3/2}} e^{-\eta t} + \frac{3}{2} \frac{\alpha}{t^{5/2}} e^{-\eta t}, \tag{6.22}$$

which in the long-time limit reduces to

$$\psi(t) \propto \frac{\text{const}}{t^{3/2}} e^{-\eta t}. \tag{6.23}$$

Note that the truncated inverse power law is reminiscent of the waiting-time distribution density discussed earlier, where we found that the inverse power law, with $\mu = 1.5$,

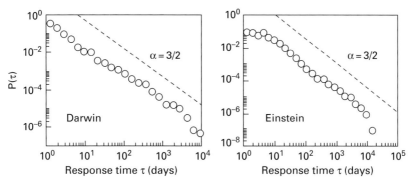

Figure 6.3.   The graphs indicate the distributions of response times to letters by Darwin (left) and Einstein (right). These distributions have the same meaning as the waiting-time distribution density introduced earlier. This figure was adapted from [27] with permission.

holds only for a limited time regime, whereafter it is followed by an exponential decay. More remarkable is the fact that Oliveira and Barabási [27] found that the time interval between the receipt of a letter by either Darwin or Einstein and the sending of their replying letters is satisfactorily described by waiting-time distribution densities with $\mu = 1.5$, as depicted in Figure 6.3. Both Darwin and Einstein began writing letters as teenagers and the volume of letters sent and received tended to increase as a function of time.

Barabási numerically studied the limiting case of $p = 1$, namely the case in which at any time step the highest-priority customer is served, and determined in this case that the waiting-time distribution is the inverse power law

$$\psi(\tau) \propto \frac{1}{\tau}. \tag{6.24}$$

Note that $p = 1$ is a singularity that on the one hand makes the exponential cutoff diverge and on the other hand generates a $\psi(\tau)$ of vanishing amplitude. Thus, Vásquez [33] and Gabrielli and Caldarelli [17] developed special methods to derive the power-law index $\mu = 1$ without these limitations.

On the basis of the fact that on moving from the totally random to the totally deterministic condition the waiting-time distribution changes its structure from the exponential form (6.19) to the inverse power law with index $\mu = 1$ (that Barabási found to be a proper description of the email communication [8]), Barabási decided to refine his model so as to study the cases intermediate between random choice of the task and the highest-priority-task criterion. He adopted for the probability that a task with priority x is chosen for execution per unit time the prescription

$$\Theta(x) = \frac{x^{\gamma}}{\displaystyle\sum_{i=1}^{N} x_i^{\gamma}}. \tag{6.25}$$

Note that in this section we are identifying Barabási's tasks with the N lines moving in lanes of equal width L in a post office, in the normalized interval $[0, 1]$; see Figure 6.4.

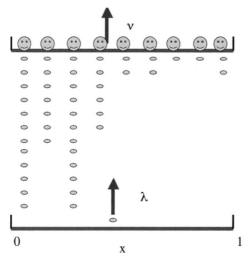

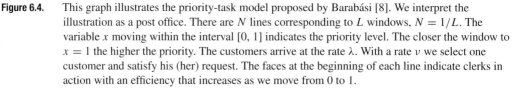

Figure 6.4.　This graph illustrates the priority-task model proposed by Barabási [8]. We interpret the illustration as a post office. There are N lines corresponding to L windows, $N = 1/L$. The variable x moving within the interval $[0, 1]$ indicates the priority level. The closer the window to $x = 1$ the higher the priority. The customers arrive at the rate λ. With a rate ν we select one customer and satisfy his (her) request. The faces at the beginning of each line indicate clerks in action with an efficiency that increases as we move from 0 to 1.

Thus, we convert a problem of the human mind, with its prioritizing of tasks, into an issue of social organization. In fact, the clerks of the post office have a different efficiency and the smiling face of Figure 6.4 closest to the maximum priority $x = 1$ is the most efficient clerk in the office.

A merely random choice corresponds to $\gamma = 0$ in (6.25), whereas setting $\gamma = 1$ is equivalent to always selecting the customer in line closest to $x = 1$. Note that here we set $\nu > \lambda$. Thus, there is no delay caused by more customers arriving per unit time than customers being satisfied by the clerks of this post office per unit time.

The probability that a task with priority x waits a time interval τ before execution is

$$\psi(x, \tau) = [1 - \Theta(x)]^{\tau - 1} \, \Theta(x). \tag{6.26}$$

Note that we have again adopted the notation of the waiting-time distribution and used the product of the probability of not having a priority x in each of the first $\tau - 1$ time intervals with that of having a priority x in the last interval. This is a discrete version of the following:

$$\psi(x, \tau) \approx \exp[-\Theta(x)\tau]\Theta(x) = -\frac{d\Psi(x, \tau)}{d\tau}, \tag{6.27}$$

which is the definition of the waiting-time distribution density in terms of the time derivative of the survival probability. Here we have generalized the survival probability such that the rate carries the priority index

$$\Psi(x, \tau) = \exp[-\Theta(x)\tau], \tag{6.28}$$

namely the survival probability related to a task of priority x. The dependence of ψ and Ψ on the priority index x is due to the fact that the waiting time depends on the priority level.

The average waiting time of a task with priority x is obtained by averaging over the time t weighted with the priority-indexed waiting-time density $\psi(x, t)$, giving

$$\tau(x) = \Theta(x) \sum_{t=1}^{\infty} t[1 - \Theta(x)]^{t-1}. \tag{6.29}$$

In the continuous-time notation the sum is replaced with an integral and we obtain

$$\tau(x) = \int_0^{\infty} t\, dt\, \psi(x, t) = \int_0^{\infty} t\, dt\, \Theta(x)e^{-\Theta(x)t}, \tag{6.30}$$

so that introducing the new integration variable $\theta = t\Theta(x)$ allows us to write

$$\tau(x) = \frac{1}{\Theta(x)} \int_0^{\infty} \theta\, d\theta\, e^{-\theta} = \frac{1}{\Theta(x)}. \tag{6.31}$$

Therefore the average waiting time of a task of priority x is inversely related to the probability that a task with that priority is carried out. Moreover, from the form of the probability of a task of priority x being chosen (6.25) we have the inverse power-law form for the average waiting time

$$\tau(x) \propto \frac{1}{x^{\gamma}}. \tag{6.32}$$

According to (6.32) the probability density of waiting a time τ, $\psi(\tau)$, is related to the probability $\pi(x)$ of selecting a given line of Figure 6.4 by means of the equality of probabilities

$$\pi(x)dx = \psi(\tau)d\tau. \tag{6.33}$$

Here we select the priority by randomly selecting a line in Figure 6.4 and therefore choose $\pi(x) = 1$ to obtain

$$\psi(\tau) = \left| \frac{dx}{d\tau} \right|. \tag{6.34}$$

Consequently, from (6.32) we have

$$\psi(\tau) \propto \frac{1}{\tau^{1+1/\gamma}}, \tag{6.35}$$

so that the properties in the Einstein and Darwin correspondence in Figure 6.3 can be explained by setting

$$\gamma = 2. \tag{6.36}$$

We note that this inverse power law (6.32) with an index of 3/2 was obtained by Barabási [8] with the deterministic protocol $p = 1$ and is recovered by setting $\gamma = \infty$.

Nothing in the above argument determines the roles of the arrival parameter λ and the servicing parameter ν in the post-office lines of Figure 6.4. Using queuing theory Abate and Whitt [1] proved that the average waiting time is given by

$$\tau_0 = \frac{1}{\eta} = \frac{1}{v(1 - \sqrt{r})} \tag{6.37}$$

from (6.23), where

$$r \equiv \frac{\lambda}{v}. \tag{6.38}$$

We note that in the case $r > 1$ we can express the length l of a line as

$$l(t) = (\lambda - v)t. \tag{6.39}$$

In fact, there are more customers arriving than there are being served. On the other hand it is evident that due to the stochastic nature of the process, when $\lambda = v$, the line will temporarily become large and from time to time it will shrink back to the condition $l = 0$. The distance between two consecutive returns to the origin $l = 0$ is the time that a customer has to wait to be served. Earlier we determined that the time interval between two consecutive returns to the origin is given by $\psi(\tau) \propto \tau^{-3/2}$. This is the prediction of (6.23) when $r = 1$, in which case the average waiting time diverges as seen in (6.37).

When $r < 1$, the arrival rate of the tasks is smaller than the execution rate. Thus, very frequently the line will vanish and the task will be executed almost immediately on the arrival of a customer. We see that in the case $r \to 0$ the waiting time becomes identical to the execution time.

When $r > 1$, the arrival rate of the tasks is greater than the execution rate. Thus, very frequently the line becomes infinitely long and the task will not be executed almost immediately on the arrival of a customer. We see that in the case $r \to \infty$ there are customers who are never served or letters that go unanswered.

What about the index 5/2 of the inverse power law of (6.22)? Techniques have been developed to obtain analytic solutions for priority-based queuing models that are biased, but the arguments depend on phase-space methods for the development of probability densities, particularly the first-passage-time probability density. In the last section we discussed a model of the time intervals between letter responses of Darwin and Einstein developed by Barabási. We now generalize that formalism to determine the first-passage-time pdf. With this pdf in hand we apply (3.183) to the first-passage-time pdf for a diffusion process with drift

$$a(q_0) = v > 0 \tag{6.40}$$

and strength of the fluctuations

$$b(q_0) = D, \tag{6.41}$$

resulting in the second-order differential equation of motion with constant coefficients

$$D\frac{\partial^2 \widehat{W}}{\partial q_0^2} + v\frac{\partial \widehat{W}}{\partial q_0} - u\widehat{W} = 0. \tag{6.42}$$

The general solution to this equation satisfying the vanishing boundary condition as $q_0 \to \infty$ is

$$\widehat{W}(q_a, u; q_0) = \exp\left[\left(v - \sqrt{v^2 + 4uD}\right)\frac{q_a - q_0}{2D}\right] \tag{6.43}$$

and here again the first-passage-time pdf is obtained by taking the inverse Laplace transform. The form is given by

$$W(q_a, t; q_0) = \exp\left[\frac{v}{2D}(q_a - q_0)\right]\mathcal{LT}^{-1}\left[\exp\left[-\frac{(q_a - q_0)}{D}\sqrt{\theta}\right]; t\right], \tag{6.44}$$

where the inverse Laplace transform is taken with respect to θ,

$$\theta = v^2 + 4uD. \tag{6.45}$$

Thus, using the inverse Laplace transform for $\exp[-a\sqrt{u}]$, we obtain

$$W(q_a, t; q_0) = \frac{q_a - q_0}{\sqrt{4\pi D}}\frac{1}{t^{3/2}}\exp\left[-\frac{(q_a - q_0 - vt)^2}{4Dt}\right] \tag{6.46}$$

for a biased Gaussian distribution.

However, the task-priority model of Barabási is not universally accepted. Hong *et al.* [21] analyzed the time interval between two consecutive short messages (SMs) via cell phone. They used eight volunteers (denoted by the letters A through H). These volunteers consisted of one company manager (C) and seven university students (else). The overall time spans of those phone records ranged from three to six months. As shown in Figure 6.5 the inter-event distribution density $\psi(\tau)$ is well fit by the inverse power-law function

$$\psi(\tau) \propto \frac{1}{\tau^\mu}, \tag{6.47}$$

where the exponent μ is in the interval $1.5 < \mu < 2.1$. Each curve in Figure 6.5 has an obvious peak appearing at approximately ten hours. This peak is related to the physiologic need for humans to sleep. When it is time to sleep, individuals postpone their response to phone messages for the duration of the sleep interval.

The time statistics of SM communications are similar to those observed in email [8] and surface-mail [27] communications. However, the SM communication is directly perceived through the senses. We cannot invoke the priority-task principle. We may reply soon to urgent and important letters, and the ones that are not so important or require a difficult reply may wait for a longer time before receiving a response. The same ideas apply to emails. By contrast, we usually reply to a short message immediately. The delay in replying here might be caused not by the fact that the received message is not important, but by other motivating factors.

6.2 Graphic properties of webs

In this section we further compare random webs with real webs. We find that both random and real webs are characterized by small length scales. The small-length-scale

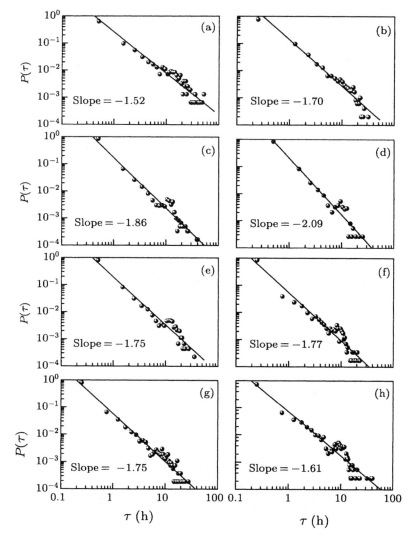

Figure 6.5. The log–log plots of the inter-event time distributions. The black dots represent the empirical data, and the straight lines are the linear fitting to the data. The panels (a)–(h) correspond to the records of the participants A through H, respectively. This graph is reproduced from [21] with permission.

property is easily accounted for by a small-world model proposed by Watts and Strogatz [35] in 1998 that shows that real webs exist in a condition intermediate between order and randomness. However, this model does not account for another important property of real webs, *the scale-free condition*. The model of preferential attachment discussed in the previous sections yields the inverse power-law coefficient $\alpha = 3$, whereas for real webs α ranges from 2 to 3. In the latter part of this section we present a concise review of the work done to generate values of α smaller than 3. We conjecture that the large clustering observed in real webs serves the purpose of facilitating the attainment of local consensus and that the scale-free distribution density of links is related to the transmission of information from one cluster to another so as to transform local consensus into global synchrony.

In the literature of complex networks there is a widely shared conviction that webs with large clustering coefficients that are scale-free are generated for the main purpose of reaching the highest possible degree of efficiency. In this section we are interested in determining whether the concept of topological efficiency corresponds to the perhaps more intuitive notion of dynamical efficiency, interpreted as the condition favoring the realization of global consensus. From an intuitive point of view, a large clustering coefficient corresponds to the existence of local tight-knit communities that are expected to generate local consensus. However, for the consensus to become global, it is necessary to have links connecting different clusters. This raises the question of whether or not the scale-free nature of real webs is generated by the communication between clusters. If this is the case, however, we have to establish whether it can be realized without recourse to a hierarchical prescription. We impose this constraint because according to Singer [31] the brain is an orchestra without a conductor, suggesting that hierarchical networks are not a proper model to describe the most complex of networks, the human brain.

6.2.1 Random webs

Random-web models are similar in spirit to random-walk models in that their mathematical analysis does not capture the full richness of reality; they are not intended to, but they provide a solvable mathematical setting in which to answer specific questions. Consequently, completely random webs provide a useful starting point for investigating how elements arbitrarily distributed in space influence one another. This model is not the real world, but becoming facile with it can provide a tool with which to better understand the real world. Where real-world data yield parameters that deviate markedly from the model, the model mechanism on which that parameter is based can be discarded or modified. Measures of such things as clustering, betweenness and web size all serve to distinguish real-world from random webs.

Random webs are an example of totally disordered systems. According to a widely accepted perspective, random webs are at one end of a continuum with completely ordered webs at the other, locating complex webs somewhere in between. Consequently a totally random web is not complex. However, we have to establish full command of their properties, so as to properly address issues such as those discussed in the previous section.

Some definitions
Random networks were widely studied by Erdös and Rényi in 1959, 1960 and 1961. For an accurate reference to the work of these two outstanding mathematicians we refer the readers to the review paper [2]. In this subsection we develop some elementary notions that will make life easier as we proceed. A graph is a pair of sets $G \equiv \{P, E\}$; P is a set of vertices (points, nodes) $P_1, P_2, \ldots, P_N$ and E is a set of edges (links or lines) that connect two elements of P. Figure 6.6 shows a graph with five points and two edges.

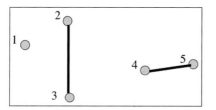

Figure 6.6.   A graph with five points (nodes) and two edges (links).

Thus, the nodes might be individuals within a group of people and a link is the friendship relationship. So if two people are friends they are linked; otherwise they are not.

If N is the number of nodes, the maximum possible number of edges connecting nodes is

$$E_{\text{max}} = \frac{N(N-1)}{2}. \tag{6.48}$$

Note that we divide $N(N-1)$ by 2 because the order of the pairs is not taken into account. The edge $i-j$ is equivalent to the edge $j-i$. However, in some of the webs the coupling is asymmetric, so the direction becomes important. For such webs the 2 is not present in (6.48). Family relations may be one-way as in "father to" or "son of" or two-way as in "cousin to."

Random graphs are established by assigning to any of the E_{max} pairs an edge with probability p. The probability that one pair does not have an edge is $1 - p$. The probability that two pairs do not have any edge is $(1 - p)^2$, and so on. The probability that the whole graph does not have any edge is

$$\Psi(E_{\text{max}}) = (1 - p)^{E_{\text{max}}}. \tag{6.49}$$

Note that this probability can be written in exponential form

$$\Psi(E_{\text{max}}) = e^{E_{\text{max}} \ln(1-p)}. \tag{6.50}$$

When the probability of there being a pair with an edge is very small $p \ll 1$ we make the approximation

$$\ln(1 - p) \approx -p, \tag{6.51}$$

so that the probability that no edges are formed simplifies to the exponential

$$\Psi(E_{\text{max}}) \approx e^{-pE_{\text{max}}}. \tag{6.52}$$

As a consequence, even if p is very small, the probability undergoes a rapid decay for large N, thereby revealing that many edges appear as we increase the number of nodes N. Said differently, the fact that the probability that no edge is formed goes to zero with increasing N implies that it becomes certain that edges are formed.

In Figure 6.7 we sketch the case in which $N = 10$. Consequently the maximum number of edges is $E_{\text{max}} = 45$ and, using (6.49), the probability of no edge being formed is $\Psi(E_{\text{max}}) \approx 0.01$ for $p = 0.1$ and is even smaller, $\Psi(E_{\text{max}}) \approx 0.0001$, for $p = 0.2$. The estimated number of links between nodes is given by the formula

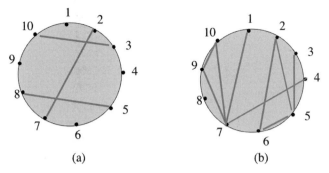

Figure 6.7. The elements in the web are evenly spaced on a circle and the probability of connecting any two nodes is determined by the probability p: (a) $p = 0.1$; (b) $p = 0.2$.

$$L_N(p) = E_{\max} p,$$ (6.53)

which becomes more and more exact as $E_{\max} \to \infty$. For $p = 0.1$ the estimated number of links is between four and five, and for $p = 0.2$ it is about nine. In Figure 6.7 we assign three and nine links to the cases $p = 0.1$ and $p = 0.2$, respectively, for illustrative purposes. It is interesting to note that for $p = 0.2$ two triangles appear. These triangles constitute our first meeting with the clustering phenomenon, but it is by no means our last.

The Poisson distribution of links

Let us now evaluate the probability that a given node of a random network with probability p has k links. The average number of links z is given by

$$z = (N - 1)p.$$ (6.54)

We have $N - 1$ possible links to this specific node. Let us imagine that all these links are contained in a box and that we decide to draw k of them to assign to a single node. This can be done in a number of ways given by the combinatorial expression

$$\binom{N-1}{k} \equiv \frac{(N-1)!}{(N-1-k)!k!}.$$ (6.55)

The probability of getting k links is p^k and the probability of not realizing $N - 1 - k$ links is $(1 - p)^{N-1-k}$. Consequently, the probability of assigning k links to a given node is determined by the product of these three expressions:

$$p_k = \binom{N-1}{k} p^k (1 - p)^{N-1-k}.$$ (6.56)

Let us express (6.56) in terms of the average number of links z given by (6.54) to obtain by direct substitution

$$p_k = \binom{N-1}{k} \left(\frac{z}{N-1}\right)^k \left(1 - \frac{z}{N-1}\right)^{N-1-k}$$ (6.57)

and as $N \to \infty$ by simple factoring we obtain

$$p_k = \binom{N-1}{k} \left(\frac{z}{N-1-z} \right)^k e^{-z}. \tag{6.58}$$

Let us rewrite (6.58) as

$$p_k = F(N) \frac{z^k}{k!} e^{-z}, \tag{6.59}$$

where the function to be simplified is

$$F(N) \equiv \frac{(N-1)!}{(N-1-k)!(N-1-z)^k} = \frac{(N-1)!}{(N-1-k)!(N-1)^k(1-p)^k}. \tag{6.60}$$

Under the condition $p \ll 1$, we neglect the p-dependence and write

$$F(N) \approx \frac{(N-1)!}{(N-1-k)!(N-1-z)^k} = \frac{(N-1)!}{(N-1-k)!(N-1)^k}. \tag{6.61}$$

On implementing the lowest order of the Stirling approximation $\ln N! = N \ln N - N$, we find, after some algebra, that for $N \to \infty$

$$\ln F(N) = 0, \tag{6.62}$$

thereby yielding $F(N) \approx 1$. We conclude that the distribution of k links is therefore given by

$$p_k = \frac{z^k}{k!} e^{-z}, \tag{6.63}$$

the Poisson distribution.

Note that z of (6.54) is identical to the mean number of links per node:

$$z = \langle k \rangle. \tag{6.64}$$

Let us double check this important property. The mean value $\langle k \rangle$ is defined by

$$\langle k \rangle = \sum_{k=1}^{\infty} k p_k. \tag{6.65}$$

By inserting (6.59) into (6.65) we obtain

$$\langle k \rangle = \sum_{k=1}^{\infty} \frac{1}{(k-1)!} z^k e^{-z} = z e^{-z} \sum_{k=1}^{\infty} \frac{z^{k-1}}{(k-1)!} = z. \tag{6.66}$$

Note the normalization condition

$$\sum_{k=0}^{\infty} \frac{z^k}{k!} e^{-z} = 1. \tag{6.67}$$

Recall our earlier derivation of the Poisson distribution and specifically how to generalize it beyond the exponential survival probability.

Giant clusters

We have seen that increasing p has the effect of establishing a larger number of links and also that of creating clusters, the triangles found in Figure 6.7 being, in fact, a kind of lowest-order cluster. Is it possible to find a critical value of p, much smaller than 1, that allows a walker to move freely from one border to the opposite border of the network? In Figure 6.8 we sketch a giant cluster making it possible for the walker to realize a complete North–South or East–West trip. What is the critical value of p that we have to assign to the network for this giant cluster to appear?

Let us denote by s the fraction of the nodes of the graph that *do not belong* to the giant cluster. When the number of nodes is very large, the quantity s can also be interpreted as the probability that a node randomly chosen from the network does not belong to the giant component. The fraction S of the graph occupied by the giant cluster is therefore

$$S = 1 - s. \tag{6.68}$$

Now consider a node not belonging to the giant cluster and assume that the node has degree k, namely it has k links. If the node we are considering does not belong to the giant cluster, neither do its k nearest neighbors. In fact, if one of these neighbors belonged to the giant cluster, then the node under consideration would belong to it as well. Thus, we obtain

$$s = \sum_{k=0}^{\infty} p_k s^k. \tag{6.69}$$

In fact, since s is a probability it follows that the probability that the k neighbors do not belong to the giant cluster is s^k. The probability that the node under consideration has k neighbors and the probability that the neighbors do not belong to the giant component are independent of one another. This explains the product $p_k s^k$. The sum from $k = 0$

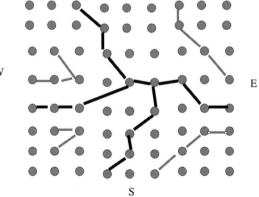

Figure 6.8. When the probability p reaches a critical value $p_c = 1/(N-1)$ a walker can travel from the Southern to the Northern border and from the Western to the Eastern border. The black links indicate a giant cluster spanning the world from East to West and from North to South. The gray links indicate local clusters.

to $k = \infty$ takes into account all possible numbers of links. By inserting the Poisson distribution (6.63) into (6.69) we obtain the transcendental equation for the probability

$$s = e^{z(s-1)}. \tag{6.70}$$

Using the definition (6.68) we transform (6.70) into the transcendental equation for the fraction of the graph occupied by the giant cluster,

$$S = 1 - e^{-zS}. \tag{6.71}$$

In this equation z is a control parameter, or, using (6.53), it is proportional to the probability p. To establish the critical value of the control parameter in this case, we assume that p_c is the critical value of the probability compatible with the emergence of a very small, but finite, value of S. For S very small we expand the exponential and (6.71) becomes

$$S = zS, \tag{6.72}$$

which indicates that the critical value is $z_c = 1$, thereby yielding, on the basis of (6.53),

$$p_c = \frac{1}{N-1}. \tag{6.73}$$

This result is interesting, because it indicates that for very large numbers of nodes N the critical value of p or z may be extremely small. In Figure 6.9 we sketch the shape of this phase-transition process.

The length of the random web

Let us now define another important property of random webs. This is the length l of the web. How many nodes do we have to encounter to move from a generic node i to another generic node j? An intuitive answer to this question can be given by assuming that each node has a very large number of links that we identify with the mean value $z \gg 1$. This is equivalent to making the assumption that we are far beyond the threshold value $z_c = 1$.

Each node of a network has z neighbors ($z = 5$ in the case of Figure 6.10). Thus, in the first layer there are z points connected to P. In the second layer there are z^2 points. In the lth layer we have for the number of points ultimately connected to P

$$N_l = z^l. \tag{6.74}$$

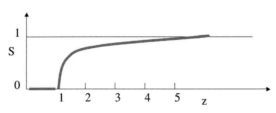

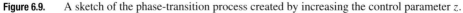

Figure 6.9.   A sketch of the phase-transition process created by increasing the control parameter z.

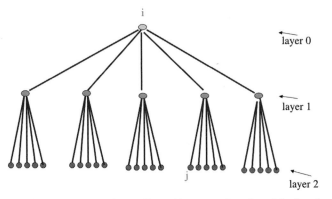

Figure 6.10. This figure illustrates the small-world concept. In spite of the fact that the node denoted by j is a unit of a very large population, it is located at a distance $l = 2$ from the node denoted by i.

Let us now make the assumption that l is sufficiently large as to ensure that all of the nodes are connected, $N_l \approx N$. We have

$$N = z^l, \tag{6.75}$$

where l defines the network length.

Note that (6.75) is an approximation, or rather an estimate, of the number of nodes necessary to characterize the length of the web, because in the earlier layers there are points that we are not counting. To convince the reader that the error corresponding to this approximation is small, consider the quantity R defined by

$$R = \frac{z^{l-1} + z^{l-2} + \cdots + z}{z^l} = \frac{1}{z} + \cdots \frac{1}{z^{l-2}} + \frac{1}{z^{l-1}}. \tag{6.76}$$

This quantity is the ratio of the neglected population to the total population. The condition $z \gg 1$ yields

$$R \approx \frac{1}{z} \ll 1. \tag{6.77}$$

If the total population is $N = 300{,}000{,}000$, namely the population of the United States, corresponding to the sixth layer $l = 6$, we have $z \approx 26$, that is, $26^6 \approx 3 \times 10^8$ and the population of the fifth layer is about $12{,}000{,}000$ people.

In summary, we can use (6.75) to define the length of the network, thereby obtaining

$$l = \frac{\ln N}{\ln z}. \tag{6.78}$$

Owing to the logarithmic dependence on N, we reach the conclusion that the length of even a random web is small. Consequently, we live in a small world.

The clustering coefficient

Let us now study another important property of random webs: *the clustering coefficient*. This coefficient is also of value in the general setting of complex networks. Let us consider a generic node P and assume that the node P in Figure 6.11 is linked to two

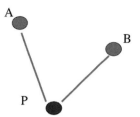

Figure 6.11.   This figure illustrates the concept of a triple. P is a friend of both A and B. Is A a friend of B? If he/she is a friend, then A is linked to B and the triple is a triangle. If P is an actor, for instance Marilyn Monroe, and she plays in a picture with both A, Tony Curtis, and B, Jack Lemmon, does A play in the picture with B? In that case all three actors played in the same movie and the triple is a triangle.

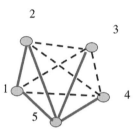

Figure 6.12.   This figure illustrates the clustering definition proposed by Newman [26]. The full lines denote the realized links. The dashed lines denote the potential links that are not realized.

nodes, A and B. The nodes are assumed to represent agents A and B, who are friends of P. We are interested in the probability that A is a friend of B. It is evident that, when the probability of A being a friend of B is high, this is the signature of a community. How can we measure the value of this property called *clustering*?

Newman [26] proposes the method illustrated by Figure 6.12. The formula for the clustering coefficient he proposed is

$$C = \frac{3\Xi}{\Pi},$$ (6.79)

where Ξ denotes the *number of triangles in the network* and Π is the *number of connected triples of vertices*. A triangle is a set of three nodes that are completely inter-connected. In Figure 6.12 we see only one triangle, that constituted by nodes 1, 2 and 5. A triple is a set of three nodes with at least two links. If A is not linked to B (see Figure 6.11) these three nodes are a triple, but not yet a triangle. If A is linked to B, we have a triangle. In a triangle there are three triples. This explains the factor of 3 in (6.79). If all the possible triples are associated with a triangle, we get for C the maximum value of 1. In Figure 6.12 we note that nodes 1 and 2 each correspond to a triple. The node 3 generates six triples, with (12), (14), (15), (24), (25) and (45). Thus we obtain for the clustering coefficient for the graph in Figure 6.12

$$C = \frac{3}{8}.$$ (6.80)

Another prescription for the determination of the parameter C was given by Watts and Strogatz [35]. This prescription reads

$$C = \frac{\sum_{i=1}^{N} C_i}{N}, \tag{6.81}$$

where N denotes the number of nodes, the index i runs from 1 to N, and C_i is the local clustering coefficient for the node i defined by

$$C_i = \frac{\Xi}{\Pi}, \tag{6.82}$$

where Ξ is the *number of triangles connected with the ith node* and Π denotes the *number of triples connected with the ith node*. In the case of Figure 6.12 we see that node $1 \Rightarrow C_1 = 1$, node $2 \Rightarrow C_2 = 1$, node $3 \Rightarrow C_3 = 0$, node $4 \Rightarrow C_4 = 0$ and node $5 \Rightarrow C_5 = 1/6$. Thus we obtain from (6.81)

$$C = \frac{1 + 1 + 1/6}{5} = \frac{13}{30}. \tag{6.83}$$

An equivalent, but more intuitive, way of defining the clustering coefficient C_i is illustrated in Figure 6.13. We focus our attention on the ith node and consider all k_i nodes connected with it. If all the friends of the ith node are friends with one another, the environment of the ith node has the number of connections N_C given by

$$N_C = \frac{k_i(k_i - 1)}{2}. \tag{6.84}$$

Actually, for the ith node there will be a smaller number of connections, Δ_i. Thus, the connectivity of the ith node is given by

$$C_i = \frac{2\Delta_i}{k_i(k_i - 1)}. \tag{6.85}$$

In the case of a random web each node has $\langle k \rangle = p(N - 1)$ links. The approximate number of links of its environment is given by

$$\Delta_i \approx \frac{\langle k \rangle(\langle k \rangle - 1)p}{2} \tag{6.86}$$

and, consequently, in this case we find, assuming $C = C_i$,

$$C = p = \frac{\langle k \rangle}{N - 1} \approx \frac{\langle k \rangle}{N}. \tag{6.87}$$

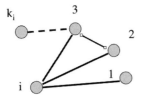

Figure 6.13. This graph illustrates an intuitive way of defining C_i. The full lines correspond to the realized links. The dashed line corresponds to a generic set of nodes from node 4 to node k_{i-1}.

We see that a random graph for N very large is characterized by a small clustering coefficient.

Web efficiency

Following [22], let us define the concept of *web efficiency*. Consider the shortest distance between two arbitrary nodes i and j and denote it by the symbol L_{ij}. In the case of Figure 6.13 the shortest distance is $L_{ij} = 3$. The communication efficiency between i and j is defined by

$$\epsilon_{ij} = \frac{1}{L_{ij}}. \tag{6.88}$$

When there is no path in the graph between i and j, $L_{ij} = \infty$ and consequently $\epsilon_{ij} = 0$, namely we have the condition of minimum efficiency and no communication. The global efficiency of the web is defined by

$$E = \frac{1}{N(N-1)} \sum_{i,j} \frac{1}{L_{ij}}. \tag{6.89}$$

In the case of all-to-all coupling, $L_{ij} = 1$ and

$$\sum_{i,j} \frac{1}{L_{ij}} = N(N-1). \tag{6.90}$$

Thus, the all-to-all coupling case corresponds to that of maximum efficiency.

The question is how do real webs depart from the condition of weak clustering that is typical of random webs?

6.2.2   Real webs

Watts and Strogatz (WS) [35] have shown that the statistical properties of real webs significantly depart from those of random webs. This observation captured the attention of the scientific community, resulting in their paper becoming very popular; as of March 3, 2009 it had over 4,000 citations. This number of citations is particularly significant because the distribution of citations is another inverse power law and the average number of citations to a scientific paper in a year is a little over 3.

WS studied the following real webs.

(a) *Film actors.* The first web they considered was one motivated, at least in part, by the party game determining the distance between two actors in terms of the movies in which they have acted. For example, if they acted in a movie together the separation distance is zero. If, however, they have each acted in a movie with a third party then the distance between them is one and so on. Two actors are considered to be joined by an edge or link if they have acted in a film together; see, for example, Figure 6.14. So, in terms of the web, the above party game can be cast in terms of the links between actors. In an earlier section we saw that random webs may have giant clusters, where the linked elements form closed groups. Although the web

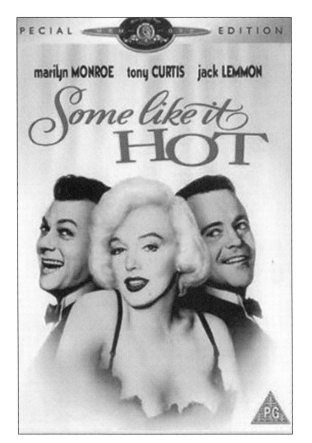

Figure 6.14. In the famous movie *Some Like It Hot*, Marilyn Monroe, Tony Curtis and Jack Lemmon play together. This yields a triangle and fleshes out the content of Figure 6.11.

of movie actors is not random (the complex process by which actors are selected for parts in a movie is very mysterious to the authors), it does seem to have giant clusters. A given actor often plays a certain kind of character and consequently there is correlation between each actor and their partners. However, movie-actor networks also have a giant component, which includes approximately 90% of all actors listed in the Internet Movie Database (available at http://us.imdb.com/). The observation made in [35] uses data ending in April 1977. WS found that the total number of actors is $N = 222,526$ and the average connectivity is $\langle k \rangle = 61$.

(b) *Power grids*. The next web of interest is physical and is the power grid in the United States. Electrical power transmission on the grid distributes electricity from power plants across vast geographical regions to substations and, from substations, across smaller geographical regions to consumers. It is evident that in order to perform its function the power grid forms a web that cannot be random. In fact, the power grid is determined by several conditions, such as the population density, the geographical and historical nature of the territory, and so on (see Figure 6.15). WS studied a web with a total number of stations and substations $N = 4,941$, which they determined

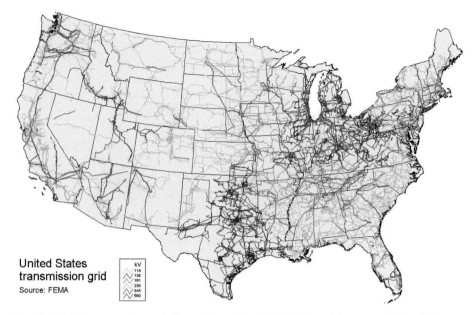

Figure 6.15.   The United States power-transmission grid consists of 300,000 km of lines operated by 500 companies. From http://en.wikipedia.org/wiki/Electric_power_transmission.

Figure 6.16.   *Caenorhabditis elegans* is a free-living, transparent nematode (round-worm), about 1 mm in length, which lives in temperate soil environments. From http://en.wikipedia.org/wiki/Caenorhabditis_elegans.

to have an average connectivity of $\langle k \rangle = 2.67$. Of course, we learned in Chapter 2 that blackouts have a size–frequency distribution that is an inverse power law.

(c) *C. elegans*. The final exemplar is a biological web, *Caenorhabditis elegans*, a worm. In the wild this worm feeds on bacteria that develop on decaying vegetable matter (see Figure 6.16). *C. elegans* is the most primitive organism that exists that shares many of the biological characteristics of mammals. The nervous system of *C. elegans* is quite remarkable, having about 300 neurons and 4,000 connections. Because of its small size this worm is often used to simplify the analysis of such entities as a mouse brain. WS found that the total number of nodes is $N = 282$, and the average connectivity is $\langle k \rangle = 14$. Research into the molecular and developmental biology of *C. elegans* was begun in 1974 by Sydney Brenner and it has since been used extensively as a model organism.

WS decided to evaluate the statistical properties of these three different kinds of real webs. They evaluated the degree of clustering in each web according to their definition of a clustering coefficient C (6.81). They also evaluated the property

$$D = \langle L_{ij} \rangle, \tag{6.91}$$

where L_{ij} measures the minimal separation distance between two vertices in the graph. D is the average over all possible pairs.

WS compared the results of their observations on the three real networks with the theoretical predictions of the random-graph theory for webs with the same numbers of nodes. The mean number of links per node is given by

$$\langle k \rangle = 2 \frac{L_{\text{random}}}{N}, \tag{6.92}$$

where L_{random} is the number of links that are realized in a random web with probability p. We have

$$p = \frac{L_{\text{random}}}{L_{\text{max}}}, \tag{6.93}$$

where

$$L_{\text{max}} = \frac{N(N-1)}{2}. \tag{6.94}$$

As pointed out earlier, the factor of 2 in (6.92) takes into account that each edge contributes to the number of edges of the node i and to the number of edges of the node j at the same time. From (6.93)

$$L_{\text{random}} = p L_{\text{max}}, \tag{6.95}$$

so that inserting (6.95) into (6.92) yields the average number of connections

$$\langle k \rangle = \frac{2p L_{\text{max}}}{N}. \tag{6.96}$$

Finally, by inserting (6.94) into (6.96), we obtain

$$\langle k \rangle = p(N-1), \tag{6.97}$$

in accordance with what was found earlier.

In summary, then, using (6.97) we determine the values of the probability p of the three webs under the assumption that they are random. We have for the film actors

$$p = \frac{61}{222,525} = 0.000274127, \tag{6.98}$$

for the power-grid networks

$$p = \frac{2.67}{4,940} = 0.00054 \tag{6.99}$$

and for the *C. elegans* network

$$p = \frac{14}{281} = 0.0498. \tag{6.100}$$

Table 6.1. Derived from [35], with the value of 0.005 for C_{random} of the power grid probably being a misprint, which is here replaced with the value 0.0005

	L_{actual}	L_{random}	C_{actual}	C_{random}
Film actors	3.65	2.99	0.79	0.00027
Power grid	18.7	12.4	0.08	0.0005
C. elegans	2.65	2.25	0.28	0.05

Insofar as the distance L is concerned, according to (6.78) we can write

$$L_{random} = \frac{\ln N}{\ln(\langle k \rangle)}, \tag{6.101}$$

which yields for the film-actor web

$$L_{random} = \frac{\ln(222{,}526)}{\ln(61)} = \frac{12.31}{4.11} = 2.99, \tag{6.102}$$

for the power-grid web

$$L_{random} = \frac{\ln(4{,}941)}{\ln(2.67)} = \frac{8.51}{0.98} = 8.68 \tag{6.103}$$

and for the C. elegans web

$$L_{random} = \frac{\ln(282)}{\ln(14)} = \frac{5.64}{2.63} = 2.14. \tag{6.104}$$

Note that there seems to be an ordering from the most human-designed web with the highest clustering to the least human-designed, or, said differently, the most naturally evolved web has the lowest clustering. Insofar as C is concerned, using (6.87) we expect

$$C_{random} = p. \tag{6.105}$$

From Table 6.1 we get some important information. Both random and real webs have short distances. In other words, the small-world property is shared by both actual and random webs. The most significant difference between theory and data is given by the clustering coefficient C. The real webs have large clustering coefficients, whereas the random webs have small clustering coefficients.

WS [35] claim that real webs are complex in the sense that they exist in a condition intermediate between randomness and order. They also propose a model to create networks with the same statistical properties as real networks. With probability p they rewire some of the links established according to the deterministic prescription. Note that the probability p adopted by them to rewire their links should not be confused with the probability p used in Section 6.2.1. In fact, when WS rewire a link, they adopt the criterion of uniform probability that is equivalent to setting p of Section 6.2.1 equal to the inverse of the number of possible new links.

Note that when $p = 0$ and the *regular* condition of Figure 6.17 applies, all the triples are triangles. Thus, $C = 1$. When $p = 1$, all the links are established randomly.

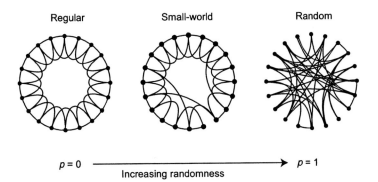

Figure 6.17. The random rewiring procedure for interpolating between a regular ring lattice and a random web, without altering the number of vertices or edges in the graph. WS start with a ring of N vertices, each connected to its k nearest neighbors by undirected edges. (For clarity, $N = 20$ and $k = 4$ in the schematic examples shown here.) WS choose a vertex and the edge that connects it to its nearest neighbor in a clockwise sense. With probability p, they reconnect this edge to a vertex chosen uniformly at random over the entire ring, with duplicate edges forbidden; otherwise they leave the edge in place. They repeat this process by moving clockwise around the ring, considering each vertex in turn until one lap has been completed. Next, WS consider the edges that connect the vertices to their second-nearest neighbors clockwise. As before, they randomly rewire each of these edges with probability p, and continue this process, circulating around the ring and proceeding outward to more distant neighbors after each lap, until each edge in the original lattice has been considered once. (Since there are $Nk/2$ edges in the entire graph, the rewiring process stops after $k = 2$ laps.) Three realizations of this process are shown, for different values of p. For $p = 0$, the original ring is unchanged; as p increases, the graph becomes increasingly disordered until, for $p = 1$, all edges are rewired randomly. One of the main results of the WS paper [35] is that, for intermediate values of p, the graph is a small-world network: highly clustered like a regular graph, yet with small characteristic path length, like a random graph [35]. Reproduced with permission.

The probability p of Section 6.2.1 is given by the ratio of the number of links established according to the deterministic rule, which is $2N$, to the maximum number of links, which is $N(N - 1)/2$. The ratio of these two quantities is

$$R = \frac{4}{N - 1}. \tag{6.106}$$

This formula applies for $N > 5$, and, in the case of $N = 1,000$ considered in Figure 6.18, yields $R = 0.004$. Note that this value is small, but larger than the critical probability $p_c = 1/(N - 1) = 0.001$, which corresponds, according to (6.73), to the emergence of a giant cluster. According to (6.105), with p replaced by R, we expect that $C(1) = 0.004$, which corresponds in fact to the value plotted in Figure 6.18.

Beyond the model of preferential attachment

Earlier we found that real webs are characterized by the scale-free property. We have justified the observation that the distribution of links is an inverse power law by pointing out that this corresponds to the prescriptions of the laws of Pareto and Zipf. Because of

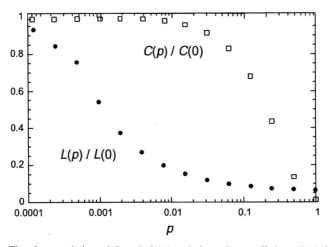

Figure 6.18. The characteristic path length $L(p)$ and clustering coefficient $C(p)$ for the family of randomly rewired graphs described in Figure 6.17. Here L is defined as the number of edges in the shortest path between two vertices, averaged over all pairs of vertices. The clustering coefficient C is defined according to the proposal of (6.105). The data shown in this figure are averages over twenty random realizations of the rewiring process described in Figure 6.17, and have been normalized by the values $L(0)$ and $C(0)$ for a regular lattice. All the graphs have $N = 1,000$ vertices and an average degree of $k = 10$ edges per vertex. We note that the logarithmic horizontal scale has been used to resolve the rapid drop in $L(p)$ corresponding to the onset of the small-world phenomenon. During this drop, $C(p)$ remains almost constant at its value for the regular lattice, indicating that the transition to a small world is almost undetectable at the local level [35]. Reproduced with permission.

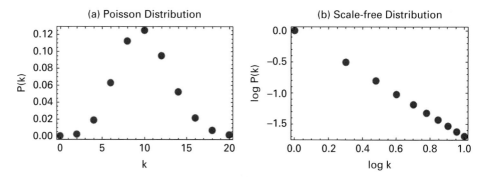

Figure 6.19. (a) A random graph is characterized by a Poisson distribution. The typical scale of this web is given by the average node degree $\langle k \rangle$. The probability of finding nodes with high or low node degree is statistically very small because the degree distribution $P(k)$ decays exponentially. (b) By contrast, scale-free webs are characterized by a power-law degree distribution $P(k) \propto k^{-\alpha}$. In this web, there is no typical scale; that is, it is scale-free and it is possible to find highly connected nodes (hubs). In the log–log plot, the power law is characterized by a straight line of negative slope.

these empirical inverse power laws, one expects that real webs do not have the Poisson structure of the left panel of Figure 6.19 but rather correspond to the structure sketched in the right panel of that figure. However, it seems that $\alpha = 3$ generated by the model of preferential attachment is the upper limit of the power-law indices observed experimentally. The experimental observation of biological networks reveals [25] that α is scattered between 2.1 and 3. For this reason, AB [3] proposed a modified version of the model of preferential attachment. They start with m_0 isolated nodes, and at each step they perform one of the following three operations.

(i) With probability p they add m ($m \leq m_0$) new links. For each link one node is selected randomly and the other in accord with the criterion of preferential attachment. This process is repeated m times.

(ii) With probability q AB rewire m links. They select randomly a node i and a link l_{ij} connected to it. Then they replace it with a new link $l_{ij'}$ that connects node i with node j' selected in accord with the preferential attachment criterion. This process is repeated m times.

(iii) With probability $1 - p - q$ AB add a new node. The new node has m new links that are connected to the nodes already present in the web that had been selected in accord with the preferential attachment criterion.

Using this model AB realize a distribution $\theta(k)$ that can be either an inverse power law or an exponential function. In Figure 6.20 examples of inverse power-law and exponential distributions are depicted for the various conditions. It is important to point out that AB prove that, in the scale-free regime, α changes from 2 to 3. Figure 6.20 depicts the comparison between the numerical simulations and the predictions made using continuum theory. The dashed lines in Figure 6.20(a) illustrate the theoretical prediction in the scale-free region. In Figure 6.20(b) the exponential region is depicted; here the distribution is graphed on semilogarithmic paper, converging to an exponential in the $q \to 1$ limit.

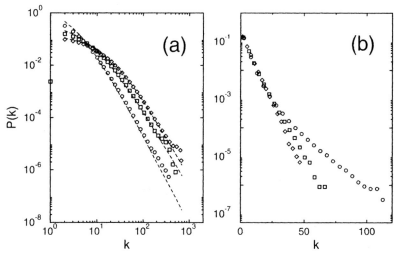

Figure 6.20. (a) In the simulations $t = 10,000$ and $m_0 = m = 2$. Circles, $p = 0.3, q = 0$; squares, $p = 0.6, q = 0.1$; and diamonds, $p = 0.8, q = 0$. The data were logarithmically binned. (b) Circles, $p = 0, q = 0.8$; squares, $p = 0, q = 0.95$; and diamonds, $p = 0, q = 0.99$. From [29] with permission.

It is worth stressing that there exists another proposal regarding how to create a scale-free network with $\alpha < 3$ [29]. The authors of that model propose that the number of new links arriving is a fluctuating rather than a fixed quantity.

6.2.3   Web failures

We have all experienced failure in grades, in sporting events, or in our love life. This is part of the unpredictability of the world's complex social webs and how we humans interact with and within them. These setbacks are part of what we learn as we grow older; love is lost when our overtures no longer match the needs of the pursued and tests are blotched when the interests of the professor are not taken into account during preparation. These losses are devastating when they occur, but they rarely have long-term negative effects, and they prepare youth for the challenges of future responsibility. Of course, there are more elaborate social networks in which failure has more dire outcomes than a bruised ego, such as failure due to attacks on computer networks by malicious hackers and on social organizations by terrorists. The latter can unravel the very fabric of society, unless thoughtful people can determine how complex webs respond to attacks or even to unintentional forms of failure.

We discussed failure and survival previously in a manufacturing context and learned that the probability of failure of a given fraction of a large number of manufactured widgets is given by an exponential. A single parameter characterized the fraction of widgets lost and that was the constant rate of failure during the manufacturing process. In this simple manufacturing model the exponential predicts that the same fraction of widgets fail in every equal interval of time. Knowing this failure pattern enables the industrialist to plan for the future. Why do simple networks respond to failure in this way and how are complex networks different?

There are two general classifications of internally generated failure, namely failure that is local in space and time, and failure that is catastrophic. Local failure can be the loss of a single circuit in a large switching network that has essentially no effect on the functioning of the web. Catastrophic failure spreads over vast physical regions such as the failure of the power grid we discussed earlier and/or extends over long times such as an economic depression. The extent of web damage caused by failure is a consequence of how the elements of the web are interconnected. A random web has all the nodes more or less connected so that the loss of any particular node causes the same damage as the loss of any other node. On the other hand, in a scale-free network the connectivity of the nodes can be quite different, so the aggregated damage due to failure of a particular node depends on the properties of that node. Not all nodes are equal.

Another kind of failure is caused by external disruption, whether it is the human body or a computer web being invaded by viruses, or social/political organizations being attacked by terrorists. We want to know how robust the biological or social web is with respect to such disruption. One measure of the degree of breakdown of the web is the loss of ability to support and transport information. Put somewhat differently, the disruption is a reflection of how tolerant a web is to error and a web's error tolerance is not a fixed quantity. It is the intrinsic structure of the web, its topology, that often determines its robustness, which is to say, the web's ability to absorb error and still function. Part of what we examine in this subsection is how the heterogeneity of complex webs influences their robustness.

Figure 6.19 depicts two classes of networks. One has a uni-modal distribution of connections indicative of a random network peaking at the average connectivity $\langle k \rangle$. The other web has an inverse power-law distribution of connections indicative of a scale-free network. The error tolerance of these two kinds of webs is very different since the probability of a node in the random web having a connectivity much greater than the average $k \gg \langle k \rangle$ decreases exponentially, whereas the probability of there being highly connected nodes can be substantial in scale-free webs. It is the difference between these two mathematical forms that enables us to quantify the robustness of the web. The random web is like the world of Gauss, with all the elements in the web being well represented by the mode of the distribution. In the scale-free web the connectivity of various nodes can be significantly different, with some being the airport hubs of Dallas and Atlanta, and others being the smaller airports in Buffalo and Cleveland. The former are the well-connected hubs of the web and the latter are members of the sparsely connected majority.

Albert *et al.* [4] were among the first to discuss how to quantify the error or attack tolerance of webs. They examined the changes in the web diameter, the diameter being the average distance between any two nodes on the web, when a small fraction of the number of elements is removed. A general result is that the diameter of the web increases with the loss of nodes. The node loss translates into a loss in the number of paths traversing the web and it is the loss of paths that produces an increase in the web's diameter. Erdös and Rényi [15] determined that a random network has Poisson statistics and consequently the number of nodes with high connectivity is exponentially small. As Albert *et al.* [4] explain, in a random web the diameter increases monotonically with the fraction of nodes that fail as a consequence of the exponential nature of the tail. As pointed out above, the exponential has the property of equal fractional decreases for equal incremental increases in the dynamical variable. This property of the exponential is reflected in the reduced ability of the various parts of the web to communicate with one another, since removing each node produces the same amount of damage in the web. This is very different from the tolerance expressed in a scale-free web.

The calculations done by Albert *et al.* [4] indicate that as many as 5% of the nodes can be removed from the web with the transmission of information across the web unaffected. This is the robust behavior often discussed with regard to scale-free webs and is a consequence of the inhomogeneous nature of the nodal connectivity. Recall that in an inverse power-law distribution there are a few nodes with a great many connections, corresponding to the number of very wealthy in the Pareto distribution of income, but the vast majority of nodes have very few links, corresponding to the much lower income of the much greater number of people in the middle and lower classes. Consequently, if nodes fail at random, or are attacked at random for that matter, the probability is high that the failing nodes have only a few links and are of no consequence to the functioning of the web. Moreover, the diameter of the web is only weakly dependent on these low-degree nodes, so their removal, even in large numbers, does not change the overall structure of the web, including its diameter. Thus, following Albert *et al.* [4], we can say that a scale-free web is more robust against random attacks than is a random web.

However, modern terrorist networks do not attack at random, nor does the malicious hacker. These people seek out the weakness in their target and attack where they believe the web is the most vulnerable, because they want their act to have maximum impact. In a random network this is of no consequence since all the nodes are equivalent. Therefore, irrespective of whether an attack on a random network is random or carefully planned, the result is the same. The degradation produced by the attack will be small and of little consequence. In the case of the scale-free network the result is dramatically different. The hacker inserting a virus into a high-degree server can efficiently distribute damage throughout the web using the web's own connectivity structure to bring it to its knees. The same is true for terrorists, who attack the highest-degree node in the network and use the connectivity to amplify the effect of the damage. These attacks make use of the small diameter of scale-free webs; for example, the diameter of the World Wide Web, with over 800 million nodes, is approximately 19, and social networks with over six billion people are believed to have a diameter of around six, as pointed out by Albert *et al.* [4]. It is the small diameter of scale-free webs that produces the extreme fragility with respect to attack on a hub. This small-world property is a consequence of the slow increase in the average diameter of the web with the number of nodes; it is essentially logarithmic, $\langle L \rangle = \ln N$.

Of course, understanding the robustness of a web can be used to counter-attack terrorist webs and to strategically defend one's own web. Gallos *et al.* [18] study the tolerance of scale-free networks under systematic variation of the attack strategy, irrespective of whether the attack is deliberate or unintentional. Since the web is scale-free the probability of a node being damaged through failure, error or external attack depends on the degree k of the node. It is not clear how to characterize the probability of damaging a node because the characteristics of different webs can be distinct. They explain that in computer networks the hubs usually serve many computers and are built more robustly than other computers in the web, and consequently their probability of failure is less than that of the other computers for a given attack. However, one can also argue from the other point of view and conclude that the nodes with more links are less robust. The highly connected nodes on a computer web have more traffic and this may make them more susceptible to failure.

As mentioned above, the intentional attack may be quite efficient by virtue of disrupting the highest-degree nodes first. This implies, of course, that the ordering of the node connectivity is apparent, as it is for many social networks. However, in terrorist and other criminal organizations the individual higher in the hierarchy may have more connections, but they may also be less well known. Thus, the probability of removing them in an attack is significantly lower than that of removing individuals lower in the food chain, that is, those with fewer links. In order to proceed beyond this point it is necessary to postulate a form for the probability $W(k_j)$ that the node j of degree k_j is deactivated by either failure or attack [18]:

$$W(k_j) = \frac{k_j^\beta}{\sum\limits_{j=1}^{N} k_j^\beta}, \qquad (6.107)$$

where the superscript β can be used to represent external attacks, internal failures or the superposition of the two.

Using the probability (6.107) Gallos *et al.* [18] studied the web tolerance under various attack strategies. They determined that, in an intentional attack, only a little knowledge about the highly connected nodes, the Don in the above crime scenario, is necessary in order to dramatically change the threshold relative to the random case. They point out that the Internet, for example, can be significantly damaged when only a small fraction of hubs is known to the malicious hacker. The news is not all negative, however. On the positive side, in the immunization of populations, even if we know very little about the virus spreaders, that is, the probability of identifying them is low, the spreading threshold can be reduced significantly with very little knowledge.

6.3 Deterministic web models

The search for the power-law index $\alpha < 3$ motivated many researchers to refocus their investigations away from random processes with preferential attachment to deterministic processes. The first model that we illustrate here is that of [7].

6.3.1 Hierarchal webs

In Figure 6.21 we show the prescription adopted to generate a hierarchal model web. We can interpret the sketch as follows. At $n = 0$ we have very poor resolution and perceive the web as a single black dot. At $n = 1$ the resolution is a little bit better and we see that the black dot is actually a set of three black dots, with the central black dot linked to the two side black dots by a link for each of them. The better resolution of $n = 2$ shows that each of the three black dots of $n = 1$ is, in turn, a set of three black dots and that the central set is linked to the two side sets by two links per set. In other words, the single links of $n = 1$ are a coarse-grained representation of two links. The better resolution of $n = 3$ shows that the four long-range links of $n = 3$ correspond to eight long-range links, and so on. All the bottom points of the graph are connected to the top point, which is called the root.

For instance, with $i = 1$, we obtain $k_R = 2$, which is in fact the degree of the root at the first step of Figure 6.21. If we keep proceeding with the next steps, we see that there are hubs, namely nodes with a large number of links that are equivalent to the root at $i = 1$. It is easy to assess that after $n - i$ more steps, namely at the nth step, the number of nodes with k_R links is

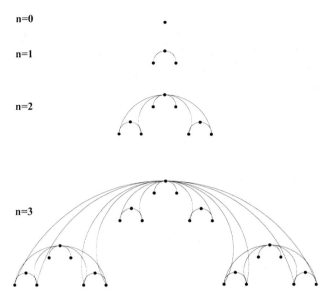

n=0

n=1

n=2

n=3

Figure 6.21.   Construction of the deterministic scale-free network, showing the first four steps of the iterative process. Adapted from [7].

$$N_R = \left(\frac{2}{3}\right) 3^{n-i}.$$ (6.108)

For instance, at $n = 3$ the number of nodes with degree $k_R = 2$ is six, in accordance with (6.108). Let us imagine that we iterate the procedure up to time $n \gg 1$. Let us focus on $i \gg 1$. Thus (6.111) yields

$$k = 2^i,$$ (6.109)

from which we get

$$i = \frac{\ln k}{\ln 2}.$$ (6.110)

It is easy to assess that the root, namely the hub with the highest degree, after i iterations attains the degree

$$k_R = 2^{i+1} - 2.$$ (6.111)

Taking the logarithm of (6.108) yields

$$\ln N_R = \ln \frac{2}{3} + (n - i) \ln 3.$$ (6.112)

On inserting (6.110) into (6.112) we have

$$\ln N_R = \ln \frac{2}{3} + n \ln 3 - \frac{\ln 3}{\ln 2} \ln k = \ln \left(\frac{2}{3} 3^n\right) + \ln \left(\frac{1}{k^\alpha}\right),$$ (6.113)

with the power-law index

$$\alpha = \frac{\ln 3}{\ln 2}.$$ (6.114)

We write (6.113) as

$$\ln N_R = \ln \left(\frac{A(n)}{k^\alpha} \right), \tag{6.115}$$

with the coefficient

$$A(n) \equiv 2 \cdot 3^{n-1}. \tag{6.116}$$

In conclusion the probability density for connectivity of the nodes is, using the number of nodes with k_R links,

$$\theta(k) \propto \frac{1}{k^\alpha}, \tag{6.117}$$

with α defined by (6.114), the numerical value of which is

$$\alpha \approx \frac{1.099}{0.6939} \approx 1.59. \tag{6.118}$$

The above argument is similar in spirit to that given for determining the length of a fractal curve connecting two points separated by a finite Euclidean distance. Let us construct the curve obtained by adding a bit of length to the line segment connecting the two points at each step of an iteration process. Consider a line segment of unit length and partition it into three pieces of equal size. Now we replace the center piece with a section that is twice the original length of that segment, thereby making a sharp cap in this interval as shown in Figure 6.22. The length of the curve therefore increases from unity at $n = 0$ to $4/3$ at $n = 1$. Now carry this process a step further and partition each of these four line segments into thirds and place a cap in the center third of each one of these as shown in Figure 6.22 for $n = 2$. The total length of the curve at $n = 2$ is $16/9$, where each of the sixteen line segments is of length $1/9$. In general, if η is the length of the ruler and $N(\eta)$ is the number of times the ruler is laid end to end to measure the length of the curve, the length $L(\eta)$ is given by

$$L(\eta) = N(\eta)\eta. \tag{6.119}$$

This relation is, of course, true in general.

In the construction process the size of the ruler is determined by the iteration number and in the example given the ruler size is $\eta = 1/3^n$ at generation n. Consequently the length of the curve at generation n is

$$L(n) = \frac{N(n)}{3^n}. \tag{6.120}$$

The number of times the ruler is laid down to measure a curve of fractal dimension D is given by $N(n) = \eta^{-D}$, so the length given by (6.120) is

$$L(n) = \left(\frac{1}{3^n} \right)^{1-D}. \tag{6.121}$$

Moreover, the curve becomes incrementally longer with each iteration and the lengths of the curve at successive generations are related by

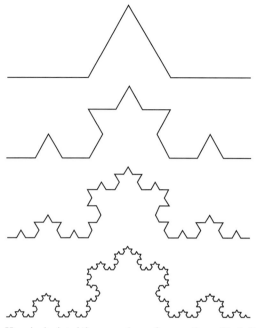

Figure 6.22. Here is depicted the procedure of generating a Koch line segment resulting from expanding the middle third of each line segment in going from iteration n to iteration $n + 1$. In the limit $n \to \infty$ the resulting line segment is one that is infinitely crinkled and has a fractal dimension $D = \ln 4 / \ln 3$.

$$L(n) = \frac{4}{3}L(n - 1).$$ (6.122)

Inserting (6.121) into both sides of (6.122) yields the relation

$$3^{n(D-1)} = \frac{4}{3}3^{(n-1)(D-1)},$$ (6.123)

which simplifies to the value for the fractal dimension

$$D = 2\frac{\ln 2}{\ln 3} \approx 1.26;$$ (6.124)

this is twice the dimension of the fractal dust found in an earlier chapter. The original line segment has been crinkled to such a degree that the distance along the curve between the end points becomes infinite. This divergence in length occurs because $1 - D < 0$, so that in the limit $n \to \infty$ the length of the line segment given by (6.121) becomes infinite, that is, $\lim_{\eta \to 0} L(\eta) \to \infty$. A more extensive discussion of the consequences of this scaling is given by West and Deering [37].

Unfortunately, the web in Figure 6.21 has little clustering [11]. The authors of [11] propose an alternative deterministic web yielding $\alpha \approx 1$ and large clustering. The web is realized according to the prescription illustrated in Figure 6.23. In this web the root (a university president) interacts with two vice presidents, one for science and one for the humanities. The two vice presidents, in turn, work with the help of two subordinate

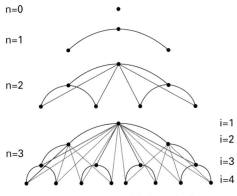

n=0

n=1

n=2

i=1
i=2
n=3
i=3
i=4

Figure 6.23. Construction of the deterministic scale-free network, showing the first four steps of the iterative process. Adapted from [11].

co-workers. We expect that the network becomes more efficient if the root interacts directly also with the subordinate co-workers of his/her two vice presidents. The same remarks apply to his/her vice presidents. They do not interact only with their two direct co-workers but also with the co-workers of the co-workers. The straight-line segments at the level $n = 2$ indicate that the root interacts also with the co-workers of his/her direct co-workers. The straight-line segments emanating from $i = 1$ to $i = 4$ at the level $n = 3$ show that the two vice presidents organize their interaction with their cascade of co-workers according to the rules of this university. They interact directly also with the co-workers of their co-workers. Note that the straight-line segments at the level $n = 3$ indicate that the president has a direct connection with all the nodes of the bottom layer.

The authors of this model prove numerically that

$$\alpha \approx 1. \tag{6.125}$$

They also prove analytically, but we omit the proof here, that the length of this web is

$$L = \frac{2^{2N+3} - (2N+4)2^{N+1}}{(2^{N+1} - 1)(2^{N+1} - 2)}. \tag{6.126}$$

Thus,

$$\lim_{N \to \infty} L(N) = \lim_{N \to \infty} 2\frac{2^{2N+2} - (N+2)2^{N+1}}{2^{2N+2}} = 2. \tag{6.127}$$

As a consequence, it is possible to move from any node i to any other node j in two steps or more. This is really a very small world and next we argue why this is the smallest possible world.

6.3.2 From small to ultra-small networks

Bollobás and Riordan [9] used rigorous mathematical arguments to derive the results of BA, the case of preferential attachment illustrated earlier, and reached the conclusion that the length of the BA network is

$$L = \frac{\ln N}{\ln(\ln N)}. \tag{6.128}$$

Using (6.78) we may conclude that the mean number of links is given by $z = \ln N$. The same value for the length of the preferential attachment network has more recently been derived by Chen *et al.* [12], who confirmed that (6.128) holds for any value of m (the rigorous demonstration of (6.128) was given by [9] only for $m = 1$).

In 2003 Cohen and Havlin [13] studied the problem of the relation between α and L, and found that for $\alpha = 3$, which is the boundary between the region where the mean-square number of connections is finite ($\alpha > 3$),

$$\langle k^2 \rangle < \infty,$$

and the region where the mean-square number of connections diverges ($\alpha < 3$),

$$\langle k^2 \rangle = \infty,$$

the web has the following length:

$$L \approx \frac{\ln N}{\ln(\ln N)}. \tag{6.129}$$

They also found that for

$$2 < \alpha < 3 \tag{6.130}$$

the web length is given by

$$L \approx \ln(\ln N). \tag{6.131}$$

For this reason these webs are considered to be ultra-small.

It is evident that the smallness of a web is important in order for it to carry out its function. This is especially important for the Internet. If the length of the web increased linearly with the addition of new sites, the excess information would make the Internet useless. The fact that the Internet is scale-free yields $L \approx \ln N$, thereby strongly reducing the size increase with increasing number of nodes N. In the special case $\alpha < 3$, the increase of L with N is further reduced.

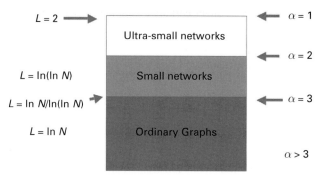

Figure 6.24. This figure shows the relation between the scale-free property of the web and the web size L.

The web proposed by the authors of [11] with $\alpha \approx 1$ is expected to generate the smallest size possible ($L = 2$). Figure 6.24 illustrates the relation between α and L in these various domains.

6.4 Reflections

In this section we considered whether there exists a connection between the topology of the web and its efficiency. The *small-world experiment* by Milgram [24, 32] suggested that the structure of a social network is such as to make it very efficient for the transport of information. In the experiment, Milgram sent several packages to 160 random people living in Omaha, Nebraska, asking them to forward the package to a friend or acquaintance whom they thought would bring the package closer to a particular final individual, a stockbroker in Boston, Massachusetts. The letter included this specific condition: "If you do not know the target person on a personal basis, do not try to contact him directly. Instead, mail this folder to a personal acquaintance who is more likely than you to know the target person." The outcome of the experiment revealed that, without any global network knowledge, letters reached the target recipient using, on average, 5.2 intermediate people, demonstrating that social acquaintance networks were indeed small worlds.

Here we point out that the scale-free structure may be hidden and that finding the shortest path connecting two generic sites, i and j, may require a special algorithm. After all, from [32] we learn that the participants were given, in addition to his name and address with the prescription of not sending the letter directly to the target (or of doing that only if the participant knew the target), his occupation and place of employment, his college and year of graduation, his military service dates, and his wife's maiden name and hometown. Then the participants were asked to make sure that each of the following recipients of the letter recorded their name. This was to prevent looping. If we activate random walkers of the same kind as those described earlier, the instruction of not visiting nodes that have already been explored, except for the purpose of avoiding localization, is probably easy to implement. The choice of the nodes, on the basis of their social condition, is, on the contrary, something much more difficult to assign to the random walker, being closely connected to what we define as human intelligence. Anyway, this experiment certainly established that a given social web is complex.

There are many open questions. What is the relation between the topological efficiency and the dynamical efficiency?

In the last chapter we discussed dynamical efficiency, with a special focus on the issue of social consensus. Here we limit ourselves to noticing that both random and complex webs have a small length, but the random webs have small clustering coefficients, whereas the real webs have large clustering coefficients. We think that this fact can be explained by our conjecture that the topology of a real web is created so as to facilitate global consensus. In fact, the large clustering corresponds to communities of strongly interacting individuals, who debate issues and eventually reach local consensus. The small length of the world has the ensuing effect of making this consensus spread throughout the globe. The small length of random networks does not serve any purpose, insofar as in this case there is no information on the decision-making issue to transmit from the local to the global level.

What is the role of the scale-free condition in this context? In the literature we have not found any discussion of this specific issue. In the absence of results of research work aiming at addressing this specific issue, we have to limit ourselves to making a further conjecture. We conjecture that the scale-free behavior favors the transmission of local consensus to the global level, turning the local consensus into a global consensus. This assumption seems to be supported by the diagram of Figure 6.24, which shows that the increase of N yields no significant increase of L and that for $\alpha < 2$ the world becomes impressively small.

However, this hypothesis is not unproblematic. We have seen, in fact, that the deterministic models are hierarchal. A hierarchal web is vulnerable, and it is plausible to accept the Singer view [31] that the brain is not hierarchal. Therefore, we have to refine our search for models that may ensure the fast attainment of global consensus through a non-hierarchal structure, without ruling out the possibility that the global consensus is essentially determined by the cooperative action of local communities and that the transformation of local into global consensus may in principle occur without triggering the scale-free distribution of nodal connections. It would be highly desirable, though, to find a scale-free model compatible with the laws of Zipf and Pareto, and with a high clustering coefficient that does not rest on any hierarchal prescription. There are indications in the study of Internet dynamics that this ideal condition may be possible [10].

6.5   Problems

6.1 Power laws

Discuss at least four empirical webs that are observed to have an inverse power-law structure in the context of the model presented in Section 6.1.1. Identify the empirical mechanisms that satisfy the properties assumed to be true in the derivation of the model.

6.2 The first-passage time

In Section 6.1.1 we introduced the Kolmogorov backward equation to calculate the first-passage-time distribution density for a diffusive process. Fill in the details of the analysis resulting in the biased Gaussian distribution.

6.3 Fractal curves

Redo the calculation depicted in Figure 6.22. However, instead of inserting two sides of length 1/3 in the middle third section of the line, replace the middle third section with three lines of length 1/3 that form a top hat. Iterate this process and note where the discussion differs from the one given in the text. What is the fractal dimension of the asymptotic curve?

References

[1] J. Abate and W. Whitt, "Asymptotics for M/G/1 low-priority waiting time tail probabilities," *Queuing Systems* **25**, 173 (1997)

[2] R. Albert and A.-L. Barabási, "Statistical mechanics of complex networks," *Rev. Mod. Phys.* **74**, 48 (2002).

[3] R. Albert and A.-L. Barabási, "Topology of evolving networks: local events and universality," *Phys. Rev. Lett.* **85**, 5234 (2000).

[4] R. Albert, H. Jeong and A.-L. Barabási, "Error and attack tolerance of complex networks," *Nature* **406**, 378–382 (2000).

[5] A. Arenas, A. Diaz-Guilera, J. Kurths, Y. Moreno and C. Zhou, "Synchronization in complex networks," *Phys. Rep.* **469**, 93 (2008).

[6] A.-L. Barabási and R. Albert, "Emergence of scaling in random walk networks," *Science* **286**, 509 (1999).

[7] A.-L. Barabási, E. Ravasz and T. Vicsek, "Deterministic scale-free networks," *Physica A* **299**, 559 (2001).

[8] A.-L. Barabási, "The origin of bursts and heavy tails in human dynamics," *Nature* **435**, 207 (2005).

[9] B. Bollobás and O. Riordan, "The diameter of a scale-free random graph," *Combinatorica* **24**, 5–34 (2004).

[10] A. Caldas, R. Schroeder, G. S. Mesch and W. H. Dutton, "Patterns of information search and access on the World Wide Web: democratizing expertise or creating new hierarchies?," *J. Computer-Mediated Commun.* **13**, 769–793 (2008).

[11] M. Chen, B. Yu, P. Xu and J. Chen, "A new deterministic complex network model with hierarchical structure," *Physica A* **385**, 707 (2007).

[12] F. Chen, Z. Chen, X. Wang and Z. Yuan, "The average path length of scale free networks," *Commun. Nonlinear Sci. Numerical Simulation* **13**, 1405–1410 (2008).

[13] R. Cohen and S. Havlin, "Scale-free networks are ultrasmall," *Phys. Rev. Lett.* **90**, 058701 (2003).

[14] A. de Moivre, *The Doctrine of Chances: A Method of Calculating the Probability of the Events of Play*, London (1718); 3rd edn. (1756); reprinted by Chelsea Press (1967).

[15] P. Erdös and A. Rényi, "Random graphs," *Publ. Math. Inst. Hungarian Acad. Sci.* **5**, 17 (1960).

[16] G. T. Fechner, *Elemente der Psychophysik*, Leipzig: Breitkopf und Härtel (1860).

[17] A. Gabrielli and G. Caldarelli, "Invasion percolation and critical transient in the Barabási model of human dynamics," *Phys. Rev. Lett.* **98**, 208701 (2007).

[18] L. K. Gallos, R. Cohen, P. Argyrakis, A. Bunde and S. Havlin, "Stability and topology of scale-free networks under attack and defense strategies," *Phys. Rev. Lett.* **94**, 188701 (2005).

[19] M. S. Granovetter, "The strength of weak ties," *Amer. J. Sociol.* **78**, 1360–1380 (1973).

[20] E. J. Gumbel, *Statistics of Extremes*, New York: Columbia University Press (1958).

[21] W. Hong, X.-P. Han, T. Zhou and B.-H. Wang, "Heavy-tailed statistics in short-message communication," *Chinese Phys. Lett.* **26**, 028902 (2009).

[22] V. Latora and M. Marchiori, "Efficient behavior of small-world networks," *Phys. Rev. Lett.* **87**, 198701 (1–4) (2001).

[23] R. K. Merton, "The Matthew effect in science," *Science* **159**, 56–63 (1968).

[24] S. Milgram, "The small world problem," *Psychol. Today* **1**, 60–67 (1967).

[25] J. C. Nacher and T. Akutsu, "Recent progress on the analysis of power-law features in complex cellular networks," *Cell. Biochem. Biophys.* **49**, 37–47 (2007).

[26] M. E. J. Newman, "The structure and function of complex networks," *SIAM Rev.* **45**, 47 (2002).

[27] J. C. Oliveira and A.-L. Barabási, "Darwin and Einstein correspondence pattern," *Nature* **437**, 1251 (2005).

[28] D. J. de Sola Price, "A general theory of bibliometric and other cumulative advantage processes," *J. Amer. Soc. Inform. Sci.* **27**, 292–306 (1976).

[29] P. Sheridan, Y. Yagahara and H. Shimodaira, "A preferential attachment model with Poisson growth for scale-free networks," *Ann. Inst. Statist. Math.* **60**, 747–761 (2008).

[30] H. A. Simon, "On a class of skew distribution functions," *Biometrika* **42**, 425–440 (1935).

[31] W. Singer, "The brain – an orchestra without a conductor," *Max Planck Res.* **18**, 3 (2005).

[32] J. Travers and S. Milgram, "An experimental study of the small world problem," *Sociometry* **32**, 425–443 (1969).

[33] A. Vásquez, "Exact results for the Barabási model of human dynamics," *Phys. Rev. Lett.* **95**, 248701 (2005).

[34] D. J. Watts, *Six Degrees*, New York: W. W. Norton (2003).

[35] D. J. Watts and S. H. Strogatz, "Collective dynamics of "small-world" networks," *Nature* **393**, 440–442 (1998).

[36] B. J. West, *Where Medicine Went Wrong*, Singapore: World Scientific (2006).

[37] B. J. West and W. Deering, *The Lure of Modern Science*, Singapore: World Scientific (1995).

[38] G. U. Yule, "A mathematical theory of evolution based on the conclusions of Dr. J. C. Willis," *Philos. Trans. R. Soc. London B* **213**, 21–87 (1925).

7 Dynamics of chance

In this chapter we explore web dynamics using a master equation in which the rate of change of probability is determined by the probability flux into and out of the state of interest. The master equation captures the interactions among large numbers of elements each with the same internal dynamics; in this case the internal dynamics consist of the switching of a node between two states. A two-state node may be viewed as the possible choices of an individual, say whether or not that person will vote for a particular candidate in an election. This is one of the simplest dynamical webs which has been shown mathematically to result in synchronization under certain well-defined conditions. We focus on the intermittent fluctuations emerging from a phase-transition process that achieves synchronized behavior for the strength of the interaction exceeding a critical value. This model provides a first step towards proving that these intermittent fluctuations, rather than being a nuisance, are important channels of information transmission allowing communication within and between different complex webs. The crucial power-law index μ discussed earlier and there inserted for mathematical convenience is here determined by the web dynamics. This observation on the inverse power-law index leads us to define the network efficiency in a form that might not coincide with earlier definitions proposed through the observation of the network topology.

Both the discrete and the continuous master equation are discussed for modeling web dynamics. One of the most important aspects of the analysis concerns the perturbation of one complex network by another and the transfer of information between complex clusters. A cluster is a network with a uniformity of opinion due to a phase transition to a given state. This modeling strategy is not new, but in fact dates back to Yule [71], who used the master-equation approach, before the approach had been introduced, to obtain an inverse power-law distribution. We investigate whether the cluster's opinion is robust or whether it can be easily changed by perturbing the way members of the cluster interact with one another.

A general theme of this chapter is to develop the point of view (already briefly introduced in Section 3.2.2) that linear response theory (LRT) suitably generalized can be considered a universal principle. Specifically, we generalize the LRT of non-equilibrium statistical physics to faithfully determine the average web response to a perturbation and we examine the evidence for such a generalization. The latter part of the chapter is devoted to proving this remarkable theorem. This would be the first such principle that is apparently independent of the specific type of complex web being considered. This principle is of more than theoretical interest and may well explain the fading behavior of complex webs in response to simple stimuli manifest in multiple phenomena, including perturbed liquid crystals, as we demonstrate.

The main purpose of Section 7.3 is to determine whether a universal LRT does in fact exist, even when the stationary condition does not. To support our arguments we adopt an idealization of reality and study the case when the regression to equilibrium, after a suitable initial preparation, has an infinitely long duration. We show that in this case it is also possible to make theoretical predictions on the basis of the LRT structure. If the non-equilibrium condition is not perennial, the non-stationary LRT has to be valid both in the initial out-of-equilibrium condition and in the final equilibrium regime. We shall see that it is possible to satisfy this condition, even if at the present time we are not aware of any theoretical treatment joining the initial out-of-equilibrium process to the final equilibrium.

Generalized LRT refers to a condition in which dynamics are determined by events, and the non-equilibrium condition corresponds, in fact, to making the number of events per unit time decrease with increasing time. Our arguments do not require a Hamiltonian formalism, which is a significant extension of the LRT property from physical to neurophysiological and sociological webs [68]. This theory, in the ordinary ergodic and stationary case, yields results that apparently coincide with the prediction of conventional LRT. We present applications of these ideas to the well-known phenomenon of habituation in the last part of the chapter. We developed a model to explain the psychophysical phenomena of habituation and dishabituation. Habituation is a ubiquitous and extremely simple form of learning through which animals, including humans, learn to disregard stimuli that are no longer novel, thereby allowing them to attend to new stimuli. As Breed [17] points out, one of the more interesting aspects of habituation is that it can occur at different levels of the nervous system. Cohen *et al.* [20] show that sensory networks stop sending signals to the brain in response to repeated stimuli; for example, such an effect occurs with strong odors. But odor habituation has been shown in rats to also take place within the brain by Deshmukh and Bhalla [25], not just at the sensor level. We hypothesize that $1/f$ noise, which is characteristic of complex networks, arising as it does both in single neurons [63] and in large collections of neurons [22, 31], is the common element that may explain both these observations by virtue of suppressing signals being transmitted to the brain and inhibiting signals being transferred within the brain.

7.1   The master equation

There is a variety of ways to introduce the dynamics of discrete probabilities, depending on the phenomenon being modeled. The chapter on random walks illustrates one such technique. Another favorite method of statistical physics is the *master equation*, which describes the temporal behavior of singlet probabilities, all multivariate one-time probabilities for *all* stochastic processes and the conditional probabilities for Markov processes [51]. A singlet probability is the probability that an event occurs at a given discrete location on a lattice at a given time; a multivariate one-time probability is the probability of multiple events occurring at various sites but at a specific time; a conditional probability depends on two lattice sites and two times, the occurrence of an event at the latter location and time being conditional on the occurrence of an event at the former location and earlier time. All these various technical definitions of probabilities were given full voice in Markov's theory at the turn of the last century, but, other than acknowledging the existence of this body of theory, we find that much of what we need to understand about the statistics of complex webs lies outside the domain of Markov theory. To clearly label this difference the complex webs discussed fall under the heading of non-Markov phenomena.

7.1.1   Discrete dynamics

The master equation was first written under that name as a differential-difference equation to describe gain–loss processes [50]. In its simplest representation the master equation can be written in matrix form, where $\mathbf{p} = (p_1, p_2, ..., p_N)$ is a probability vector and $\mathbf{K}$ is an $N \times N$ matrix of time-independent transfer coefficients,

$$\frac{d\mathbf{p}(t)}{dt} = \mathbf{K}\mathbf{p}(t), \tag{7.1}$$

which clearly has the formal solution

$$\mathbf{p}(t) = e^{\mathbf{K}t}\mathbf{p}(0). \tag{7.2}$$

The behavior of the probability density is therefore completely determined by the eigenvalue spectrum of the transition matrix. Suppose that $\mathbf{S}$ is the matrix that diagonalizes the transition matrix, yielding

$$\mathbf{SKS}^{-1} = \Lambda = \begin{pmatrix} \lambda_0 & 0 & 0 & 0 & \bullet \\ 0 & \lambda_1 & 0 & & \\ 0 & 0 & \bullet & 0 & \\ & & 0 & \bullet & \\ \bullet & & & & \lambda_{N-1} \end{pmatrix}, \tag{7.3}$$

where the eigenvalues are ordered $|\lambda_0| < |\lambda_1| < ... < |\lambda_{N-1}|$ for a finite lattice of size N. The general solution (7.2) can then be written as

$$\mathbf{p}(t) = \mathbf{C}_0 + \sum_{j=1}^{N-1} \mathbf{C}_j e^{\lambda_j t}, \tag{7.4}$$

where the $\mathbf{C}$-vectors are expressed in terms of the similarity-transform elements and the initial condition on the probability vector. In order for the asymptotic probability density to be a constant vector, that is, a steady-state solution to the master equation

$$\lim_{t \to \infty} \mathbf{p}(t) = \mathbf{C}_0, \tag{7.5}$$

the lowest eigenvalue must vanish, $\lambda_0 = 0$, and the real parts of the remaining eigenvalues must be negative. These are the constraints on the eigenvalues and consequently on the elements of the transition matrix that make the master equation an acceptable model of the phenomenon being investigated. Of course there are mathematical subtleties that confound this straightforward interpretation, such as the situation when more than one eigenvalue is zero, in which case the exponential relaxation of the web to an asymptotic steady state might not occur. Other, more complex, forms of relaxation may arise; see Oppenheim *et al.* [51] for a more rigorous mathematical discussion of solutions to the master equation and a presentation of the background literature.

For illustrative purposes, it is convenient to consider the case in which the unit time is $\Delta t = 1$. Thus, we write a discrete version of (7.1) in the form of an iterative equation

$$\mathbf{p}(n+1) = (\mathbf{I} + \mathbf{K})\mathbf{p}(n), \tag{7.6}$$

where $\mathbf{I}$ is the $N \times N$ unit matrix. Let us assume that at any time step n we toss a coin to select the state occupied by the network at time $n + 1$. It is evident that such a process reduces the dimensionality of the master equation to two, changing the generic vector to $\mathbf{p}(n) = (p_1(n), p_2(n))$ into the equilibrium condition $\mathbf{p}(n) = \left(\frac{1}{2}, \frac{1}{2}\right)$ since $p_1 + p_2 = 1$. Consequently, the coefficient of the preceding state can be written as

$$(\mathbf{I} + \mathbf{K}) = \begin{pmatrix} \frac{1}{2} & \frac{1}{2} \\ \frac{1}{2} & \frac{1}{2} \end{pmatrix}, \tag{7.7}$$

where $\mathbf{I}$ is the 2×2 unit matrix and the transition matrix becomes

$$\mathbf{K} = \begin{pmatrix} -\frac{1}{2} & \frac{1}{2} \\ \frac{1}{2} & -\frac{1}{2} \end{pmatrix}. \tag{7.8}$$

This form of the transition matrix is typical of the classical master equation.

7.1.2   A synchronized web of two-level nodes

Consider the master equation to describe the dynamics of the two-level node obtained from the equations with which we ended the last subsection:[1]

$$\frac{dp_1}{dt} = -g_{12}p_1 + g_{21}p_2,$$
$$\frac{dp_2}{dt} = -g_{21}p_1 + g_{12}p_2. \tag{7.9}$$

Here the elements of the 2×2 coupling matrix $\mathbf{K}$ are the gs, which in the simplest case of a coin flip are all 1/2. When the matrix elements are all equal $g_{21} = g_{12} = g$ and this equation can be derived from the subordination to the coin-tossing prescription discussed earlier, by using the subordination function

$$\psi(\tau) = re^{-r\tau}, \tag{7.10}$$

so that $g_{21} = g_{12} = g = r/2$. Now we generalize the model and embed this single two-state node into a web of such nodes as indicated schematically in Figure 7.1.

---

[1] Much of the discussion of this subsection is taken from [15].

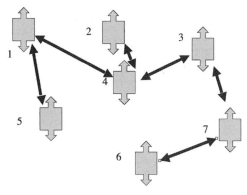

Figure 7.1. A set of interacting two-state nodes forms a small web. The two-headed arrows indicate the interactions among the two-state nodes.

Let us imagine that each node of the given network is a dynamical unit described by the master equation (7.9). To establish a dynamical connection among the nodes of this network imagine that a given unit is influenced by all the other units to which it is directly connected. In Figure 7.1 we illustrate this condition, where node 1 interacts directly with nodes 4 and 5 and indirectly with nodes 2 and 3. As a consequence of these interactions one possible choice for the coupling coefficients in (7.9) is to use the time-dependent transition probabilities per unit time given by [15, 64]

$$g_{ij}(t) = g \exp(K[\pi_j(t) - \pi_i(t)]). \tag{7.11}$$

Note that in the Ising model mentioned earlier, a model used extensively in condensed-matter physics, all the coupling coefficients are constants and we emphasize that this is not the case being considered here. The weighting factors in the interaction coefficient (7.11) are given by the relative number of sites in the indicated state, that is,

$$\pi_r = \frac{M_r}{M}, \tag{7.12}$$

where M_r is the number of sites of that specific subset that are in the state $r = 1, 2$ and M denotes the total number of nodes connected to the site which is being considered. For instance $M = 2$ in the case in which the site we are considering is site 1 of Figure 7.1. The parameter K is the control parameter in the exponential (7.11) generating the coupling factor

$$C_{ij} = \exp\left(K\left[\pi_j(t) - \pi_i(t)\right]\right). \tag{7.13}$$

To illustrate what we mean, let us assume that the site we are considering, namely site 1 in our example, is in the state $r = 1$. Assume that both sites linked to it are in the same state. In this situation the coupling factor is e^{-2K}. Thus, site 1 tends to remain in the state $r = 1$ for a more extended time than it would in the absence of any coupling. If both linked sites are in the state $r = 2$, the coupling factor is e^{2K}, and site 1 makes a transition from state $r = 1$ to state $r = 2$ earlier than it would in the absence of coupling. When site 4 is in the state $r = 1$ ($r = 2$) and site 5 is in the state $r = 2$ ($r = 1$), site 1 has to make a decision on its own, since in this case the coupling factor is one and does not exert any influence on the site.

The key point is that the master equation for each site is perfectly well defined, but it is a fluctuating master equation. In fact, the transition coefficients $g_{ij}(t)$ depend on the quantities π_r, which have random values depending on the stochastic time evolution of the environment at a specific site. As a matter of fact, we may define another ratio

$$\Pi_r = \frac{J_r}{J}, \tag{7.14}$$

where J denotes the total number of sites, namely $J = 7$ in the example of Figure 7.1. It is evident that the quantity Π_r is also an erratic function of time, even if it is expected to be smoother than π_r. In fact, Π_r is a global property, obtained from the observation of the whole network, whereas π_r is a property of the environment at a given site. The smaller the cluster, the more erratic the quantity π_r.

One of the web properties of interest is the cooperation among the nodes and the conditions under which we can expect *perfect consensus*. In the case of perfect consensus all the nodes are in either the state $r = 1$ or the state $r = 2$ and do not jump from one state to the other. The condition for *imperfect consensus*, by contrast, corresponds to the collection of nodes making random jumps from one state to the other. The condition for a lack of consensus corresponds to the coupling parameter being below some specified value $K < K_c$, where K_c is the critical value of the control parameter below which the flipping of the nodes between states is uncoordinated. In this case there are uncorrelated fluctuations around the vanishing mean value. For $K > K_c$, on the other hand, there are fluctuations around two distinct non-vanishing values and the web after a long sojourn in one of the two states makes an abrupt transition to the other state.

Imagine that a web topology necessary to produce perfect consensus exists, in which case the probability of being in a given state is

$$p_r(t) = \pi_r(t) = \Pi_r(t); \quad r = 1, 2. \tag{7.15}$$

With this assumption we also set the condition of there being no difference from site to site in the network. In a real case, however, $p_r(t)$ does change from site to site. Consequently, when the condition of perfect consensus is realized all the sites obey the same master equation (7.9) with the transition rates given by

$$g_{ij}(t) = g \exp[-K(p_i(t) - p_j(t))]. \tag{7.16}$$

On introducing the difference variable

$$\Pi(t) = p_1(t) - p_2(t) \tag{7.17}$$

the master equation simplifies to

$$\frac{d\Pi}{dt} = -(g_{12} + g_{21})\Pi - (g_{12} - g_{21}). \tag{7.18}$$

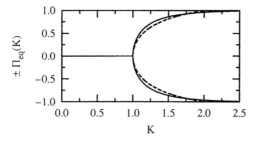

$\pm \Pi_{eq}(K)$ — vertical axis, values 1.0, 0.5, 0.0, −0.5, −1.0; horizontal axis K values 0.0, 0.5, 1.0, 1.5, 2.0, 2.5

Figure 7.2. The equilibrium positions as a function of the coupling parameter K, redrawn from [15] with permission.

By inserting the definition of the transition rates given by (7.16) into (7.18) we obtain the mean-field equation for the web of two-state nodes,

$$\frac{d\Pi}{dt} = -2g\Pi \cosh(K\Pi) - 2g \sinh(K\Pi) = -\frac{\partial U(\Pi)}{\partial \Pi}, \qquad (7.19)$$

which corresponds to the dynamics of a particle in a double-well potential introduced in Chapter 3.

Equation (7.19) describes the overdamped motion of a particle, whose "position" is Π within the symmetric potential $U(\Pi)$, and the values of its minima depend only on the coupling constant K. It is straightforward to show that there is a critical value of the coupling parameter given by $K = K_c = 1$, such that (1) if $K \leq K_c$ the potential has only one minimum located at $\Pi = 0$; or (2) if $K > K_c$ the potential is symmetric and has two minima located at $\pm \Pi_{min}$ separated by a barrier with the maximum centered at $\Pi = 0$. The potential has the same form as the quartic potential shown in Figure 3.4 and the height of the potential barrier is a monotonic function of the coupling parameter K.

The time evolution of the dynamical variable Π is determined by the extrema of the potential and consequently two kinds of dynamical evolution are possible: (1) if $K \leq K_c$, the dynamical variable $\Pi(t)$ will, after a brief transient, settle to an asymptotic value $\Pi(\infty) = 0$ independently of the initial condition $\Pi(0)$; (2) if $K > K_c$, the dynamical variable $\Pi(t)$ will, after a transient, reach the asymptotic value $\Pi(\infty) = \Pi_{min} \neq 0$ for an initial condition $\Pi(0) > 0$, $\Pi(\infty) = -\Pi_{min} \neq 0$ for an initial condition $\Pi(0) < 0$ and finally $\Pi(t) = 0$ for all time given an initial condition $\Pi(0) = 0$. In Figure 7.2 we compare the minima $\pm \Pi_{min}$ and the numerical evaluation of $\Pi(\infty)$ for various values of the coupling constant K.

The location of the potential minima as a function of the coupling parameter K is given in Figure 7.2, which shows that a phase transition occurs at $K = K_c = 1$. For a single two-state node the condition $\Pi(\infty) = \pm \Pi_{min} = p_1 - p_2 \neq 0$ corresponds to the statistical preference of the particle to be in either the state $r = 1$ or the state $r = 2$. This is a consequence of the transition rates being different if the coupling parameter is greater than the critical value. By inserting the mean-field value of the probabilities (7.15) into the expression for the transition rates (7.16) and allowing the dynamical variable to reach its asymptotic value we obtain

$$g_{12} = g e^{-K\Pi(\infty)} \neq g_{21} = g e^{K\Pi(\infty)}. \qquad (7.20)$$

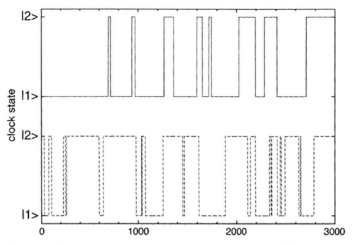

Figure 7.3. The typical time series of a single two-state node in a web of the ensemble. The solid line refers to a positive initial state and the dashed line to a negative initial state. The number of nodes in the single web of the ensemble is 10^5, the unperturbed transition rate is $g = 0.01$ and the coupling parameter is $K = 1.05$; redrawn from [15] with permission.

Figure 7.3 confirms the prediction of (7.20), showing that if $\Pi(\infty) = \Pi_{min}$ the single two-state node spends on average more time in the state $r = 1$ and if $\Pi(\infty) = -\Pi_{min}$ the single node spends on average more time in the state $r = 2$. The probability density functions for the sojourn times both in the preferred and in the non-preferred states are exponential functions with different mean sojourn times.

7.1.3 Stochastic approximation to a synchronized web

Let us now explore the collective behavior of a single web of N two-state nodes under the all-to-all coupling condition. In order to do this Bianco *et al.* [15] introduced a familiar collective phase variable. The global web variable is defined as

$$\xi(t) \equiv \frac{1}{N} \sum_{j=1}^{N} e^{i\Phi_j(t)} = \frac{N_1(t) - N_2(t)}{N};$$
(7.21)

Φ_j is the phase of the jth node and has the value of 0 if the node is in state $r = 1$ and π if the node is in the state $r = 2$, $N_1(t)$ is the number of nodes in the former state at time t and $N_2(t)$ is the number of nodes in the latter state at time t. In the decision-making model this is the difference between those voting for a candidate and those voting against. In the mean-field case, when $N \to \infty$, the single web becomes a statistical ensemble of identical nodes. In this case the master equation for the network is (7.9), where p_r is the probability that any node in the ensemble is in the state r. Consequently, from the definition of the web variable, we obtain in the mean-field limit

$$\xi(t) = p_1(t) - p_2(t) = \Pi(t).$$
(7.22)

Thus, for a web with infinitely many nodes, the temporal behavior of the global web variable $\xi(t)$ is identical to that of the dynamical variable $\Pi(t)$.

When the mean-field limit is not taken, there is a finite number of nodes and the dynamical picture stemming from the above master equation is changed. Bianco *et al.* [15] pointed out that in the finite-number case the master equation is that for the infinite-number situation except that the transition rates fluctuate,

$$g_{ij} \rightarrow g_{ij} + \varepsilon_{ij}, \tag{7.23}$$

where the fluctuations are on the order of $1/\sqrt{N}$ as we found for the random-walk process. If the number of nodes is very large, but still finite, we consider the mean-field approximation to be nearly valid and are able to replace the deterministic equation (7.19) with the stochastic equation

$$\frac{d\Pi(t)}{dt} = -\frac{\partial U(\Pi)}{\partial \Pi} - \eta(t)\Pi(t) + \varepsilon(t), \tag{7.24}$$

where the multiplicative fluctuation has the form

$$\eta(t) = \varepsilon_{12}(t) + \varepsilon_{21}(t) \tag{7.25}$$

and the additive fluctuation is given by

$$\varepsilon(t) = \varepsilon_{12}(t) - \varepsilon_{21}(t). \tag{7.26}$$

The random fluctuations induce transitions between the two states of the potential well. Thus, for a web with a finite number of nodes the phase synchronization of (7.19) is not stable.

A simpler stochastic equation is obtained if the fluctuations are anti-symmetric; that is, if $\varepsilon_{12}(t) = -\varepsilon_{21}(t)$. In this simpler case the multiplicative coefficient vanishes, $\eta(t) = 0$, and (7.24) reduces to the Langevin equation with additive random fluctuations,

$$\frac{d\xi(t)}{dt} = -\frac{\partial U(\xi)}{\partial \xi} + \varepsilon(t), \tag{7.27}$$

using (7.22) for the web variable. Equation (7.27) represents the Smoluchosky approximation to the physical double-well potential discussed in Chapter 3 with the addition of a random force. The fluctuations drive the particle from one well of the potential to the other by giving it sufficient amplitude to traverse the barrier between the wells. However, here the fluctuations arise from the finite number of nodes in the web rather than from the thermal behavior of a heat bath.

The global variable fluctuates between the two minima as described by (7.27) for values of the coupling parameter greater than the critical value as depicted in Figure 7.4. The single node follows the fluctuations of the global variable, switching back and forth from the condition where the state $r = 1$ is preferred statistically to that where the state $r = 2$ is preferred statistically. In physics this is similar to what occurs in the Ising model of magnetization, where each element has either up or down spin and spins are allowed to interact with one another. For constant-strength interactions the material undergoes a phase transition, going from a state of no magnetization with each spin fluctuating

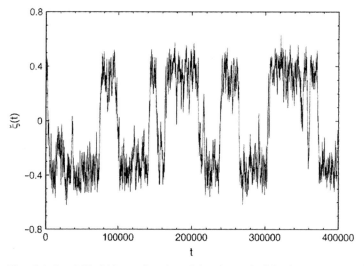

Figure 7.4.   The global variable $\xi(t)$ as a function of time is graphed for the parameter values $K = 1.05$ and $g = 0.01$ for a web with 10^5 nodes; redrawn from [15] with permission.

between its two states to a condition in which all the spins are aligned and the material is magnetized. In the social domain the phase transition is the process of reaching consensus or agreement.

The probability density functions of the sojourn times in the states $\xi > 0$ and $\xi < 0$ as shown in Figure 7.4 are identical since the potential is symmetric. Thus, we denote them by $\psi(\tau)$. In the simplest case, where the coupling parameter is on the border of criticality, $K = K_c = 1$, and the barrier between potential wells vanishes, the probability density for the web variable at time t starting initially at the origin is given by

$$p(\xi, t) = \frac{1}{\sqrt{4\pi Dt}} e^{-\frac{\xi^2}{4Dt}}, \qquad (7.28)$$

using a random-walk argument. The probability density of crossing the origin at time t is therefore given by

$$p(0, t) = \frac{1}{\sqrt{4\pi Dt}}. \qquad (7.29)$$

We observe that the population at the origin is composed of particles that left the origin and went back to it an arbitrarily large number of times. As a consequence, after summing over all the possible number of zero-crossings we obtain

$$R(t) = \sum_{n=0}^{\infty} \psi_n(t). \qquad (7.30)$$

The function $R(t)$ has been calculated and found to be proportional to $1/t^{2-\mu}$ for renewal processes. On comparing this result with (7.29) we find that the power-law index is given by

$$\mu = 1.5. \qquad (7.31)$$

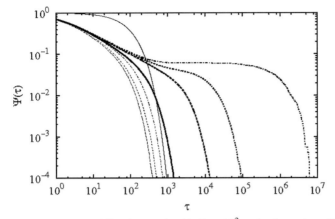

$\Psi(\tau)$ versus τ

Figure 7.5. The survival probability for a web with $N = 10^3$ nodes for various K values.

Consequently, the probability density for the length of time the web remains in a given global state is an inverse power law of the form $\psi(\tau) \propto \tau^{-3/2}$ for an extended interval of sojourn times.

The survival probability corresponding to the above argument is one having the inverse power-law form $\Psi(\tau) \propto \tau^{-1/2}$ for any fixed unperturbed value of the coupling parameter g. In Figure 7.5 we see that as the number of nodes in the web increases the time over which the survival probability has the predicted inverse power-law form increases. However, in each case the distribution is eventually truncated by an exponential and the dynamics are determined by solving the Langevin equation (7.27). This behavior is predicted by Kramers' theory, the predicted shape being an inverse power law, followed by a shoulder ending in an exponential.

7.1.4 The Yule process

In our earlier presentation of the BA model we pointed out that the preferential-attachment mechanism central to their discussion had been identified previously under a variety of names by a number of investigators, the first of whom was the mathematician Yule. As a matter of fact the BA model is a limiting case of the more general argument presented by Yule in 1925 [71] to explain the hyperbolic relation discovered by Willis in his experimental investigations of speciation of flowering plants published with Yule in 1922 [70]. The Yule process is reviewed by Newman [49] using a master equation, a technique that had not been developed at the time of Yule's analysis. Although Yule and subsequently Newman concluded the argument in terms of new species being added to a genus or family, we follow Simon [58] and make the presentation in terms of words in order to avoid confusion over biological jargon.

Simon presented the stochastic model in terms of word frequency. Consider the writing of a book that has reached n words in length; the double-indexed quantity $p_{k,n}$ is the fraction of *different* words that have occurred exactly k times in these first n words. The stochastic process was defined by the following two assumptions.

I. The probability that the $(n + 1)$st word is a word that has already appeared exactly k times is proportional to $np_{k,n}$.

II. There is a constant probability κ that the $(n + 1)$st word is a new word.

The stochastic process describing the probability that a particular word will be the next one written depends on what words have already been written. The master equation for the new number $(n + 1)p_{k,n+1}$ of words occurring exactly k times is given by employing the two assumptions:

$$(n + 1)p_{k,n+1} = np_{k,n} + \theta[(k - 1)p_{k-1,n} - kp_{k,n}], \tag{7.32}$$

where $\theta = 1 - \kappa$ is the probability that the $(n + 1)$st word is not a new word. The form of this master equation is given by Newman using a somewhat longer argument involving speciation and is a version of the equation given by Simon. As both Newman and Simon point out, the only exception to (7.32) is for $k = 1$, which obeys the equation

$$(n + 1)p_{1,n+1} = np_{1,n} + 1 - \theta p_{1,n}. \tag{7.33}$$

Note that the number of words in this formulation corresponds to the discrete time index, so that in this master equation, as observed by Newman, the analogue of the passage of time is the incidence of new events, which in (7.32) is the writing of words.

We can now determine the asymptotic form of the solution to the master equations for the number of words being generated in this mythical book. The asymptotic or number-independent form of the probability is given by taking the limit

$$p_k = \lim_{n \to \infty} p_{k,n}, \tag{7.34}$$

so that taking this limit in (7.33) yields

$$p_1 = 1 - \theta p_1 \Rightarrow p_1 = \frac{1}{1 + \theta}. \tag{7.35}$$

Applying the same limit to (7.32) yields

$$p_k = \theta[(k - 1)p_{k-1} - kp_k],$$

which can be rearranged to provide the iteration equation

$$p_k = \frac{k - 1}{k + 1/\theta} p_{k-1}. \tag{7.36}$$

The solution to the iteration equation is provided in both Newman [49] and Simon [58] by using (7.35) as the initial value to be

$$p_k = \frac{1}{\theta} \frac{\Gamma(k)\Gamma(1 + 1/\theta)}{\Gamma(k + 1 + 1/\theta)} = \frac{1}{\theta} B(k, 1 + 1/\theta). \tag{7.37}$$

Here the ratio of gamma functions is denoted by $B(a, b)$, the beta function, which Simon called the Yule distribution, since this is the distribution obtained by Yule in his mathematical argument for the distribution of new species [71]. As is well known in classical analysis, the beta function has a power-law tail $B(a, b) \sim a^{-b}$, from which we can immediately conclude that the probability p_k has an asymptotic power-law distribution with exponent

$$\mu = 1 + 1/\theta, \tag{7.38}$$

thereby restricting the domain of the power law to $\mu \geq 1$. The empirical power-law index obtained by Willis and Yule [70] for all flowering plants is $\mu = 2.5 \pm 0.1$, implying that the constant probability of a new species being generated is $\kappa = 1/3$.

It is worth pointing out that the inverse power-law index obtained here is not restricted to the value $\mu = 3$ as it was in the BA model. It would be a useful exercise for the student to reinterpret the above argument in terms of the graph-theory parameters introduced in our review of the BA model. Newman [49] also generalized the above argument to networks composed of a collection of objects, such as genera, cities, papers published, citations, web pages and so forth. He maintains that the Yule process may well explain the reputed power laws of all these phenomena.

7.2 General modeling strategy

The solution to the master equation has been written in general and the eigenvalue spectrum determines the dynamics of the probability in the state space for the network of interest. The question of how the web's dynamics responds to external perturbations now arises. This is an issue of fundamental importance that has been widely studied in the field of ordinary statistical physics, particularly in the non-Markov case of the generalized master equation (GME). More specifically, the neural web of the human brain is continually responding to external stimulation, as do the physiologic networks of the human body to a changing environment and the economic markets to world news. The theory of complex webs, concerning as it does non-ergodic networks, has forced us to address the same issue from a more general perspective that should reduce to the traditional prescriptions when we recover the ordinary equilibrium condition.

7.2.1 From crucial events to the GME

Let us now make the assumption that the events do not occur at regular times, but instead the interval between two consecutive events is determined by the hyperbolic distribution density $\psi(t)$ given by (3.239). It is important to notice that, in the case in which the time fluctuations are characterized by a well-defined time scale, the departure from the condition of event occurrence at regular times will not produce significant effects, resulting in only a change of time scale. It is evident that interesting effects emerge when the events are crucial, and especially when $\mu < 2$. The reasons for this will become clear subsequently. For the time being let us limit ourselves to noticing that a set of interacting units located on the nodes of a complex network generates global dynamics characterized by abrupt changes corresponding to those crucial events, which are incompatible with ergodic theory, namely the crucial events occur with $\mu < 2$.

We follow the CTRW prescription discussed in a previous chapter and make the assumption that there exists a stochastic connection between the discrete time n and the physical (continuous) time t. In the natural time scale, namely when we set the condition that the elementary time step $\Delta t = 1$, the classical master equation has the discrete form given by (7.6). However, now unlike the argument preceding (7.6), where the statistics are determined by the toss of a coin, we want to incorporate the waiting-time distribution into each occurrence of the n events. We assume that the time interval $\tau(n) = t(n+1) - t(n)$ is derived from a waiting-time distribution density (3.239). The state $\mathbf{p}(n)$ lasts from $t(n)$ to $t(n+1)$. The event occurring at time $t(n)$ is an abrupt change from the state $\mathbf{p}(n-1)$ to the state $\mathbf{p}(n)$. At a generic time t we can write the probability as

$$\mathbf{p}(t) = \sum_{n=0}^{\infty} \int_0^t dt' \, \psi_n(t') \Psi(t - t')(\mathbf{I} + \mathbf{K})^n \mathbf{p}(0). \tag{7.39}$$

Note that $\psi_n(t)dt$ is the probability that n events have occurred, the last one occurring at time t. The function $\Psi(t)$ denotes the probability that no event occurs up to time t and is given by (3.218). The occurrence of an event corresponds to activating the matrix $(\mathbf{I} + \mathbf{K})$, so that activating n events transforms the initial condition $\mathbf{p}(0)$ into $(\mathbf{I} + \mathbf{K})^n \mathbf{p}(0)$. This form in (7.39) is kept from t', at which time the last event occurs, up to time t, the time interval from t' to t being characterized by no event occurring. Of course, the expression (7.39) takes into account that the number of possible events may range from the no-event case to the case of infinitely many events. For this mathematical idealization to be as realistic as possible, we have to assume that the waiting time τ may be arbitrarily small, as small as we wish. This is not quite realistic, insofar as there may be a shortest truncation time. In real networks there exists also a largest truncation time. We assume that the latter is so extended that it is possible to accommodate a very large (virtually infinite) number of events.

Note that the renewal nature of the events driving the dynamics of the network of interest S makes it possible to establish a hierarchy coupling the successive events:

$$\psi_n(t) = \int_0^t dt' \, \psi_{n-1}(t') \psi_1(t - t'). \tag{7.40}$$

In fact, if at time t the nth event occurs, the $(n-1)$th event must have occurred at an earlier time $0 < t' < t$. We make the assumption that at $t = 0$ an event certainly occurs, namely

$$\psi_0(t) = \delta(t). \tag{7.41}$$

The waiting-time distribution density $\psi(t)$ is identified with the probability of the first event's occurrence at a time $t > 0$,

$$\psi_1(t) = \psi(t), \tag{7.42}$$

and consequently the convolution form of (7.40) reduces to the product of Laplace transforms

$$\widehat{\psi}_n(u) = \widehat{\psi}_{n-1}(u)\widehat{\psi}_1(u),$$

so that substituting from (7.42) and carrying out the iteration yields (3.193)

$$\widehat{\psi}_n(u) = \left[\widehat{\psi}(u)\right]^n. \tag{7.43}$$

The Laplace transform of the survival probability $\Psi(t)$ is

$$\widehat{\Psi}(u) = \frac{1 - \widehat{\psi}(u)}{u}, \tag{7.44}$$

so that inserting (7.43) and (7.44) into the Laplace transform of (7.39) and again using the convolution theorem yields

$$\widehat{\mathbf{p}}(u) = \frac{1 - \widehat{\psi}(u)}{u} \sum_{n=0}^{\infty} \left[\widehat{\psi}(u)(\mathbf{I} + \mathbf{K})\right]^n \mathbf{p}(0). \tag{7.45}$$

The eigenvalues of the matrix $(\mathbf{I} + \mathbf{K})$ are ≤ 1 and $\widehat{\psi}(u) < 1$, having the value $\widehat{\psi}(u) = 1$ only in the limiting case $u = 0$. Thus, we evaluate the geometric sum appearing in (7.45) according to the well-known rule

$$\sum_{n=0}^{\infty} \left[\widehat{\psi}(u)(\mathbf{I} + \mathbf{K})\right]^n = \frac{1}{1 - \widehat{\psi}(u)(\mathbf{I} + \mathbf{K})}. \tag{7.46}$$

Consequently, inserting this form of the series into (7.45) gives us for the Laplace transform of the probability density

$$\widehat{\mathbf{p}}(u) = \frac{1 - \widehat{\psi}(u)}{u - u\widehat{\psi}(u)(\mathbf{I} + \mathbf{K})}\mathbf{p}(0) = \frac{1 - \widehat{\psi}(u)}{u(1 - \widehat{\psi}(u)) + u\widehat{\psi}(u) - u\widehat{\psi}(u)(\mathbf{I} + \mathbf{K})}\mathbf{p}(0)$$

and, using the form of the unit matrix, this reduces to

$$\widehat{\mathbf{p}}(u) = \frac{1 - \widehat{\psi}(u)}{u(1 - \widehat{\psi}(u)) - u\widehat{\psi}(u)\mathbf{K}}\mathbf{p}(0). \tag{7.47}$$

By dividing both the numerator and the denominator of (7.47) by $(1 - \widehat{\psi}(u))$ we obtain

$$\widehat{\mathbf{p}}(u) = \frac{1}{u - \widehat{\Phi}(u)\mathbf{K}}\mathbf{p}(0), \tag{7.48}$$

where

$$\widehat{\Phi}(u) = \frac{u\widehat{\psi}(u)}{1 - \widehat{\psi}(u)}, \tag{7.49}$$

namely the typical memory kernel that emerged in CTRW theory discussed previously. With a little algebra (7.48) can be rewritten as

$$u\widehat{\mathbf{p}}(u) - \mathbf{p}(0) = \widehat{\Phi}(u)\mathbf{K}\widehat{\mathbf{p}}(u), \tag{7.50}$$

the lhs of which is the Laplace transform of the time derivative of the probability and the rhs of which is the convolution of the product of the memory kernel and the probability density. Consequently, taking the inverse Laplace transform of (7.50) yields the generalized master equation (GME)

$$\frac{d\mathbf{p}(t)}{dt} = \int_0^t dt' \, \Phi(t - t') \mathbf{K} \mathbf{p}(t'). \tag{7.51}$$

Note that the GME is non-Markov except when the memory kernel is a delta function in time. The derivation of the GME rests on the hypothesis that the dynamics of complex webs are determined by the action of events, and that it is necessary to assume that these events are crucial to account for the deviation from ordinary statistical physics revealed by real experiments; see for example [5, 57].

7.2.2   Relaxation of the two-level GME

Many continuous phenomena are well modeled by two states; for example we have the on and off states of blinking quantum dots, a form of physical network; in biology there is the respiratory network with the inhalation and exhalation of the lungs; in information theory there is the connectivity of the Internet; in sociology there is the two-state decision-making process, say voting for a candidate; we can use two-state models to capture the dynamics of any web characterized by deciding between two alternatives. In the situation where $N = 2$ we can reduce the GME given by (7.51) to two coupled integro-differential equations,

$$\frac{dp_1(t)}{dt} = -\frac{1}{2} \int_0^t dt' \, \Phi(t - t')[p_1(t') - p_2(t')],$$

$$\frac{dp_2(t)}{dt} = +\frac{1}{2} \int_0^t dt' \, \Phi(t - t')[p_1(t') - p_2(t')], \tag{7.52}$$

where we have used the form of the transition matrix given by (7.8). On taking the difference between these two equations and again introducing the difference variable the two-state GME reduces to the compact form

$$\frac{d\Pi(t)}{dt} = -\int_0^t dt' \, \Phi(t')\Pi(t - t'). \tag{7.53}$$

An interesting general relation results from the solution to the equation for the difference variable. The Laplace transform of (7.53) yields

$$u\widehat{\Pi}(u) - \Pi(0) = -\widehat{\Phi}(u)\widehat{\Pi}(u),$$

which, after some rearrangement, gives

$$\widehat{\Pi}(u) = \frac{\Pi(0)}{u + \widehat{\Phi}(u)}.$$

However, replacing the memory kernel in this equation using (7.49) results in

$$\widehat{\Pi}(u) = \frac{1 - \widehat{\psi}(u)}{u}\Pi(0) = \widehat{\Psi}(u)\Pi(0), \tag{7.54}$$

where we have also used the relation between the Laplace transform of the survival probability and the waiting-time distribution density. Taking the inverse Laplace transform of (7.54) yields the remarkable result

$$\frac{\Pi(t)}{\Pi(0)} = \Psi(t). \tag{7.55}$$

A network that is initially prepared to be out of equilibrium decays back to the equilibrium condition in a manner determined by the survival probability. But why is this result of interest?

Let us go back to the equation for the elements of the transition matrix (7.7). According to this prescription the web reaches equilibrium in one step in the natural time representation. Thus, it is reasonable that in the continuous-time representation the regression to equilibrium of $\Pi(t)$ must coincide with the probability that in this time scale no event occurs up to time t. The waiting-time distribution $\psi(\tau)$ serves the purpose of creating in the continuous time scale t a process subordinated to that occurring in the natural time scale n. Therefore, we call the waiting-time distribution the *subordination function*. When we adopt the waiting-time distribution density of the hyperbolic form given by (3.239) as a subordination function we obtain the regression to equilibrium to be

$$\frac{\Pi(t)}{\Pi(0)} = \Psi(t) = \left(\frac{T}{T+t}\right)^{\mu-1}, \tag{7.56}$$

which is much slower than the typical exponential regression that is assumed in much of the dynamical modeling of physical networks.

Nearly a century ago Lars Onsager proved that physical systems that are out of equilibrium relax back to their equilibrium state by means of a variety of transport mechanisms. Reichl [55] explains that one assumption necessary to prove Onsager's relations is that fluctuations about the equilibrium state in a physical network, on average, decay according to the same laws that govern the decay of macroscopic deviations from equilibrium. Typically this relaxation is exponential with a macroscopic rate of relaxation. Consequently, it is the non-exponential relaxation back to the equilibrium state given by (7.56) that is so interesting.

What this form of relaxation means in the present context is that one ought not to expect complex webs to relax exponentially in time from excited configurations back to their equilibrium condition. The complexity of the network slows the process from an exponential to an inverse power-law form when that process is determined by crucial events.

7.2.3 Perturbing the GME

The time-convoluted form of the GME (7.51) is by now familiar. It should be borne in mind that this is the form of the master equation that arises when the network's dynamics are driven by events, including crucial events, of course, when the waiting-time distribution $\psi(\tau)$ is not exponential. Recall that memory depends on the fact that the bath dynamics have a long time scale, as we saw in the discussion of booster dynamics. On the other hand, for very fast dynamics, the memory kernel undergoes a rapid regression to equilibrium and we recover the typical memoryless structure of the classical master equation. To see this more clearly, let us suppose that the waiting-time distribution density is exponential,

$$\psi(t) = ge^{-gt}, \tag{7.57}$$

with the Laplace transform

$$\widehat{\psi}(u) = \frac{g}{g+u}. \tag{7.58}$$

On inserting the Laplace transform for the waiting-time density (7.57) into the Laplace transform for the memory kernel (7.49) we obtain

$$\widehat{\Phi}(u) = g,$$

which implies, using the inverse Laplace transform, that the memory kernel is a delta function in time,

$$\Phi(t) = g\delta(t). \tag{7.59}$$

The delta-function memory kernel restricts the GME (7.51) to the simpler form

$$\frac{d\mathbf{p}(t)}{dt} = g\mathbf{K}\mathbf{p}(t). \tag{7.60}$$

This master equation has the same Markov structure as does the original, only the transition elements are scaled by the constant factor g.

So now we know that the subordination function is not exponential and we address the question of how to describe the effect of a perturbation on the complex network. It is reasonable to assume that the perturbation affects the parameters of the transition matrix since these elements model the interactions between the states of the network. Sokolov [59] has recently proposed that in the general case the GME should read

$$\frac{d\mathbf{p}(t)}{dt} = \int_0^t dt' \, \Phi(t - t')\mathbf{K}(t)\mathbf{p}(t') \tag{7.61}$$

and this somewhat arbitrary generalization of the GME leads to some interesting results. Here the perturbation is assumed to occur after the last event, that is, at time t and no earlier. In the two-state case we write

$$\mathbf{K}(t) = \begin{pmatrix} -K_+(t) & +K_-(t) \\ +K_+(t) & -K_-(t) \end{pmatrix} \tag{7.62}$$

and weakly perturb the matrix elements

$$K_\pm(t) = \frac{1}{2}[1 \mp \epsilon\xi_p(t)]. \tag{7.63}$$

Going back to the difference variable and inserting the perturbed matrix elements into (7.61) gives the perturbed GME

$$\frac{d\Pi(t)}{dt} = \varepsilon \xi_\mathrm{p}(t) R(t) - \int_0^t dt'\, \Phi(t - t')\Pi(t'),\qquad(7.64)$$

where we see that the perturbation appears as a coefficient of the previously introduced event-generation function

$$R(t) = \int_0^t dt'\, \Phi(t') = \sum_{n=1}^{\infty} \psi_n(t).\qquad(7.65)$$

The inhomogeneous term in (7.64) gives a time-dependent coefficient to the perturbation strength that depends on the number of events that occur in the time interval $(0, t)$.

It is useful to note that the Poisson condition (7.58) inserted into the last term in (7.65) generates

$$R(t) = g.\qquad(7.66)$$

Thus, in the ordinary Poisson case the number of events generated per unit time is constant, in spite of adopting the preparation prescription at time $t = 0$, which is a form of the out-of-equilibrium condition. That is not the case when $2 < \mu < 3$, namely the case in which the network S generates crucial events. Here we see that the condition $\mu > 2$ is less dramatic than the condition $\mu < 2$. In fact, when $\mu > 2$, the network tends towards the Poisson condition

$$R(\infty) = \frac{1}{\langle \tau \rangle},\qquad(7.67)$$

but it takes an infinitely long time to do so. Note that the constant in (7.67) can be associated with the constant in (7.66) and consequently the statistics become Poisson asymptotically. It is reasonable to imagine that the response to perturbation of the network during this infinitely extended regression to equilibrium departs from the ordinary linear response prescription.

The condition $\mu < 2$ implies an even stronger departure from the conventional condition. In fact, in this case the network lives in a perennial out-of-equilibrium condition. In the case $\mu > 2$ it is possible to prepare the network in a stationary condition. In the case $\mu < 2$ the stationary condition is not possible.

Let us again consider the memoryless structure of the classical master equation by assuming that the memory kernel is given by the delta function (7.59) so that the perturbed GME (7.64) reduces to

$$\frac{d\Pi(t)}{dt} = \varepsilon g \xi_\mathrm{p}(t) - g\Pi(t),\qquad(7.68)$$

which has the exact solution

$$\Pi(t) = e^{-gt}\Pi(0) + \varepsilon g \int_0^t dt'\, e^{-g(t-t')}\xi_\mathrm{p}(t').\qquad(7.69)$$

We assume without loss of generality that the initial value vanishes, $\Pi(0) = 0$, and define the variable $\xi_s(t)$ that has the value 1 when the web is in the "on" state and the value -1 when the web is in the "off" state to replace $\Pi(t)$. Thus, the solution (7.69), which is the average response to the perturbation, can be written

$$\langle \xi_s(t) \rangle = \varepsilon \int_0^t dt' \, \chi(t, t') \xi_p(t'). \tag{7.70}$$

In this case the response function is defined by the exponential

$$\chi(t, t') = g e^{-g(t-t')}. \tag{7.71}$$

We note that the autocorrelation function of the fluctuation response has the form

$$\langle \xi_s(t) \xi_s(t') \rangle = \Psi_s(t, t') = \Psi_s(t - t') = e^{-g(t-t')}, \tag{7.72}$$

which is a direct consequence of the Poisson statistics. On comparing (7.72) with (7.71) it is clear that the response function is the time derivative of the autocorrelation function

$$\chi(t, t') = \frac{d \Psi_s(t - t')}{dt'}. \tag{7.73}$$

The structure of (7.70) is the general form of the response to perturbations derived by Kubo [48] when the response function is defined by (7.73). A stochastic process $\xi_s(t)$, which in the absence of perturbation would vanish, in the presence of a stimulus with the time dependence $\xi_p(t)$ generates a non-vanishing average value $\langle \xi_s(t) \rangle$. The kernel $\chi(t, t')$ is the time derivative of the unperturbed autocorrelation function. This prescription from statistical physics holds in general and it is always valid, provided that the network is at or near equilibrium compatible with the existence of stationarity for the autocorrelation function $\Psi_s(t)$.

7.2.4   Towards a new LRT

In this section we present a solution to the perturbed GME using an ansatz, that is, we assume a form of the solution and then verify that it satisfies the GME (7.64). Note that this is the strategy often used to solve differential equations; we assume a general form for the solution with parameters that are eventually adjusted to satisfy certain specified conditions. Consider an equation for the assumed response of the two-level GME to a time-dependent perturbation $\xi_p(t)$:

$$\Pi(t) = \varepsilon \int_0^t dt' \, R(t') \Psi(t - t') \xi_p(t'). \tag{7.74}$$

The time derivative of (7.74) is given by

$$\frac{d \Pi(t)}{dt} = \varepsilon R(t) \Psi(0) \xi_p(t) + \varepsilon \int_0^t dt' \, R(t') \frac{d \Psi(t - t')}{dt} \xi_p(t'), \tag{7.75}$$

but $\Psi(0) = 1$ by definition for the survival probability. The waiting-time distribution density is the negative time derivative of the survival probability so that (7.75) simplifies to

$$\frac{d\Pi(t)}{dt} = \varepsilon R(t)\xi_p(t) - \varepsilon \int_0^t dt' \, R(t')\psi(t-t')\xi_p(t'). \tag{7.76}$$

Equation (7.76) ought to be exactly the same as the perturbed GME (7.64) if, in fact, (7.74) is a solution to the GME. Assume (7.74) to be true and equate the integral terms of the two equations to obtain

$$\varepsilon \int_0^t dt' \, R(t')\psi(t-t')\xi_p(t') = -\int_0^t dt' \, \Phi(t-t')\Pi(t'), \tag{7.77}$$

the validity of which must be established. Inserting the assumed form for the solution $\Pi(t)$ into (7.77) yields the rather awkward-looking expression

$$\varepsilon \int_0^t dt' \, R(t')\psi(t-t')\xi_p(t') = \varepsilon \int_0^t dt' \, \Phi(t-t') \int_0^{t'} dt'' \, R(t'')\Psi(t'-t'')\xi_p(t''). \tag{7.78}$$

The Laplace transform of (7.78) together with judicious use of the convolution theorem leads to the expression

$$\varepsilon \widehat{\psi}(u) \mathcal{LT}\left[R(t)\xi_p(t); u\right] = \varepsilon \widehat{\Phi}(u) \frac{1 - \widehat{\psi}(u)}{u} \mathcal{LT}\left[R(t)\xi_p(t); u\right] \tag{7.79}$$

and cancelling out common terms on the right- and left-hand sides of this equation leads to

$$\frac{u\widehat{\psi}(u)}{1 - \widehat{\psi}(u)} = \widehat{\Phi}(u). \tag{7.80}$$

Note that (7.80) is the expression we found earlier for the Laplace transform of the memory kernel in terms of the waiting-time distribution density. Consequently we have established the veracity of (7.74). This indirect method of establishing that (7.74) is the solution to the perturbed GME is actually much easier than solving the original equation directly.

7.2.5 Stochastic resonance and conventional LRT

Stochastic resonance is a phenomenon that has attracted the attention of many investigators, too many to list here, since its introduction in the early 1980s [14, 30] as a possible explanation of the 100,000-year cycle in the climate record [12, 13, 47]. The intent was to develop a simple model of the alterations of the Earth's climate between ice ages and periods of warm weather due to the eccentricity of the Earth's orbit. It is currently believed by the climatology community that this direct effect is not strong enough to determine such a large change in climate. Benzi *et al.* [12, 13] coincidently with Nicolis and Nicolis [47] conjectured that there may exist a cooperative mechanism between the weak change of eccentricity in the Earth's orbit and the natural random fluctuations in the average global temperature that might account for the periodic and strong climate changes. The theory developed to support this conjecture turned out to be very successful and as of August 2009 the original papers have over a thousand citations.

However, the popularity of stochastic resonance is due not to its intended explanation of climate change, but instead to its neurophysiologic applications [66]. The pioneering work of Neiman *et al.* [48] established that an external noise can enhance neuron synchronization and consequently improve the transmission of information when that information is in the form of a harmonic signal [52].

Stochastic resonance has to do with the propagation of a signal in a stochastic environment with the surprising result that the signal-to-noise ratio increases with increasing noise intensity, rather than decreasing as one intuitively expects. In this mechanism the fluctuations act to facilitate the signal rather than inhibit it under certain conditions. To establish the essence of this phenomenon, let us consider the equation of motion

$$\frac{dQ}{dt} = -\frac{\partial}{\partial Q}\Phi(Q) + f(t) + k\xi_{p}(t).\tag{7.81}$$

The variable $Q(t)$ is the coordinate of a particle (the signal of interest), moving under the overdamped condition within a double-well potential $\Phi(Q)$. Recall that this equation can be derived from Newton's force law in the Smoluchosky approximation discussed in Section 3.2.3. The particle is also under the influence of white noise $f(t)$ and an external stimulus $\xi_{p}(t)$. In the original formulation and most subsequent discussions, the external stimulus is assumed to have the periodic form

$$\xi_{p}(t) = \cos(\omega t).\tag{7.82}$$

The potential is assigned the double-well form

$$\Phi(Q) = \frac{Q_0}{a^4}(Q^2 - a^2)^2.\tag{7.83}$$

To move from $Q > 0$ to $Q < 0$ the particle must overcome a barrier of intensity Q_0 separating the two wells. At the bottom of the two wells, $Q = \pm a$, the potential $U(Q)$ vanishes. Owing to the overdamped condition, the function $\Phi(Q)$ and the white noise $f(t)$ are the true potential $U(Q)$ and the true stochastic force $F(t)$, divided by the friction Γ, respectively. The parameter Γ is the friction coefficient of the phenomenological equation $dv(t)/dt = -\Gamma v(t)$, with $v = dQ/dt$.

It is straightforward to prove that the time-dependent potential

$$\Phi'(Q, t) = \frac{U(Q)}{\Gamma} - kQ \cos(\omega t)\tag{7.84}$$

generates two wells with different depths, and that their depths decrease or increase with the unperturbed displacement Q_0 by the quantity

$$\Delta Q \approx \mp ak \cos(\omega t).\tag{7.85}$$

Using Kramers' theory [37] the transition rates between the wells in the GME can be written

$$g_{\pm}(t) = R \exp\left[-\frac{Q_0}{D}(1 \mp ka \cos(\omega t))\right].\tag{7.86}$$

Note that Kramers' theory affords insight into how to express the pre-factor R, which in the case of the double-well potential can be shown to be the product of the potential frequency at the bottom of the well and that at the top of the barrier divided by the friction Γ. The diffusion coefficient D divided by the dissipation parameter Γ is determined by the Einstein relation to be the thermodynamic temperature $k_B T$. Of course, this is only the case in discussing a physical web and is not true in general, as we shall show. Here we expand the perturbation piece of the exponential and replace (7.86) with

$$g_\pm(t) = g_0(1 \mp \varepsilon \cos(\omega t)), \tag{7.87}$$

where we limit ourselves to stressing that $\varepsilon \ll 1$,

$$\varepsilon = \frac{Q_0}{D} ka, \tag{7.88}$$

and the unperturbed coupling parameter is

$$g_0 = R \, \exp\left[-\frac{Q_0}{D} \right]. \tag{7.89}$$

We are naturally led to replace (7.60) with

$$\frac{d}{dt}\mathbf{p(t)} = g_0 \mathbf{K}(t) \mathbf{p}(t), \tag{7.90}$$

where the time-dependent matrix $\mathbf{K}(t)$ now has the form

$$\mathbf{K}(t) = \begin{pmatrix} -[1 - \varepsilon \cos(\omega t)]/2 & +[1 + \varepsilon \cos(\omega t)]/2 \\ +[1 - \varepsilon \cos(\omega t)]/2 & -[1 + \varepsilon \cos(\omega t)]/2 \end{pmatrix}. \tag{7.91}$$

Using the perturbed form of the coupling matrix (7.91) in (7.90) yields the equation of motion for the difference variable

$$\frac{d}{dt}\Pi(t) = -g_0\Pi(t) + g_0\varepsilon \, \cos(\omega t), \tag{7.92}$$

where $p_1 + p_2 = 1$ is used in the second term. The exact solution to this equation is

$$\Pi(t) = \exp(-g_0 t)\Pi(0) + \varepsilon g_0 \int_0^t ds \, \exp(-g_0(t - s)) \cos(\omega s). \tag{7.93}$$

Let us assume that initially the web is in equilibrium so the probability of being in either of the wells is the same and consequently $\Pi(0) = 0$. We define the variable $\xi_s(t)$, which has the value 1 when the network S is in the right well and the value -1 when the network S is in the left well. Thus, we write for the average web response (7.93) in the familiar LRT form

$$\langle \xi_s(t) \rangle = \varepsilon \int_0^t \chi(t, t') \xi_p(t') dt', \tag{7.94}$$

where the response function is

$$\chi(t, t') = g_0 \exp[-g_0(t - t')] \tag{7.95}$$

and $\xi_p(t)$ is the harmonic perturbation. Note that the autocorrelation for the network response is given by (7.72) and the linear response function is given by (7.73).

In the specific case of a harmonic perturbation (7.82), after some straightforward algebra, we obtain from (7.93) for $t \to \infty$

$$\Pi(t) = \varepsilon \frac{g_0}{g_0^2 + \omega^2} [g_0 \cos(\omega t) + \omega \sin(\omega t)], \tag{7.96}$$

which can be written in the more compact form

$$\Pi(t) = \varepsilon \frac{g_0}{(g_0^2 + \omega^2)^{1/2}} \cos(\omega t - \phi), \tag{7.97}$$

with the phase defined as

$$\phi = \arctan\left(\frac{g_0}{\omega}\right). \tag{7.98}$$

The measure of stochastic resonance is typically the signal-to-noise ratio. We note that the coefficient D in (7.86) is the measure of the noise intensity. In the unperturbed case the coupling coefficient is given by (7.89). Let us assume, for simplicity's sake, that $R = 1$ and $Q_0 = 1$, which yields for the strength of the noise

$$D = \frac{1}{\ln(1/g_0)}. \tag{7.99}$$

One measure for the visibility of the signal in the noise background is given by the signal-to-noise intensity ratio, which in this case is given by the amplitude in (7.97):

$$\frac{S}{D} = \frac{\varepsilon g_0}{(g_0^2 + \omega^2)^{1/2}} \ln\left(\frac{1}{g_0}\right). \tag{7.100}$$

Consequently, when the intensity of the noise vanishes, $D = 0$, then so does the coupling coefficient $g_0 = 0$, and, consequently the signal-to-noise ratio does too, $S/D = 0$.

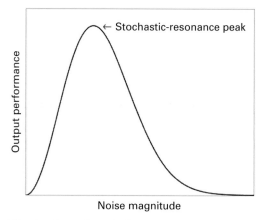

Figure 7.6.   The signal-to-noise ratio for a generic stochastic-resonance process is depicted as a function of the noise magnitude. It is clear that the signal-to-noise ratio is non-monotonic and achieves a maximum for some intermediate value of the noise magnitude.

With increasing noise D, g_0 increases and tends to its maximum value, which, in this idealized representation, is $g_0 = 1$. The logarithmic contribution to the signal-to-noise ratio (7.100), which is negligible for $g_0 \to 0$, becomes active in the case of large values of g_0, and makes the signal-to-noise ratio decrease to zero after a broad maximum, in the region of g_0 of the order of ω. This non-monotonic behavior of the signal-to-noise ratio with increasing noise is why this process is called stochastic resonance. In this subsection we have followed the approach to stochastic resonance proposed by McNamara and Wiesenfeld [46].

7.3 The Universality of LRT

In this section we discuss LRT from a perspective outside the stochastic differential equations used in Chapter 3. The new approach extends the formulation and application from physical processes driven by classical or quantum Liouville equations to the dynamics of complex networks in a variety of venues. Such extensions are necessary because the mechanisms that can be identified in physical webs might not be acting in the webs of society or in those of life; the interactions between organizations or organs may be less tangible than those in physical webs. This is an ambitious undertaking; however, there are preliminary indications that the non-stationary LRT proposed in this section may turn out to be a universal principle. This would make the more general LRT one of the few principles applicable to the dynamics of complex webs that is not mechanism-specific; another would be the generalized Onsager principle.

No general theory exists that successfully extends LRT from the condition of an equilibrium web to that of a non-equilibrium web [21]. Although the question is interesting, little progress has been made to explain some particular aspects of non-equilibrium web response to perturbation; no theorem as general as the equilibrium fluctuation–dissipation theorem (FDT) is yet available [21]. Furthermore, the recent theoretical prediction on the lack of a steady response to external stimuli [61] may generate the conviction that in the non-stationary case it is not possible to have any form of LRT.

The recent literature on non-Poisson renewal processes has increased the interest regarding the action of non-ergodic renewal events [19, 45, 54], with a wide set of applications, ranging from quantum mechanics [18] to brain dynamics [32], thereby casting significant doubt on the possibility of using conventional LRT to study the effects of perturbation in these cases. This limitation is a consequence of two things: (i) it is very difficult, if not impossible [16], to describe the time evolution of event-driven webs by means of Hamiltonian operators (the classical or quantum Liouville equation); and (ii) it is not yet well understood how to use the linear response structure of (7.70) when a stationary autocorrelation function is not available, in spite of the fact that some prescriptions for the non-stationary situation already exist [4, 7]. These are probably the reasons why Sokolov and Klafter [60, 61] coined the term "death of linear response" to denote the fading response of a complex web to a harmonic stimulus. This interesting phenomenon piqued the interest of other researchers and also inspired a debate on the best way to generate this response in surrogate sequences [35, 43, 44].

The event-based theory, in the ordinary ergodic and stationary case, yields results that apparently coincide with the predictions of conventional LRT.

7.3.1   The Hamiltonian approach to LRT

In this section we discuss LRT in the context of Hamiltonian webs following the concise form proposed by Zwanzig [73]. In the presence of a perturbation the classical Liouville equation can be written in terms of two Liouville operators,

$$\frac{\partial}{\partial t}\rho(t) = L\rho(t) \equiv [-L_0 + L_1 E(t)]\rho(t), \tag{7.101}$$

where $\rho(t)$ denotes the trajectory density in phase space and its average value yields a probability density. The quantity L_0 is the unperturbed Liouville operator and $L_0\rho$ is the Poisson bracket of H_0, the unperturbed Hamiltonian, and the density ρ will be as shown explicitly below. The perturbation enters through the second Liouville operator L_1, which is the Poisson bracket of $-M$, the web variable used to couple the web to the external perturbation. The intensity of the time-dependent perturbation is proportional to the function $E(t)$. In other words, the total Hamiltonian H of the web is time-dependent and expressed by

$$H(t) = H_0 - M E(t). \tag{7.102}$$

Let us separate the phase-space distribution into two pieces, $\rho(t) = \rho_0(t) + \rho_1(t)$, and split the Liouville equation into the lowest-order piece

$$\frac{\partial}{\partial t}\rho_0(t) = -L_0\rho_0(t) \tag{7.103}$$

and the first-order piece

$$\frac{\partial}{\partial t}\rho_1(t) = -L_0\rho_1(t) + L_1\rho_0(t)E(t). \tag{7.104}$$

The first-order solution to the Liouville equation (7.101) is given by solving the inhomogeneous equation (7.104) to obtain

$$\rho_1(t) = \int_0^t dt' \exp[-(t-t')L_0]L_1 E(t')\rho_0(t'). \tag{7.105}$$

The Liouville operator L_1 acts on the unperturbed distribution $\rho_0(t)$ as follows:

$$L_1\rho_0(t) = -\left(\frac{\partial M}{\partial \mathbf{p}}\frac{\partial \rho_0(t)}{\partial \mathbf{q}} - \frac{\partial M}{\partial \mathbf{q}}\frac{\partial \rho_0(t)}{\partial \mathbf{p}}\right). \tag{7.106}$$

The vectors $\mathbf{p}$ and $\mathbf{q}$, the canonical variables for the web degrees of freedom, are multidimensional and the scalar product is assumed in (7.106). Thus, in practice this picture may apply to an arbitrarily large number of particles.

Now we assume that prior to the application of the perturbation the network is in equilibrium and is characterized by the canonical distribution

$$\rho_{\text{eq}}(\mathbf{q}, \mathbf{p}) = \frac{\exp(-\beta H_0)}{Z}, \tag{7.107}$$

where $\beta = 1/(k_B T_{abs})$ and Z is the classical partition function. Assuming that the web is originally equilibrated allows us to express the LRT in terms of an equilibrium auto-correlation function, but at the same time establishes a condition beyond which we plan to work. Assuming that $\rho_0 = \rho_{eq}$ makes it possible to turn (7.106) into

$$L_1 \rho_{eq} = \beta \rho_{eq} \left(\frac{\partial M}{\partial \mathbf{p}} \frac{\partial H_0}{\partial \mathbf{q}} - \frac{\partial M}{\partial \mathbf{q}} \frac{\partial H_0}{\partial \mathbf{p}} \right). \tag{7.108}$$

The observable time-evolution is driven by L^+, the operator adjoint to L. In the Hamiltonian case $L^+ = -L$, thus

$$L_0^+ = -L_0. \tag{7.109}$$

The time-evolution of M as an effect of the zeroth Liouville operator is given by

$$\dot{M} = \frac{d}{dt} M = L_0^+ M = -L_0 M$$
$$= \frac{\partial M}{\partial \mathbf{p}} \frac{\partial H_0}{\partial \mathbf{q}} - \frac{\partial M}{\partial \mathbf{q}} \frac{\partial H_0}{\partial \mathbf{p}}. \tag{7.110}$$

Thus, comparing (7.110) with (7.108) yields

$$L_1 \rho_{eq} = \beta \dot{M} \rho_{eq} \tag{7.111}$$

and by inserting (7.111) into (7.105) we obtain the first-order phase-space distribution

$$\rho_1(t) = \beta \int_0^t dt' \, \exp[-(t - t')L_0] \dot{M} E(t') \rho_{eq} \tag{7.112}$$

in terms of the initial equilibrium phase-space distribution.

Consider the web variable A and assume that with the web in equilibrium its mean value vanishes:

$$\langle A \rangle_{eq} = 0. \tag{7.113}$$

To establish the web response to perturbation we multiply (7.112) by the web variable A and integrate over the phase-space variables to obtain for the average response

$$\langle A(t) \rangle = \beta \int_0^t dt' \, E(t') \chi_{AM}(t - t'), \tag{7.114}$$

where the stationary linear response function $\chi_{AM}(t, t') = \chi_{AM}(t - t')$ can be formally expressed as

$$\chi_{AM}(t - t') = \int d\mathbf{q} \, d\mathbf{p} \, A(\mathbf{q}, \mathbf{p}) \exp[-(t - t')L_0] \dot{M}(\mathbf{q}, \mathbf{p}) \rho_{eq}(\mathbf{q}, \mathbf{p}). \tag{7.115}$$

We emphasize that the Hamiltonian approach to establishing the form of LRT is not confined to the exponential case. Quite the contrary, the exponential case is an idealization that is not compatible with the Hamiltonian approach. We have seen that

$$\chi_{AM}(t - t') = \langle A(t) \dot{M}(t') \rangle, \tag{7.116}$$

namely the linear response function is the cross-covariance function between the web variables A and $\dot{M}$. Years ago Lee [40] proved that the Hamiltonian approach to the velocity autocorrelation function cannot generate an exponential decay. The search for a representation that may make the exponential decay compatible with the Hamiltonian treatment has been the subject of many studies (see [65] and references therein). The discussion of this problem is closely related to the derivation of irreversible processes from a picture of either quantum or classical physics that is invariant under time reversal. We cannot rule out the possibility that this transition is realized by the presence of renewal events that are invisible in conditions very far from the exponential relaxation. These events become important to eliminate the sign of the original reversibility when the exponential form becomes the predominant contribution to relaxation as investigated by Vitali and Grigolini [65]. Let us here adopt the view that the exponential decay may be a good approximation to the relaxation process. In this sense, the linear response function

$$\chi(t, t') = \exp[-\gamma(t - t')] \tag{7.117}$$

is acceptable, in spite of its exponential character.

7.3.2   Applying the formalism

To put the formalism of the last subsection into perspective, recall the simple example of LRT given in Section 3.2.2 where we considered the linear Langevin equation for the velocity with linear dissipation, a white-noise random force and a driving force $E(t)$. Recall that the exact solution for the velocity with zero initial value and the perturbation switched on at $t = 0$ is

$$\langle V(t) \rangle = \int_0^t e^{-\lambda(t-t')} E(t') dt' \tag{7.118}$$

and the exponential is the autocorrelation function

$$e^{-\lambda(t-t')} = \frac{\langle V(t) V(t') \rangle_{eq}}{\langle V^2 \rangle_{eq}}. \tag{7.119}$$

If we again assume equipartition of energy

$$\langle V^2 \rangle_{eq} = k_B T_{abs} \tag{7.120}$$

(7.118) becomes

$$\langle V(t) \rangle = \beta \int_0^t ds \, \chi_{VQ}(t - s) E(s). \tag{7.121}$$

We note that this result fits perfectly well the prediction of (7.114) and (7.115). In fact, the Hamiltonian interacting with the external perturbation in this case reads

$$H_1(t) = -q E(t), \tag{7.122}$$

and, of course, since $dQ/dt = V$, the linear response function $\chi_{VQ}(t - t')$ coincides with the velocity–displacement cross-covariance function.

In this subsection, for simplicity's sake, we focus on the case when we observe the response of the variable that is directly perturbed by the external stimulus. This is the reason why in (7.73) we do not assign subscripts to the linear response function: we are assuming that the observed variable coincides with the perturbed variable. To make evident to the reader that in this case the ordinary LRT yields (7.70) with the linear response function (7.73), let us consider the case described by

$$\frac{dQ(t)}{dt} = V(t) \tag{7.123}$$

and

$$\frac{dV(t)}{dt} = -\lambda V(t) + F(t) + E(t) - \omega^2 Q(t). \tag{7.124}$$

This is the same Brownian particle as the one described in Section 3.2.2 with the new condition that the perturbation process occurs in the presence of the harmonic potential $U(q)$ given by

$$U(q) = \frac{1}{2}\omega^2 q^2. \tag{7.125}$$

The harmonic potential produces a confinement effect, yielding for the dynamical space variable Q the finite second moment

$$\left\langle Q^2 \right\rangle_{eq} = \frac{k_B T_{abs}}{\omega^2}. \tag{7.126}$$

Let us consider the condition

$$\omega \ll \lambda, \tag{7.127}$$

which allows us to use the Smoluchosky approximation, namely, to set $dV/dt = 0$ in (7.124), so as to express $V(t)$ in terms of $Q(t)$, $E(t)$ and $F(t)$. We then insert this expression for $V(t)$ into (7.123), thereby obtaining

$$\frac{dQ(t)}{dt} = -\gamma Q(t) + \eta(t) + f(t), \tag{7.128}$$

where

$$\eta(t) \equiv \frac{E(t)}{\lambda}, \tag{7.129}$$

$$f(t) \equiv \frac{F(t)}{\lambda} \tag{7.130}$$

and

$$\gamma \equiv \frac{\omega^2}{\lambda}. \tag{7.131}$$

Equation (7.128) is formally identical to the linear Langevin equation considered in Section 3.2.2. Thus, for a vanishing initial condition we are immediately led to the average response

$$\langle Q(t) \rangle = \int_0^t \exp[-\gamma(t - t')]\eta(t')dt'.$$

(7.132)

Apparently, this equation is different from (7.70) and (7.73) because it involves the autocorrelation function

$$\Psi_s(t - t') = \exp[-\gamma(t - t')]$$

(7.133)

rather than the derivative of the autocorrelation,

$$\frac{d}{dt'}\Psi_s(t - t') = \gamma \exp[-\gamma(t - t')].$$

(7.134)

However, for a proper comparison we must take into account that ξ_s and ξ_p are dimensionless variables, whereas Q has the dimensions of space and η has the dimensions of velocity. We are assuming that the particle's mass is unity. Therefore we introduce the transformations

$$\xi_s(t) = \frac{1}{\sqrt{\langle Q^2 \rangle}} Q(t)$$

(7.135)

and

$$\xi_p(t) = \frac{1}{\gamma\sqrt{\langle Q^2 \rangle}} \eta(t),$$

(7.136)

and, by expressing Q and η in terms of the dimensionless variables and cancelling out the common factor $\sqrt{\langle Q^2 \rangle}$, we turn (7.132) into

$$\langle \xi_s(t) \rangle = \gamma \int_0^t e^{-\gamma(t-t')} \xi_p(t')dt',$$

(7.137)

which coincides with the theoretical prediction of (7.70) together with (7.73). Note that the Langevin equation is a form of *linear* stochastic physics that makes the LRT valid with no restriction that the perturbation has to be extremely small in intensity. This explains the lack of the parameter ε in both (7.137) and (7.132).

7.3.3   The experimental approach

The result (7.97), which is widely accepted by the community studying stochastic resonance, suggests how to extend LRT using conjectures associated with the non-stationary case. Let us assume that the number of events per unit time, rather than being constant, for physical reasons becomes smaller and smaller, until asymptotically this rate reaches a very small value g_{min} that is much smaller than the frequency ω of the periodic driver. Let us imagine that after achieving this very small value the rate of event generation remains constant again as we saw in Section 7.2.2. It is evident that in the long-time regime (7.97) is replaced by

$$\Pi(t) = \varepsilon \frac{g_{\min}}{\omega} \cos(\omega t + \Phi). \tag{7.138}$$

The transition from g_0 to $g_{\min}$ also has the effect of changing the phase ϕ into the phase Φ, whose explicit value depends on the details of the transition process, and this change is not discussed here.

Let us now consider the case in which the probability of event production after an earlier event, $g(t)$, rather than being constant, is time-dependent. We use the time-dependent event generator (3.233) from Section 3.5.2 and identify the occurrence of an event with the occurrence of failure. Consider the case when the exponent in (3.233) is given by $\eta = -1$. The condition $\eta < -1$ would lead to an asymptotic time with no event at all, and, consequently, to a vanishing response. Thus, it is convenient to consider the singularity condition $\eta = -1$. Let us imagine such a collection of experiments that are of such high quality as to create a Gibbs ensemble of webs, each of them characterized by the occurrence of an event at $t = 0$. In this case the Gibbs ensemble is an out-of-equilibrium web that will try to regress to equilibrium. The signature of this process of regression to equilibrium is given by the function $R(t)$, the number of events produced per unit time by the network S prepared at $t = 0$. The concept of preparation is fundamental to understanding the non-stationary nature of the condition $\mu < 2$. We assume that our statistical analysis is based on the preparation of infinitely many realizations of the sequence $\xi_S(t)$ with the condition that all these realizations have an event at time $t = 0$.

The guidelines to presenting the new LRT in an understandable format to as wide an audience as possible are indicated by the following three steps: preparation, perturbation and experimentation.

Preparation. There are webs whose non-equilibrium nature is determined by a cascade of renewal events with a rate $R(t)$ decreasing in time, as shown in a subsequent section, according to

$$R(t) \propto \frac{1}{t^{2-\mu}}, \tag{7.139}$$

after a proper experimental failure rate. Here we use liquid crystals as exemplars of physical networks belonging to this group [57]. The time decay in the number of events per unit time is the prediction of the hyperbolic form of the waiting-time distribution function. The mathematical origin of the cascade implied by the inverse power-law decay in (7.139) is a well-known [28] property of renewal events with a survival probability $\Psi(t)$ (namely the probability of waiting longer than a time t for an event to occur) decaying as $t^{1-\mu}$, with $\mu < 2$, for $t \to \infty$. In the experiment illustrated subsequently $R(t)$ is generated by the dynamics of interacting defects that are prepared at a time preceding the application of a weak perturbation to S. In the absence of perturbation the variable ξ_S is time-independent, in spite of the perennial out-of-equilibrium condition represented by the ever-drifting quantity $R(t)$ ensuing preparation. This non-ordinary condition is incompatible with the Hamiltonian-based treatments, where events are not considered and the physical observable responds only to a direct or indirect Hamiltonian interaction with the external stimulus.

Perturbation. The external stimulus, which is not directly coupled to ξ_s, affects the event dynamics according to the value of the variable $\xi_s(t)$. This is the common situation in a complex web, where multiple variables interchange information with one another. The perturbation cascades through the web, thereby generating an indirect effect on ξ_s, and consequently a bias, which is taken to be zero in the absence of the stimulus. We demonstrate this effect in a number of psychophysical phenomena after reviewing the results of the physics experiment.

Experimentation. The experiment on liquid crystals [5] supports the following theoretical prescription for the average web response modeled on (7.138):

$$\langle \xi_s(t) \rangle = C R(t) \cos(\omega t + \phi), \tag{7.140}$$

with the amplitude C and phase ϕ as fitting parameters; and the power-law index μ is determined, as we shall see, by the experiment itself. Equation (7.140) admits an extremely simple and intuitive explanation of the web response fading in time, since it resembles the response to perturbation of Poisson processes (stochastic resonance [42, 68]) with the proviso that the cascade of events generated by preparation fades away with time.

Allegrini *et al.* [5] prepare their experiment by bringing the web into a fully developed turbulent regime corresponding to a Poisson condition and to a high rate of event production. After a few seconds (20 s) they adopt a control potential to establish cooperative interaction between the web's units (corresponding to defect-mediated turbulence and $\mu = 1.5$), and apply a weak perturbation to the web by modulating the AC voltage amplitude. Unfortunately, the first equilibrium value for transmissivity is different from the latter, corresponding to the regime of interest. Therefore, the equation they adopt must take into account also the free regression to equilibrium as well as the component given by (7.140),

$$\langle \xi_s(t) \rangle = \langle \xi(0) \rangle \Psi(t) + C R(t) \cos(\omega t + \phi). \tag{7.141}$$

To derive the form of (7.140) they perturb the web with $\cos(\omega t)$ and $-\cos(\omega t)$ and evaluate the sum response

$$S(t) = \frac{\langle \xi_s(t) \rangle_+ + \langle \xi_s(t) \rangle_-}{2} \tag{7.142}$$

and the difference response

$$D(t) = \frac{\langle \xi_s(t) \rangle_+ - \langle \xi_s(t) \rangle_-}{2}. \tag{7.143}$$

Note that $\langle \xi_s(t) \rangle_+$ and $\langle \xi_s(t) \rangle_-$ are the responses of the network to $\cos(\omega t)$ and $-\cos(\omega t)$, respectively. To obtain the ensemble averages indicated by the angle brackets they repeat the preparation and perturbation procedures approximately 200 times for each measurement and then average. $D(t)$ and $S(t)$ are shown in Figures 7.7 and 7.8, respectively. Note that the result shown in Figure 7.8 allows us to determine the power index μ ($\mu = 1.56$ in this case) from the slope of the experimental curve. This method

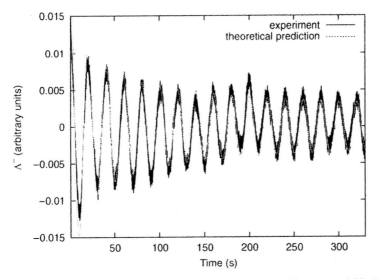

Figure 7.7. The dots are the experimental results [5] expressed as the difference variable $D(t)$ as a function of time. The theoretical prediction of (7.140) and the experimental data cannot be distinguished from one another. Redrawn with permission.

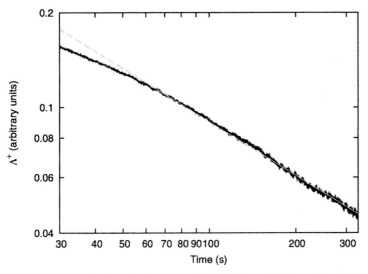

Figure 7.8. The wiggly curve is given by the sum variable $S(t)$ as a function of time. The straight-line segment is the best inverse power-law fit to the experimental data [5]. Redrawn with permission.

leads to a definition of μ that is more accurate than that determined in earlier work [62]. Thus, the excellent agreement between the fitting formula of (7.140) and the experimental results of Figure 7.7 requires only two fitting parameters, ϕ and C.

The authors of [5] afford compelling evidence that the response of the web is rigorously proportional to the perturbation intensity, as it should be if a proper LRT is applicable. To complete the experimental proof of the non-stationary LRT, they

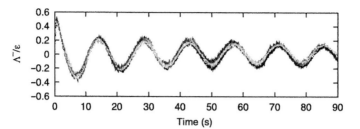

Figure 7.9.   One line is twice the response to a perturbation whose intensity is half that of the perturbation generating the other [5]. It is clear that the three data sets coincide. Redrawn with permission.

repeat the experiment, generating several perturbations of different intensity, and then rescale the results with the perturbation amplitude. The scaled results are illustrated by Figure 7.9. This experiment superimposes one curve on top of the other, thereby establishing that *for small perturbations* the response of the web is linear in the perturbation intensity.

7.3.4   Theoretical interpretation

The experimental realization of a non-stationary LRT given in the last subsection compels us to address this problem from a theoretical point of view. The first theoretically interesting question is how to derive (7.140) from LRT. On the basis of (7.73) we may make the conjecture that the new LRT is realized by

$$\chi(t, t') = \frac{d}{dt'} C(t, t'), \tag{7.144}$$

in terms of the two-time autocorrelation function. However, it is also possible to make the choice

$$\chi(t, t') = -\frac{d}{dt} C(t, t'). \tag{7.145}$$

In the stationary case the two choices are equivalent, but in the non-stationary case they are not. A discussion of which is the most convenient choice from the two alternatives was presented by the authors [4, 7], and in the earlier work [8] as well. Here we show how to relate these choices to the rate of event generation $R(t)$ given by (7.140), which is an essential step in our experiment-based assessment of the non-stationary LRT. The LRT of a process whose dynamics are dominated by events rather than being driven by the Hamiltonian of the web S, which, even when it exists, is unknown to us, must be based on the information on the events produced by the ensemble. The central prescription is given by the age-specific rate of event production [23]:

$$g(t) = \frac{\psi(t)}{\Psi(t)}. \tag{7.146}$$

In Poisson processes $g(t) = g$ is time-independent, implying that $\Psi(t) = \exp(-gt)$. The fully turbulent regime produces a high rate of events of this kind, but the transition to the defect-mediated turbulent regime makes this high g decrease in time.

The experimental result shown in Figure 7.8 forces us make Pandora's choice for the time-dependent rate:

$$g(t) = \frac{r_0}{1 + r_1 t}, \tag{7.147}$$

which, in fact, as shown previously, yields the hyperbolic survival probability

$$\Psi(t) = \left(\frac{T}{t + T}\right)^{\mu - 1}. \tag{7.148}$$

This is the proper time for us to use the stochastic rate defined in Section 3.5.2. Let us adopt the more concise, and more precise, notation

$$r(t) = g(t - t_i), \tag{7.149}$$

where t_i is the random time of occurrence of the last event prior to time t. However, to make the statistics more transparent, it is convenient to notice that we can write

$$\exp\left(-\int_{t'}^{t} r(\tau) d\tau\right) = \exp\left(-\int_{t'}^{t} g(\tau) d\tau\right) = \Psi(t - t'), \tag{7.150}$$

when there are no events from time t' to t; and the last equality arises from the definition of the survival probability. Thus, the survival probability Ψ depends on the time t' at which observation begins:

$$\Psi(t, t') = \int_0^{t'} dt'' \, R(t'') e^{-\int_{t''}^{t} r(\tau) d\tau} = \int_0^{t'} dt'' \, R(t'') e^{-\int_{t''}^{t} g(\tau) d\tau}$$

$$= \int_0^{t'} dt'' \, R(t'') \Psi(t - t''). \tag{7.151}$$

In the case in which the laminar region between two consecutive events, occurring at times t_i and t_{i+1}, is filled with values ξ_s drawn from a distribution with finite width, it has been shown, see [1], that the autocorrelation function $C(t, t')$ reduces to the age-dependent survival probability $\Psi(t, t')$, and hence the response function in (7.144) becomes

$$\chi(t, t') = \frac{d}{dt'} C(t, t') = R(t) \Psi(t - t'). \tag{7.152}$$

Alternatively, the linear response function (7.145) becomes

$$\chi(t, t') = -\frac{d}{dt} C(t, t') = \int_0^{t'} dt'' \, R(t'') \psi(t - t''), \tag{7.153}$$

which shows explicitly how the choices of (7.144) and (7.145) depend on $R(t)$. The authors of [35, 43, 60, 61] followed Sokolov [59], whose theory was proved [4, 7] to yield the choice of (7.144). The rationale for this choice is that the perturbation does not influence the event-occurrence time, but only the drawing of the variable ξ_s to fill the time intervals between two consecutive events. We refer to this theory as the *phenomenological* FDT. It has been shown [4, 7] that the choice (7.153), called the *dynamical* FDT, corresponds to the network response produced by the external perturbation affecting the event-occurrence time according to the state of the web S.

It is convenient to rewrite (7.151) in the following form:

$$\psi(t, t') = \int_0^{t'} dt'' \, R(t'') \psi(t - t''), \tag{7.154}$$

where we recall that

$$R(t) = \sum_{n=0}^{\infty} \psi_n(t). \tag{7.155}$$

The quantity $R(t)$ is the number of events per unit time produced by the ensemble of complex networks prepared at time $t = 0$ and can easily be derived from $\psi(t)$ for a renewal process.

The double-indexed probability density $\psi(t, t')$ in (7.154) refers to a web that has been prepared at time $t = 0$ when an event occurs, say one generated by an external excitation. For a renewal process the probability density of an event occurring at time t given that the last event occurred at time t' has the exact form [68]

$$\psi(t, t') = \psi(t) + \sum_{n=1}^{\infty} \int_0^{t'} \psi_n(t'') \psi(t - t'') dt''. \tag{7.156}$$

In the physics literature $\psi(t)dt$ is the probability that an event occurs in the time interval $(t, t + dt)$. Since the observation process begins at time $t' > 0$, the probability of observing a new event after t' depends on the last event occurring prior to t', at time t'' as expressed by the distribution density $\psi(t - t'')$. The event occurring at $t'' < t'$ is in general the last of a sequence of n events, occurring exactly at t'', while the earlier events can occur at any earlier time $t > 0$. The probability for this last of n events is $\psi_n(t'')$, with $n \geq 1$; this is the first term in the integrand of (7.156). The renewal nature of the process is adopted to define the function $\psi_n(t)$ through the hierarchy

$$\psi_n(t) = \int_0^t \psi_{n-1}(t - t') \psi_1(t') dt' \tag{7.157}$$

starting at the preparation $t = 0$, which is established by setting the condition

$$\psi_0(t) = \delta(t). \tag{7.158}$$

The hierarchy of (7.157) allows us to express the distribution densities $\psi_n(t)$ in terms of $\psi_1(t)$, which is identified with $\psi(t)$. The time-convolution structure of (7.157) is a consequence of the renewal nature of the process and makes it convenient for us to use the Laplace-transform method. It is straightforward to prove that (7.157) using (7.158) yields the following relation among the Laplace-transformed functions:

$$\hat{\psi}_n(u) = (\hat{\psi}(u))^n. \tag{7.159}$$

From this algebraic form of the Laplace transform of the probability density for n events it is evident that the time intervals between events in a renewal process are independent of one another.

In fact the Laplace transform of (7.155) can, using (7.159), be written

$$\hat{R}(u) = \frac{1}{1 - \hat{\psi}(u)},$$
(7.160)

from which it is clear that the number of events is determined directly by the event probability density. Note that the aged event probability density $\psi(t, t')$ expressed by (7.154) requires an independent determination of $\psi(t)$, to which we now turn our attention.

To realize the phenomenological condition necessary for $R(t)$ to decrease as $t \to \infty$ requires that the reduction of response intensity may, but need not, have a lower limit, depending on whether or not the statistics of the neural network are ergodic or non-ergodic. To establish this condition, consider the relation between the rate of event production and the event probability density given in Laplace space by (7.160). Recalling the Laplace transform of the hyperbolic distribution

$$\hat{\psi}(u) = (\mu - 1)\Gamma(1 - \mu)(uT)^{\mu-1} \left[e^{uT} - E_{\mu-1}^{uT} \right],$$
(7.161)

we can write the asymptotic expansion as $u \to 0$ to obtain to lowest order in u in the case of no finite first moment

$$\lim_{u \to 0} \hat{\psi}(u) = 1 - \Gamma(2 - \mu)(uT)^{\mu-1} + \cdots; \quad 1 < \mu < 2,$$
(7.162)

and for a finite first moment but a diverging second moment

$$\lim_{u \to 0} \hat{\psi}(u) = 1 + u\langle t \rangle - \Gamma(2 - \mu)(uT)^{\mu-1} + \cdots; \quad 2 < \mu < 3,$$
(7.163)

in which the lowest-order terms in powers of u of the two functions in (7.161) have been retained in both expressions. Inserting (7.162) into the expression for the rate of event production yields

$$\lim_{u \to 0} \hat{R}(u) = \frac{(uT)^{1-\mu}}{\Gamma(2 - \mu)}; \quad 1 < \mu < 2,$$
(7.164)

and inserting (7.163) yields

$$\lim_{u \to 0} \hat{R}(u) = \frac{1}{u\langle t \rangle} - \frac{\Gamma(2 - \mu)}{\langle t \rangle^2 (uT)^{3-\mu}}; \quad 2 < \mu < 3,$$
(7.165)

respectively. Using the Tauberian theorem relating the time and Laplace domains,

$$\frac{1}{u^n} \Leftrightarrow \frac{t^{n-1}}{\Gamma(n)},$$
(7.166)

we obtain from (7.164) for the rate of event production as $t \to \infty$ for the non-ergodic case

$$R(t) \approx \frac{\sin[\pi(\mu - 1)]}{\pi T} \left(\frac{T}{t} \right)^{2-\mu}; \quad 1 < \mu < 2,$$
(7.167)

and for the ergodic case

$$R(t) \approx \frac{1}{\langle t \rangle} \left[1 + \frac{1}{3 - \mu} \left(\frac{T}{t} \right)^{\mu-2} \right]; \quad 2 < \mu < 3.$$
(7.168)

As a consequence of (7.167) and (7.168) we see that

$$R(t) - \frac{1}{\langle t \rangle} \propto D(t) = \frac{1}{t^{|2-\mu|}}.$$ (7.169)

The dynamical LRT yields the following theoretical prediction for the difference response [8]:

$$D(t) = \frac{\varepsilon}{\Gamma(\mu - 1)} \frac{\cos(\pi \mu/2 + \omega t)}{(\omega t)^{2-\mu}},$$ (7.170)

where $\varepsilon < 1$ is the perturbation strength. This result corresponds to the survival probability of (7.148) and to the strict assumption that $\xi_s(t)$ is dichotomous. We have assessed that the adoption of different forms of survival probability, either Lévy or Mittag-Leffler, and the use of a not strictly dichotomous $\xi_s(t)$, will generate different phases and different amplitudes, while leaving unchanged the structure of (7.140). The method of experimental observation of liquid crystals that we adopt [62] does not afford information accurate enough to quantitatively address this issue. For this reason we consider both ϕ and C as fitting parameters.

7.4   The statistical habituation model (SHM)

We provide an intuitive interpretation of a statistical approach to habituation by incorporating neuronal statistics into the dynamical response. This generalization is guided by the strategy Drew and Abbott [27] used to describe the closely related phenomenon of adaptation. They contend that in modeling activity-dependent adaptation of neuronal response it is impractical to describe in detail the full range of time constants suggested by experiment. Each time constant introduces a new and distinct exponential and consequently multiple exponential processes are required to model adaptation. However, these multiple exponential processes as an aggregate were determined to be well described by a power law and the latter description has been used successfully to describe adaptation in neural networks. A list of other biological phenomena that are well modeled by inverse power-law dependences on time, from single channels up to the level of human psychophysics, is given by Drew and Abbot [27], as well as by West *et al.* [68]. Herein we do not make this phenomenological replacement, but show analytically how an average over a distribution of rates (time constants) gives rise to inverse power laws in neuronal networks.

We take cognizance of the fact that multiple time scales contribute to the phenomenon of habituation. However, rather than phenomenologically replacing the multiple exponentials by an inverse power law to represent the full range of network dynamics, we assume that the dynamics consist of multiple interacting channels, each with a statistically different time constant. The output is then determined by the aggregation of the outputs of these multiple channels so as to provide an effective synaptic strength

$$\text{output} = w_{\text{eff}}(t) \times \text{input}.$$ (7.171)

So how do we model this effective synaptic strength? As Feng [29] emphasizes, physiologic data clearly show that nearby neurons usually fire in a correlated way [72] and anatomical data reveal that neurons with similar functions group together and fire together [56]. The SHM proposed by West and Grigolini [69] represents the output from the dynamics of a cluster of neurons, that is, from a tightly coupled network of parallel channels. The synaptic strength has contributions from each of the channels in a given time interval, with the time constants contributing to $w_{\text{eff}}(t)$ changing randomly from channel to channel. If such a cluster were modeled by a collection of two-state elements that strongly interact with one another, the dynamics of individual elements synchronize [64, 68] and the global behavior would be characterized by an extended sojourn in one of the two states, with abrupt random jumps from one state to the other, as shown earlier. The distribution density of sojourn times, or equivalently the distribution density of switching between states, $\psi(t)$, is a consequence of the network dynamics. In the present case of multiple channels $\psi(t)$ is determined by the distribution of time constants.

The formal analysis determining the response of a physical web S to a weak perturbation was addressed by a number of investigators [45, 59]. These investigators studied the response of a complex web characterized by the production of non-ergodic renewal events to a simple harmonic perturbation of weak intensity and established that the web response to the harmonic perturbation is transient and asymptotically fades to zero, as we discussed previously. To make use of this result in the context of habituation we cannot rely on the traditional LRT, based as it is on the assumption that the web S is stationary. The property of stationarity is often violated by complex webs such as the neuronal networks considered here. To properly address the issue of information transmission within a complex web we adopt the generalized LRT [4, 7]. We assume that the dynamics of interest are described by the event-dominated fluctuations $\xi_s(t)$ generated in the complex web and that the time dependence of the perturbation is described by the function $\xi_p(t)$. We assume that the latter is independent of the former, whereas when we switch on the interaction between the web and the environment the fluctuations $\xi_s(t)$ can be perturbed by $\xi_p(t)$. In the absence of perturbation the average over the ensemble of realizations of the web of interest vanishes, in which case the most general form of LRT is given by

$$\langle \xi_s(t) \rangle \equiv \int_0^t \chi(t, t') \xi_p(t') dt', \tag{7.172}$$

where $\xi_p(t)$ is the stimulus. The quantity $\chi(t, t')$ is the linear response function of the web S and can be written

$$\chi(t, t') = -\frac{d\Psi(t, t')}{dt}, \tag{7.173}$$

where $\Psi(t, t')$ is the aged survival probability defined by

$$\Psi(t, t') = \langle \xi_s(t) \xi_s(t') \rangle. \tag{7.174}$$

Thus, we replace (7.172) with

$$w_{\text{eff}}(t) = \int_0^t \psi(t, t') \xi_p(t') dt', \tag{7.175}$$

with no need for the thermodynamic temperature. Consequently, the effective synaptic strength is an average of the stimulus over an aged ensemble of realizations describing the rich dynamics of the neuron web. We now examine the properties of the probability density consistent with experiment and necessary to describe the phenomenon of habituation.

7.4.1   Simple stimulus and renewal theory

In the present context we now have that the probability density for a complex web of neurons is given by the observed inter-event-interval density [63] having the hyperbolic form. This form of probability density for neuronal webs is consistent with the experimental observation of $1/f$ noise in human cognition made from the errors in the recall of intervals [31] and in the variability of implicit measures of bias [22], both of which were described by a phenomenological theory of $1/f$ noise in human cognition [33]. The theory developed in the previous section can now be used to provide the effective synaptic weight of the response of the neuronal network to an external stimulus in terms of an average over the aged event probability density. This is done for both a single frequency stimulus and a sequence of Gaussian-shaped pulses (set as a problem).

To demonstrate the phenomenon of habituation using this expression, we assume a periodic signal of frequency ω with small amplitude $\varepsilon < 1$, $\xi_p(t') = \varepsilon \cos(\omega t')$, yielding

$$w_{\text{eff}}(t) = \varepsilon \int_0^t dt' \, \psi(t, t') \cos(\omega t'). \tag{7.176}$$

The solution to (7.176) is obtained from the real part of the Laplace transform [8]:

$$\widehat{w}_{\text{eff}}(u) = \varepsilon \, \text{Re} \left[\hat{E}(u) \right], \tag{7.177}$$

where $\hat{E}(u)$ is the Laplace transform of the partially Fourier-transformed aged distribution density

$$E(t) \equiv \int_0^t \psi(t, t') e^{-i\omega t'} dt'. \tag{7.178}$$

The exact expression for $\psi(t, t')$ allows us to make exact predictions for the behavioral response when the first term on the rhs in the renewal process (7.156) is a known function $\psi_1(t) = \psi(t)$. Following Barbi *et al.* [8] as well as West *et al.* [68] and using $\psi(t)$ to determine the aged waiting-time distribution density, we can write

$$\hat{E}(u) = \frac{i}{\omega} \frac{\hat{\psi}(u + i\omega) - \hat{\psi}(u)}{1 - \hat{\psi}(u + i\omega)}, \tag{7.179}$$

but in order to proceed further we need an explicit expression for $\psi(t)$. We use the hyperbolic distribution for the probability density and reduce (7.179) to algebraic manipulation.

The Laplace transform of the hyperbolic distribution is

$$\hat{\psi}(u) = (\mu - 1)\Gamma(1 - \mu)(uT)^{\mu-1}\left[e^{uT} - E_{\mu-1}^{uT}\right] \tag{7.180}$$

in terms of the generalized exponential

$$E_{\gamma}^{z} = \sum_{0}^{\infty} \frac{z^{n-\gamma}}{\Gamma(n+1-\gamma)}, \tag{7.181}$$

which reduces to the ordinary exponential when $\gamma = 0$ and $\mu = 1$. In the case $\mu < 2$ the Laplace transform of the network response reduces [8, 68] in the long-time regime where $u \to 0$ to

$$\widehat{w}_{\text{eff}}(u) = \varepsilon \, \text{Re}\left[\frac{i}{\omega}\left(1 - \left[\frac{u}{u+i\omega}\right]^{\mu-1}\right)\right]. \tag{7.182}$$

The inverse Laplace transform of (7.182) yields the behavioral response to a periodic input signal as $t \to \infty$ [8, 68]

$$w_{\text{eff}}(t) \approx \varepsilon \frac{\cos(\omega t - \mu\pi/2)}{\Gamma(\mu-1)(\omega t)^{2-\mu}}. \tag{7.183}$$

This expression clearly shows that the condition $\mu < 2$ suppresses the response to periodic stimulation over time and thus the network habituates. Note that the response does not distort the stimulus over time; it simply decreases the amplitude of the stimulus as an inverse power-law in time and shifts its phase.

Connection with power spectra

We remind the reader that the Wiener–Khintchine theorem proves that the Fourier transform of the autocorrelation function yields the power spectral density, but only under the condition that the process is stationary. This theorem is generally assumed to be valid, so that when a spectrum of the inverse power-law form given by

$$P(f) \propto \frac{1}{f^{\alpha}} \tag{7.184}$$

with the spectral index within the interval $0.5 < \alpha < 1.5$ is obtained the process is generally interpreted to have long-time correlations. However, in complex webs the autocorrelation function can be non-stationary and non-ergodic, invalidating the traditional Wiener–Khintchine theorem. It has recently been shown that cognitive fluctuations are non-stationary, so the application of the Wiener–Khintchine theorem is unwarranted, and yet the empirical power spectral density is found to have the $1/f$ noise form. We have addressed this paradox theoretically by replacing the continuous model of the underlying process with discrete renewal events.

Now we can explain more clearly the connection between the condition of habituation and $1/f$ noise. In the last few years a new approach to $1/f$ noise has been proposed

[41, 45] on the basis of the assumption that $1/f$ noise is produced by the renewal events. Historically the connection could not have been made because $1/f$ noise required the autocorrelation function to be stationary. The stationary condition is no longer necessary when the statistics are of renewal type. According to this theory of non-stationary processes the spectral index α of (7.184) is related to the inverse power-law index μ of the hyperbolic distribution by

$$\alpha = 3 - \mu. \tag{7.185}$$

We see, therefore, that the ideal condition of $1/f$ noise, $\alpha = 1$, corresponds to $\mu = 2$. On the other hand, as we mentioned earlier, the hyperbolic form of the event probability density has a diverging mean time $\langle t \rangle$ for $\mu \leq 2$, thereby generating ergodicity breakdown [68] with $\alpha \geq 1$. Thus, the ideal condition of $1/f$ noise is the boundary between the ergodic and non-ergodic regimes.

Thus, there is always habituation of simple stimuli. In the four experiments conducted by Kello *et al.* [36] the spectral index was determined to be in the range $0.53 \leq \alpha \leq 0.66$, indicating from (7.185) that the probability index is in the interval $2.47 \geq \mu \geq 2.34$. Consequently, these experiments determine that the intrinsic fluctuations of cognitive function are ergodic, assuming that the underlying process is renewal. Note also that this range of μ values is consistent with the Gerstein–Mandelbrot finding of 1.5 for the power-law index of the survival probability for single neurons. However, these values do not exhaust the full range of inverse power-law indices.

7.4.2   Information resonance and dishabituation

It is important to point out that the variable w_{eff} changes in time as a consequence of the stimulus. The abrupt switching on of the stimulus at $t = 0$ triggers the occurrence of an event, which, in turn, activates the $R(t)$ process, with the resulting habituation. It has recently been shown [33] that a web of interacting neurons generates a sequence of renewal events corresponding to an age-dependent rate $g(t)$ that is a decreasing function of time and can yield a virtually infinite mean time $\langle t \rangle$. The phenomenon of habituation to a periodic perturbation depends on the fact that the abrupt switching on of the perturbation triggers the occurrence of an event. The time of occurrence of the ensuing events is only slightly influenced by the external stimulus and weakly departs from the condition of natural occurrence. There are preliminary indications [33] that, in the special case in which the stimulus contains renewal events with the same complexity as the network, habituation is overcome and dishabituation can occur.

Aquino *et al.* [6] consider two complex networks denoted by S and P, with hyperbolic distributions having indices μ_S and μ_P, respectively, that are interacting with one another. The indices for both webs are in the interval $1 < \mu < 3$ and it is considered that web P perturbs web S. They determined that there is a singularity at the condition $\mu_S = \mu_P = 2$, see the center of Figure 7.10, where the value of the asymptotic cross-correlation function jumps from zero to one. If the perturbing web P is non-ergodic, meaning that $1 < \mu_P < 2$, and the web being perturbed, S, is ergodic, meaning that $2 < \mu_S < 3$, then asymptotically the signals generated by the two webs are completely

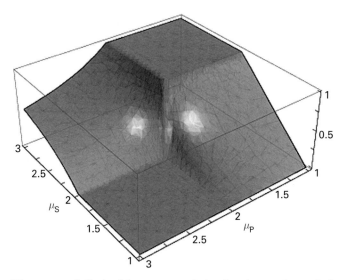

Figure 7.10.   The asymptotic limit of the cross-correlation function, on the vertical axis, is displayed for the range of power-law indices $1 < \mu_{P,S} < 3$. The vertex $\mu_S = \mu_P = 2$ marks the transition from the minimum to a condition of maximal input–output correlation. Adapted from [6].

correlated; that is, the cross-correlation function has the value one. This is the upper plateau region of the cross-correlation cube in Figure 7.10. If, on the other hand, web P is ergodic, meaning that $2 < \mu_P < 3$, and web S is non-ergodic, meaning that $1 < \mu_S < 2$, the cross-correlation of the two signals is zero. This is the lower plateau region of the cross-correlation cube. The general argument for the shape of the cross-correlation cube is given in full detail elsewhere [6, 68].

   This description of the perturbation and response of interacting complex webs suggests what happens during dishabituation. The simple stimulus used in this section is analogous to the ergodic perturbation and the effective synaptic weight is analogous to the non-ergodic response, yielding an asymptotic response in the lower plateau. The near-zero habituated response to the ergodic perturbation jumps to full strength, the upper plateau, when a disruptive non-ergodic short-time perturbation is turned on. This resetting of the effective synaptic weight is followed by the continuing ergodic perturbation and the subsequent habituation of the revitalized effective synaptic weight.

7.4.3   Discussion

From the above arguments we see that aging, an unavoidable consequence of the event probability density not having a finite average waiting time, that is, $\mu < 2$, has the effect of suppressing the asymptotic response of the complex network of neurons to coherent perturbations and consequently yields a habituated response. The intuitive explanation of this effect is that a complex web, one described by a hyperbolic distribution such as the neural web, generates signals through the interaction of many distinct neurons. These neurons without characteristic frequencies are coupled by nonlinear interactions that inhibit the formation of normal modes that would facilitate information transfer in linear single-channel webs. Consequently, the time scales in these complex neuron webs are hierarchally coupled in such a way as to promote their interdependence and synchronization over long times [68]. An excitation at a given time scale generates a cascade of perturbing multiple time scales through the multiple-channel coupling, and, because the time scale of the stimulus is only one of many, the response at that time scale is suppressed over time. However, whether the asymptotic state to which the stimulus is habituated is zero or some constant value depends on the particular kind of $1/f$ scaling of that region of the brain. Consequently, depending on whether we hear, taste, touch or smell the stimulus determines the ergodic nature of the complex neuronal webs and consequently the form of habituation.

The complexity of the time series measured herein in terms of the power-law index has an interesting interpretation. As the index approaches infinity the statistics of the process become Poisson and the arrival of new events is more or less evenly spaced in time. However, in the region of psychological interest considered here, $1 < \mu < 3$, the critical events are intermittent, occurring in clusters and having clusters within clusters, manifesting a fractal statistical structure in time. Consequently, there is no characteristic time scale in the propagation of information in neuron webs. This needs to be compared with the critical branching processes used by Beggs *et al.* [11, 53] to explain the "neuronal avalanches" measured in cortical neurons. The former effect is in time and the latter is in space. Preliminary calculations extending the decision-making model [64] to include spatial effects reveal localized "communities" with an inverse power-law distribution of sizes with slope -1, compared with the value -1.5 obtained by Beggs and Plenz [11]. These results suggest that the "neuronal avalanches," a spatial fractal effect, may be explained using either the critical branching processes [11] or the critical events introduced herein [33, 64]. However, determining the compatibility of the two approaches is an area for future investigation and is outside the scope of this book.

It should be pointed out that the habituation damping factor (7.169) is obtained using a generalization of LRT [4, 7, 8], which is one of the most robust techniques of statistical physics and follows from the fluctuation–dissipation theorem. Other researchers [35, 59, 60, 67] have adopted an approach closely related to that of our research group [4, 7, 8] and their approach is based on expressing the linear response function as $\chi(t, t') = d\Psi(t, t')/dt'$ rather than $\chi(t, t') = -d\Psi(t, t')/dt$. It is possible to show that either choice generates habituation, although it seems that networks of interacting neurons adopt the choice we made (7.173). The asymptotic convergence to zero of the response to a periodic input signal has been interpreted by physicists [35, 59, 60, 67] as the failure of LRT and in fact has been called the "death of linear response." It was expected that because the stimulus remains constant in amplitude the response should do the same rather than asymptotically vanish; when this did not occur the theory was thought to be "dead." LRT is as applicable to neural phenomena as it is to complex physical webs when suitably generalized and the so-called death of linear response is a consequence of the web's complexity.

In the critical-branching theory [11] information transfer in the web is determined by the magnitude of the branching parameter, independently of the properties of the stimulus. By contrast, in SHM the relative complexity of the web dynamics and excitation determines the efficiency of information transfer. We are convinced that the synchronization of the global fluctuations of neuron webs with a stimulus having the same degree of complexity [33, 68], the measure of complexity being the power-law index, establishes that LRT is not dead. In fact, a complex stimulus may trigger dishabituation by forcing the response from a near-zero value to one equal to its initial strength. Herein this counter-intuitive response of complex neuron webs to coherent stimuli may explain the phenomena of habituation and dishabituation.

7.5 Summary

In this chapter we investigated what is cause and what is effect in the many distributions and models discussed in the book. For example, the metrics reviewed in the previous chapter quantify some web properties resulting from web topology, whether the web is random or small-world, or any of the other choices available, and consequently the topology is interpreted as causal. We showed that the topology can be a consequence of the underlying dynamical interaction among the elements of the network and therefore the dynamics of the web are causal, not the topology, as is often assumed in the literature. The decision-making model reviewed in the present chapter uses master equations to combine uncertainty through the use of two-state probabilities for the individual elements and dynamics through the nonlinear time-dependent coupling of these probabilities. This model is shown to manifest phase transitions in which the apparently free-running behavior of the individual elements becomes synchronized, switching between the two states as a single collective unit. This consensus of opinion is seen to result in a distribution of state changes that is an inverse power law for a number of decades in time, followed by exponential behavior. The extent of the inverse power law in time depends on the number of elements in the web, as does the size of the fluctuations. In fact the fluctuations observed in the web are produced by the web having a finite number of nodes rather than by thermal excitations as they are in physical webs.

The decision-making model was used in the demonstrations of a number of general principles associated with complex webs. One was a generalization of the Onsager principle in which the traditional exponential relaxation of a spontaneous fluctuation in a physical web is replaced by the inverse power-law relaxation of a spontaneous fluctuation in complex webs. Another was the generalization of linear response theory (LRT) of non-equilibrium statistical physics, where the historical basis of LRT, namely continuous dynamical trajectories, is replaced with the discrete dynamics of critical events and renewal statistics. The familiar phenomenon of stochastic resonance was shown to be a consequence of this new LRT, as are various microscopic phenomena [5]. A final application of LRT was in the explanation of the psychophysical phenomenon of habituation, through the statistical habituation model (SHM) where a simple stimulus is shown to be attenuated as an inverse power law in time by a complex web. The form of the stimulus, whether it is heard, tasted, touched or smelled, determines the ergodic nature of the complex neuronal webs and consequently the form of habituation in the

SHM. Finally, we have an explanation as to why the uninterrupted drone of the lecturer puts even the most dedicated student to sleep.

7.6   Problems

7.1 Stochastic resonance
Show that (7.100) yields a curve that agrees qualitatively with Figure 7.6.

7.2 A harmonic oscillator in a heat bath
Consider the situation in which a simple harmonic oscillator is in contact with a heat bath so that the FP operator is determined by the potential function $U(q) = \frac{1}{2}\omega^2 q^2$. The solution to the eigenvalue equation allows an exact evaluation of the sum in (3.134) to obtain a closed-form solution to the FPE. Obtain this solution for the case in which the oscillator is initially displaced to q_0.

7.3 Linear response
It is interesting to notice that it is possible to establish a connection between the results of this and the preceding section, by assuming the web variable to be $Q(t)$ and the perturbed variable too to be $Q(t)$ so that (7.114) becomes

$$\langle Q(t) \rangle = \beta \int_0^t \left\langle Q(t)\dot{Q}(t') \right\rangle E(t')dt'.$$

Use the equipartition theorem (7.126) and the chain condition on derivatives to obtain

$$\left\langle Q(t)\dot{Q}(t') \right\rangle = \frac{d}{dt'}\langle Q(t)Q(t') \rangle = \gamma \langle Q^2 \rangle e^{-\gamma(t-t')}$$

and show that this yields

$$\langle Q(t) \rangle = \int_0^t e^{-\gamma(t-t')} \frac{\gamma}{\omega^2} E(t')dt',$$

which, using the definition (7.129), allows us to recover (7.132).

7.4 Habituation to arbitrary stimulus
The behavioral response to a single frequency is the basis for determining the response to a much richer stimulus using the principle of superposition for linear processes. For example, consider the road noise from the Interstate highway outside the window of your motel room, or the gasping air conditioning unit below that window. These signals, acoustic, olfactory, tactile and of many other types, can be represented as the superposition of M modes

$$S(t) = \sum_{k=1}^{M} A_k \cos(\omega_k t + \phi_k), \tag{7.186}$$

with real amplitudes A_k, phases ϕ_k and frequencies ω_k determined from experimental data. Redo the analysis in the text to show that the behavioral response to this stimulus is given by the exact asymptotic expression

$$w_{\text{eff}}(t) \approx \frac{1}{\Gamma(\mu-1)} \sum_{k=1}^{M} \frac{|A_k| \cos(\omega_k t + \phi_k - \mu\pi/2)}{(\omega_k t)^{2-\mu}}. \tag{7.187}$$

Graph this habituating response for a mathematical sequence of Gaussian pulses

$$|A_k| = \exp\left[-\omega_k^2/\sigma^2\right] \text{ and } \omega_k = \frac{2\pi}{T}k, \tag{7.188}$$

where T is a fundamental period and the frequencies are multiples of the fundamental frequency for ten modes.

References

[1] P. Allegrini, G. Aquino, P. Grigolini et al., "Correlation function and generalized master equation of any age," *Phys. Rev. E* **71**, 066109 (2005).

[2] P. Allegrini, F. Barbi, P. Grigolini and P. Paradisi, "Renewal, modulation and superstatistics," *Phys. Rev. E* **73**, 046136 (2006).

[3] P. Allegrini, V. Benci, P. Grigolini et al., "Compression and diffusion: a joint approach to detect complexity," *Chaos, Solitons & Fractals*, **15**, 517–535 (2003).

[4] P. Allegrini, M. Bologna, P. Grigolini and B. J. West, "Fluctuation–dissipation theorem for event-dominated processes," *Phys. Rev. Lett.* **99**, 010603 (2007).

[5] P. Allegrini, M. Bologna, L. Fronzoni, P. Grigolini and L. Silvestri, "Experimental quenching of harmonic stimuli: universality of linear response theory," *Phys. Rev. Lett.* **103**, 030602 (2009).

[6] G. Aquino, M. Bologna, P. Grigolini and B. J. West, "Beyond the death of linear response theory: $1/f$-optimal information transport," *Phys. Rev. Lett.* **105**, 040601 (2010).

[7] G. Aquino, P. Grigolini and B. J. West, "Linear response and fluctuation–dissipation theorem for non-Poissonian renewal processes," *Europhys. Lett.* **80**, 10002 (1–6) (2007).

[8] F. Barbi, M. Bologna and P. Grigolini, "Linear response to perturbation of nonexponential renewal processes," *Phys. Rev. Lett.* **95**, 22061 (2005).

[9] J. B. Bassingthwaighte, L. S. Liebovitch and B. J. West, *Fractal Physiology*, New York: Oxford University Press (1994).

[10] C. Beck, "Dynamical foundations of nonextensive statistical mechanics," *Phys. Rev. Lett.* **87**, 180601 (1–4) (2001).

[11] J. M. Beggs and D. Plenz, "Neuronal avalanches in neocortical circuits," *J. Neurosci.* **23**, 11167 (2003).

[12] R. Benzi, A. Sutera and A. Vulpiani, "The mechanism of stochastic resonance," *J. Phys. A* **14**, L453 (1981).

[13] R. Benzi, G. Parisi, A. Sutera and A. Vulpiani, "Stochastic resonance in climate change," *Tellus* **34**, 10 (1982).

[14] R. Benzi, "Stochastic resonance: from climate to biology," arXiv:nlin/0702008.

[15] S. Bianco, E. Geneston, P. Grigolini and M. Ignaccolo, "Renewal aging as emerging property of phase sychronization," *Physica A* **387**, 1387 (2008).

[16] M. Bologna and P. Grigolini, "Lévy density: the density versus trajectory approach," *J. Statist. Mech.* P03005 (2009).

[17] M. D. Breed, www.animalbehavioronline.com/habituation.html.

[18] H.-P. Breuer and B. Vacchini, "Structure of completely positive quantum master equation with memory kernel," *Phys. Rev. E* **79**, 041147 (2009).

[19] X. Brokmann, J.-P. Hermier, G. Messin *et al.*, "Statistical aging and nonergodicity in the fluorescence of single nanocrystals," *Phys. Rev. Lett.* **90**, 120601 (2003).

[20] T. E. Cohen, S. W. Kaplan, E. R. Kandel and R. D. Hawkins, "A simplified preparation for relating cellular events to behavior: mechanisms contributing to habituation, dishabituation, and sensitization of the Aplysia gill-withdrawal reflex," *J. Neurosci.* **17**, 2886–2899 (1997).

[21] F. Corberi, E. Lipiello and M. Zannetti, "Fluctuation dissipation relations far from equilibrium," *J. Statist. Mech.* P07002 (2007).

[22] J. Correll, "$1/f$-Noise and effort on implicit measures of bias," *J. Personal. and Social Psych.* **94**, 48 (2008).

[23] D. R. Cox, *Renewal Theory*, London: Methuen (1962).

[24] O. Déniz, J. Lorenzo and M. Hernández, "A simple habituation mechanism for perceptual user interfaces," *Inteligencia Artificial, Revista Iberoamericana de Inteligencia Artificial* **23**, 87 (2004).

[25] S. S. Deshmukh and U. S. Bhalla, "Representation of odor habituation and timing in the hippocampus," *J. Neurosci.* **23**, 1903–1915 (2003).

[26] V. Dragoi, "A feedforward model of suppressive and facilitatory habituation effects," *Biol. Cybern.* **86**, 419 (2002).

[27] P. J. Drew and L. F. Abbott, "Models and properties of power-law adaptation in neural systems," *J. Neurophysiol.* **96**, 826 (2006).

[28] W. Feller, *An Introduction to Probability Theory and Its Applications*, Vol. II, New York: John Wiley & Sons (1971).

[29] J. Feng, "Is the integrate-and-fire model good enough?," *Neural Networks* **14**, 955 (2001).

[30] L. Gammaitoni, P. Hänggi, P. Jung and F. Marchesoni, "Stochastic resonance," *Rev. Mod. Phys.* **70**, 223–287 (1998).

[31] D. L. Gilden, T. Thornton and M. W. Mallon, "$1/f$-Noise in human cognition," *Science* **267**, 1837 (1995).

[32] P. Gong, A. R. Nikolaev and C. van Leeuwen, "Intermittent dynamics underlying the intrinsic fluctuations of the collective synchronization patterns in electrocortical activity," *Phys. Rev.* **76**, 011904 (2007).

[33] P. Grigolini, G. Aquino, M. Bologna, M. Luković and B. J. West, "A theory of $1/f$-noise in human cognition," *Physica A* **388**, 4129–4204 (2009).

[34] C. Haldeman and J. M. Beggs, "Critical branching captures activity in living neural networks and maximizes the number of metastable states," *Phys. Rev. Lett.* **94**, 058101 (2005).

[35] E. Heinsalu, M. Patriarca, I. Goychuk and P. Hänggi, "Use and abuse of a fractional Fokker–Planck dynamics for time-dependent driving," *Phys. Rev. Lett.* **99**, 120602 (2007).

[36] C. T. Kello, B. C. Beltz, J. G. Holden and G. C. Van Orden, "The emergent coordination of cognitive function," *J. Expl. Psych.: Gen.* **136**, 551–568 (2007).

[37] H. A. Kramers, "Brownian motion in a field of force and the diffusion model of chemical reactions," *Physica A* **7**, 284–340 (1940).

[38] R. Kubo, "A general expression for the conductivity tensor," *Canadian J. Phys.* **34**, 1274–1277 (1956).

[39] Q. S. Li and Y. Liu, "Enhancement and sustainment of internal stochastic resonance in unidirectional coupled neural systems," *Phys. Rev. E* **73**, 016218 (2006).

[40] M. H. Lee, "Can the velocity autocorrelation function decay exponentially?," *Phys. Rev. Lett.* **51**, 1277 (1983).

[41] M. Luković and P. Grigolini, "Power spectra for both interrupted and perennial aging processes," *J. Chem. Phys.* **129**, 184102 (2008).

[42] M. Luković, L. Fronzoni, M. Ignaccolo and P. Grigolini, "The rate matching effect: a hidden property of aperiodic stochastic resonance," *Phys. Lett. A* **372**, 2608–2613 (2008).

[43] M. Magdziarz, A. Weron and K. Weron, "Fractional Fokker–Planck dynamical stochastic representation and computer simulation," *Phys. Rev. E* **75**, 016708 (2007).

[44] M. Magdziarz, A. Weron and J. Klafter, "Equivalence of the fractional Fokker–Planck equation and the subordinated Langevin equation: the case of a time-dependent force," *Phys. Rev. Lett.* **101**, 210601 (2008).

[45] G. Margolin and E. Barkai, "Nonergodicity of a time series obeying Lévy statistics," *J. Statist. Phys.* **122**, 137–167 (2006); G. Margolin and E. Barkai, "Nonergodicity of blinking nanocrystals and other Lévy-walk processes," *Phys. Rev. Lett.* **94**, 080601 (2005).

[46] B. McNamara and K. Wiesenfeld, "Theory of stochastic resonance," *Phys. Rev. A* **39**, 4854 (1989).

[47] C. Nicolis and G. Nicolis, "Is there a climate attractor?," *Nature* **34**, 529–532 (1984).

[48] A. Neiman, L. Schimansky-Reier, A. Cornall-Bell and F. Moss, "Noise-enhanced phase synchronization in excitable media," *Phys. Rev. Lett.* **83**, 4896–4899 (1999).

[49] M. E. J. Newman, "Power laws, Pareto distributions and Zipf's law," *Contemp. Phys.* **46**, 323–351 (2005).

[50] A. Nordsieck, W. W. Lamb and G. E. Uhlenbeck, "On the theory of cosmic-ray showers: I. The Furry model and the fluctuation problem," *Physica* **7**, 344 (1940).

[51] I. Oppenheim, K. E. Shuler and G. H. Weiss, *Stochastic Processes in Chemical Physics: The Master Equation*, Cambridge, MA: MIT Press (1977).

[52] M. Perc, "Stochastic resonance on weakly paced scale-free networks," *Phys. Rev. E* **78**, 036105 (2008).

[53] D. Plenz and T. C. Thiagarajan, "The organizing principle of neuronal avalanche activity: cell assemblies in the cortex?," *Trends Neurosci.* **30**, 101 (2007).

[54] A. Rebenshtok and E. Barkai, "Weakly nonergodic statistical physics," *J. Statist. Phys.* **133**, 565 (2008).

[55] L. E. Reichl, *A Modern Course in Statistical Physics*, Austin, TX: University of Texas Press (1980).

[56] B. R. Sheth, J. Sharma, S. C. Rao and M. Sur, "Orientation maps of subjective contours in visual cortex," *Science* **274**, 2110 (1996).

[57] L. Silvestri, L. Fronzoni, P. Grigolini and P. Allegrini, "Event-driven power-law relaxation in weak turbulence," *Phys. Rev. Lett.* **102**, 014502 (1–4) (2009).

[58] H. A. Simon, "On a class of skew distribution functions," *Biometrika* **42**, 425–440 (1935).

[59] I. M. Sokolov, "Linear response to perturbation of nonexponential renewal process: a generalized master equation approach," *Phys. Rev. E.* **73**, 067102 (2006).

[60] I. M. Sokolov and J. Klafter, "Field-induced dispersion and subdiffusion," *Phys. Rev. Lett.* **97**, 140602 (2006).

[61] I. M. Sokolov and J. Klafter, "Continuous time random walks in an oscillating field: field-induced dispersion and the death of linear response," *Chaos, Solitons and Fractals* **34**, 81–86 (2007).

[62] L. Silvestri, L. Fronzoni, P. Grigolini and P. Allegrini, "Event driven power-law relaxation in weak turbulence," *Phys. Rev. Lett.* **102**, 014502 (2009).

[63] R. G. Turcott, P. D. Barker and M. C. Teich, "Long-duration correlation in the sequence of action potentials in an insect visual interneuron," *J. Statist. Comput. & Simul.* **52**, 253–271 (1995).

[64] M. Turalska, M. Luković, B. J. West and P. Grigolini, "Complexity and synchronization," *Phys. Rev. E* **80**, 021110-1 (2009).

[65] D. Vitali and P. Grigolini, "Subdynamics, Fokker–Planck equation, and exponential decay of relaxation processes," *Phys. Rev. A* **39**, 1486–1499 (1989).

[66] L. M. Ward, "Physics of neural synchronization mediated by stochastic resonance," *Contemp. Phys.* **50**, 563–574 (2009).

[67] A. Weron, M. Magdziarz and K. Weron, "Modeling of subdiffusion in space-time-dependent force fields beyond the fractional Fokker–Planck equation," *Phys. Rev. E* **77**, 036704 (2008).

[68] B. J. West, E. L. Geneston and P. Grigolini, "Maximizing information exchange between complex networks," *Phys. Rep.* **468**, 1–99 (2008).

[69] B. J. West and P. Grigolini, "Statistical habituation model and $1/f$ noise," *Physica A* doi: 10.1016/j.physa.2010.08.033 (2010).

[70] J. C. Willis and G. U. Yule, "Some statistics of evolution and geographical distribution in plants and animals, and their signature," *Nature* **109**, 177–179 (1922).

[71] G. U. Yule, "A mathematical theory of evolution based on the conclusions of Dr. J. C. Willis," *Philos. Trans. R. Soc. London B* **213**, 21–87 (1925).

[72] E. Zohary, M. N. Shadlen and W. T. Newsome, "Correlated neuronal discharge rate and its implications for psychophysical performance," *Nature* **370**, 140 (1994).

[73] R. Zwanzig, *Nonequilibrium Statistical Mechanics*, New York: Oxford University Press (2001).

8 Synopsis

There is great value in being able to provide a brief insightful summary of an elaborate experiment, a complicated theory or a multi-faceted discussion. Suppose one of us gets on an elevator and the only other passenger is a four-star general. She smiles and asks "How is your research going?" This is one of those rare opportunities when a scientist can actually influence policy, resource allocation and the direction of science by being able to communicate effectively. Almost invariably the opportunity is lost because scientists and generals rarely speak the same language. However, there is a trick that one can employ to anticipate this rare event and that is to have something worked out in advance so as not to waste the opportunity. This is the "elevator description" of the most significant research that has been done. In another age it might have been called the "*Reader's Digest*" version.

That is the position the authors find themselves in now. We believe that we ought to provide a brief but insightful summary of the book you have just completed in a way that conveys the maximum amount of information, but without the mathematics that was necessary to make that information understandable when it was first presented. To accomplish this goal we review the high points of each of the chapters and then attempt to tie them all together into a coherent picture.

In the first chapter we argued that normal statistics do not describe complex webs. Phenomena described by normal statistics can be discussed in terms of averages. The fact that human populations are well represented by normal distributions of heights strongly influences the manufacturing of everything from the size of shoes, shirts and slacks to cars, computers and couches. The relative numbers of shirts in the small, medium and large sizes are determined by our knowledge of the average build of individuals in the population. The distribution of sizes in the manufactured shirts must match the population of buyers or the shirt factory will soon be out of business. Thus, the industrial revolution was poised to embrace the world according to Gauss and the mechanistic society of the last two centuries flourished. But as the connectivity of the various webs within society became more complex the normal distribution receded further and further into the background until it was completely gone from the data, if not from our attempted understanding.

The world of Gauss is attractive, in part, because it leads to consensus. A process that has typical outcomes, real or imagined, enables people to focus attention on one or a few quantities and, through discussion, decide what is important about the process. Physicians do this in deciding whether or not we are healthy. They measure our heart rate and blood pressure and perhaps observe us walking; all to determine whether these measurements fall within the "normal" range. The natural variability of these physiologic quantities has in the past only caused confusion and so their variation is ignored. A similar situation occurs when we try to get a salary increase and the boss brings out the charts to show that we already have the average salary for our position and maybe even a little more. Stephen Jay Gould had a gift for language and in his book *Full House* [4] he put it this way:

... our culture encodes a strong bias either to neglect or ignore variation. We tend to focus instead on measures of central tendency, and as a result we make some terrible mistakes, often with considerable practical importance.

The normal distribution is based on one view of knowledge, a view embracing the notion that there is a best outcome from an experiment, namely an outcome that is predictable and given by the average value. The natural uncertainty was explained by means of the maximum-entropy argument, where the average provides the best representation of the data and the variation is produced by the environment having the maximum entropy consistent with the average. But this interesting justification for normalcy turned out to be irrelevant, because normal statistics did not appear in any interesting data, not even in the distribution of grades in school. We did look at the distributions from some typical complex phenomena: the distribution in the number of scientific papers published, the frequency of words used in languages, the populations of cities and the metabolic rates of mammals and birds. The distributions for these complex webs were all described by an inverse power law, which from a maximum-entropy argument showed that the relevant property was the scale-free nature of the dynamical variable.

The center-stage replacement of the normal distribution with the hyperbolic distribution implies that it is the extremes of the data which dominate complex webs, not the central tendency. Consequently, our focus was shifted from the average value and the standard deviation to the variability of the process being investigated, a variability such that the standard deviation diverges. Of particular significance was the difference in the predictability of the properties of such networks, in particular their failure characteristics. No longer does the survival probability of an element or a network have the relatively rapid decline of an exponential, but instead failure has a heavy tail that extends far from the central region. Some examples were given to show the consequences of this heavy tail: the durations of power outages have this property, as do the sizes of forest fires and many other phenomena that threaten society. The why of this observation comes later.

If Chapter 1 is the embarkation point from the world of Gauss to the world of Pareto then Chapter 2 is the introduction to the ubiquity of the inverse power-law distribution in the physical, social and life sciences together with its interpretation in terms of fractals according to Mandelbrot [5]. The explanation of scaling in the observables describing these complex webs spanned the centuries from the geophysics and physiology of da Vinci to the economics of Montroll and Shlesinger [6] and the bronchial lungs of West *et al.* [8]. We also found that there are two different kinds of distributions that scale: those that follow the law of Zipf, where the relative frequency is an inverse power law in the rank of the quantity, say city size or word usage; and those that follow the law of Pareto, where the relative frequency is an inverse power law in the amount of the quantity, say income level or the grade obtained in a science class. Moreover, the two kinds of distribution are formally related and describe the scaling properties of a wide variety of physiologic networks, resulting in fractal heartbeats, fractal breaths and fractal steps.

In Chapter 2 we list nearly fifty complex webs having hyperbolic distributions, residing in disciplines from anthropology to sociology and including information theory. This list is intended to demonstrate how creative scientists have been in representing the complexity of webs in a variety of different ways that capture the intrinsic variability of the underlying dynamics. We mentioned the relative frequency in terms of rank or magnitude of a random variable, but there are also distributions in the time interval between events such as earthquakes and solar flares, the power spectrum in terms of frequency such as in music and DNA sequences, and even psychophysical distributions in terms of the stimulus intensity or the number of trials. Each distribution is a testament to the fact that the complexity of the web does not lend itself to interpretation using normal statistics. For example, there is no one number, such as the average reaction time, that can be used to characterize learning using the method of trial and error. The entire inverse power-law distribution is necessary to capture this psychophysical phenomenon; a similar effect was also observed in the cardiovascular, respiratory and motor-control webs.

In this chapter we also discussed the difference between the observed variation of the data and that which results in normal statistics. An ever-expanding set of data is accumulated with holes between regions of activity and whose overall size is determined by the inverse power-law index. The values of the web variable out in the tail of the distribution dominate the behavior of the process. In the spatial domain the inverse power law is manifest through data points that resemble clumps of sheep grazing in a meadow as opposed to the regular spacing of trees in an orchard. In the time domain the heavy-tailed phenomena give rise to a pattern of intermittency. However, even in regions of activity, magnification reveals irregularly spaced gaps between times of activity having the same intermittent behavior down to smaller and smaller time scales.

The purpose of scientific theory is to understand natural phenomena through prediction and testing, not to just describe what is observed. This is part of what made the normal distribution so attractive; even though the future could not be predicted with absolute certainty, a most probable future in terms of the average could be forecast. The quality of the forecast could then be estimated using the standard deviation or width of the distribution. As long as two successive predictions were not too far apart this seemed like a good strategy. However, when patients with heart rates in the normal range have heart attacks and die, one begins to wonder [9]. In the first two chapters the properties of data are discussed, as are methods for data analysis and description. Chapter 3 begins the description of the dynamics of webs that could possibly explain the properties of the data presented earlier and, since we should walk before we run, the beginning formalism of the dynamics is primarily linear. We started with simple linear dynamical webs and systematically increased the complexity of the web to capture various of the complicated properties observed.

The linear dynamics of Hamiltonian webs were made more complex by coupling the web of interest to the environment. The early models of classical diffusion were constructed in this way and Brownian motion was successfully explained. The dynamics were made incrementally more complex by introducing memory into the fluctuations produced by the environment, and of particular interest was the inverse power-law memory and the response of the dynamical web to such fluctuations. In this way the workhorse of statistical physics, linear response theory (LRT), was introduced, laying the groundwork for its use in nonlinear webs whose dynamics could be described perturbatively. An exemplar of such a nonlinear dynamical network is the double-well potential and Kramers' estimate of the transition rate from one well to the other. This example was used to explain the phenomenon of stochastic resonance, that is, whether a relatively small-amplitude sinusoidal signal can be amplified by internal nonlinear interactions in the presence of noise. This was subsequently shown to be the harbinger of the more general phenomenon of complexity-matching discovered by the authors [1, 3] and taken up in later chapters.

Of course, stochastic dynamical equations are only one way to describe a class of complex webs. Another is to replace the dynamical variable with a phase-space variable and the dynamical equations by a phase-space equation for the probability density, namely the Fokker–Planck equation (FPE). One particularly important solution to an FPE is the Gaussian probability distribution that results from a linear dynamical equation with additive random fluctuations (white noise). Such a strategy with the appropriate boundary conditions produced the drift-diffusion model (DDM) of decision-making, which is one of the most popular models of decision-making. We review the DDM in order to set the stage for the more general decision-making model we developed [7] involving the dynamical interaction among members of a nonlinear web.

An important asymptotic solution to a particular FPE is the inverse power law resulting from a process in which the fluctuations are multiplicative, not additive. The distribution is normalizable, but it has a diverging first moment when the power-law index is in the interval $1 < \mu < 2$. For a power-law index in this interval the statistics of the underlying multiplicative dynamical process are not ergodic.

The most significant concept introduced in Chapter 3 is the replacement of the continuous trajectories of Hamiltonian dynamics with the dynamics of discrete events. The dynamics of these events is described using renewal theory and the formalism was originally developed to solve problems in operations research. We show that this formalism gives rise to hyperbolic distributions of events, rather than the probability density of ensembles of trajectories that solve the FPE. It is natural that such human activities as decision-making lend themselves more readily to the uncertainty associated with probabilities than to the predictability of the trajectory description. The details of this difference were introduced in the next chapter.

The way physical scientists are trained to understand randomness is through random-walk theory and its generalizations. This is the starting point of Chapter 4, where the briefest of histories of random walks is used to introduce the notion of a random web. The randomness of the web provides a readily understood paradigm against which to test the more mathematically sophisticated theories that are subsequently brought to bear on understanding complex webs. The Poisson statistics of the graphical description of a web are like the normal statistics discussed earlier in that, although they do not describe any empirical data from real complex webs, they do provide insight into the inverse power law that replaces normal statistics. The random walk provides a framework in which to examine the influence of long-time memory and spatial inhomogeneity by modifying transition rates and studying how the resulting distributions are modified. In this way Lévy flights are determined to be the solution to fractional differential equations in phase space and inverse power-law spectra are determined to be the solution to fractional random walks. The fractional calculus is found to be a useful tool in modeling the potential influence of what occurred long ago and far away on what might happen in the here and now. The non-normal behavior of the world we experience is due to the fact that simple differential equations, even stochastic ones, are not sufficient to determine our future. Not all events from the past influence what we say and do in the same way. Some occurred very long ago but still burn bright, whereas what occurred yesterday has already disappeared into the forgettory. This is discussed more fully in a later chapter.

The linear dynamics from Chapter 3 were known to be inadequate for describing complex webs even when they were first discussed. However, many of the concepts introduced there bear fruit in the latter part of Chapter 4 through the notion of chaos. The notions of randomness and chaos are closely linked, particularly how the chaos resulting from the breakup of Hamiltonian orbits, nonlinear maps and strange attractors leads to unpredictability in dynamical networks. The non-integrability of Hamiltonian systems was used to define a booster, the source of statistical fluctuations in a web of interest. The discussion of the relationship between chaos and random fluctuations in stochastic equations of motion might seem like mathematical minutiae, but its importance in providing a mechanical basis for understanding randomness in physical networks cannot be over-stated. If, on the other hand, you are content with the idea that we do not have complete information about any complex web, so that in any dynamical model of the web we should always include a little randomness, that acceptance will work as well. However, one must exercise caution in introducing randomness into dynamical equations, because the way uncertainty enters the description can modify the predicted outcome. Recall that additive fluctuation led to the Gauss distribution whereas multiplicative fluctuations gave rise to the inverse power law.

In Chapter 5 we describe the dynamics of complex webs in terms of non-analytic functions. Functions of the kind that are continuous everywhere but nowhere differentiable were invented by Weierstrass and were subsequently shown to have a fractal dimension. We discussed non-analytic functions earlier in the context of fractal processes, but in Chapter 5 we showed that the fractional calculus is required in order to describe the underlying dynamics of the fractal processes described by these functions [10]. The fractional calculus was briefly reviewed and used to describe deterministic phenomena whose dynamics are non-local in time, that is, processes that depend on memory. The solutions to dissipative fractional rate equations were shown to be Mittag-Leffler functions, a Mittag-Leffler function being one that is a stretched exponential at early times and an inverse power law asymptotically. This behavior was able to generalize the Poisson distribution in which the exponential is replaced with the

Mittag-Leffler function. The solutions to fractional dissipative stochastic equations are shown to scale, that is, to display anomalous diffusion, due to this dynamical memory. More importantly, the multifractal character of certain complex physiologic webs, including migraines and cerebral blood flow, is successfully modeled with the derivative index of such fractional equations being a random variable [11].

It is true that what occurs now and in the near future may be significantly influenced by what happened in the distant past and fractional time derivatives are one way to model that influence. However, when we look at such things as earthquakes, the natural habitat of ecological webs, or the terrain over which wars are fought, one is struck by how the geography influences the propagation of information. Nothing moves in a straight line for any substantial distance, but zigs and zags to accommodate the hills and valleys. Random fields and the fractional propagation of influence on those fields are also discussed in Chapter 5, as is the first successful quantification of phenomena with diverging central moments, through continuous-time random walks (CTRWs). This theory has emerged as a systematic way to generalize the FPE to fractional form and to account for the bursting behavior in time and the clustering in space observed in many complex phenomena [10].

The promise of finding a new kind of science from the understanding of complex webs has motivated and led a great deal of the research in complex webs over the last decade. The literature on networks is much too vast for us to provide a review in one brief chapter, so what we attempted to do in Chapter 6 was to show how some of the major research themes fit into the flow of the general arguments associated with the empirical hyperbolic distributions and their mathematical descriptions in the first five chapters. The starting point is the easiest to understand since the paradigm was the random network in which the statistics of the connectivity of the members of the web are described by a Poisson distribution. The development of small-world theory and the preferential attachment used to model the growth of networks was shown to be closely related to certain historical ideas on how to understand the hyperbolic behavior of complex webs. Various metrics developed in the context of networks were reviewed and their values, such as the web size and clustering coefficient, for real webs were compared with those from a random web. One of the important applications of the scale-free modeling is the prediction of the distribution of failures both locally and catastrophically, with the discovery that inverse power-law distributions are robust against random attacks, but fragile with respect to attacks directed at hubs. The assertion made at the turn of the century that this might be a property of the Internet explains in part the subsequent explosion of interest in the science of networks. The terrorist attack of 9/11 also provides a partial explanation, because of the need to identify, disrupt and destroy terrorist webs.

Chapter 6 also introduced the idea of a deterministic web in which the complex topology gave rise to the observed scaling behavior, with no statistics involved. This geometric complexity relates back to the earlier discussion on fractals and non-analytic functions.

In Chapter 7 we attempted to determine what is cause and what is effect in the many distributions and models introduced in the book. For example, the metrics reviewed in

Chapter 6 quantify certain web properties and are a consequence of the web topology, whether the web is random or small-world, or any of the other choices available, and consequently the topology is interpreted as causal. We showed that the topology can be a consequence of the underlying dynamical interaction among the elements of the network and therefore it is the dynamics of the web which are causal, not the topology. Of course, we could have parsed the discussion into the multiple causes identified and discussed by Aristotle, but we did not. The decision-making model reviewed in Chapter 7 uses the master equation to combine uncertainty through the use of two-state probabilities for the individual elements and dynamics through the nonlinear time-dependent coupling of these probabilities in time. This model is shown to manifest phase transitions in which the apparently free-running behaviors of the individual nodes become synchronized, switching between the two states collectively. This consensus of opinion is seen to result in a distribution of state changes that is an inverse power law for a number of decades in time followed by exponential behavior. The extent of the inverse power law in time depends on the number of elements in the web, as does the size of the fluctuations. In fact the fluctuations observed in the web are produced by the web having a finite number of nodes rather than by thermal excitations as they are in physical networks.

The decision-making model was used in the demonstrations of a number of general principles associated with complex webs. One was a generalization of the Onsager principle in which the traditional exponential relaxation of a spontaneous fluctuation in a physical process is replaced by the inverse power-law relaxation of a spontaneous fluctuation in complex webs. Another was the generalization of LRT of non-equilibrium statistical physics, where the historical basis of LRT, namely continuous dynamical trajectories, was replaced with the discrete dynamics of critical events and renewal statistics. The familiar phenomenon of stochastic resonance was shown to be a consequence of this new LRT, as were various microscopic phenomena [2]. A final application of LRT was in the explanation of the psychophysical phenomenon of habituation, through the statistical habituation model (SHM) whereby a simple stimulus is shown to be attenuated as an inverse power law in time by a complex web. The form of the stimulus, whether it is heard, tasted, touched or smelled, determines the ergodic nature of the complex neuronal webs and consequently the form of habituation in the SHM. Finally, we have an explanation as to why the uninterrupted drone of the lecturer puts even the most dedicated student to sleep.

References

[1] P. Allegrini, M. Bologna, P. Grigolini and B. J. West, "Fluctuation–dissipation theorem for event-dominated processes," *Phys. Rev. Lett.* **99**, 010603 (2007).
[2] P. Allegrini, M. Bologna, L. Fronzoni, P. Grigolini and L. Silvestri, "Experimental quenching of harmonic stimuli: universality of linear response theory," *Phys. Rev. Lett.* **103**, 030602 (2009).

[3] G. Aquino, P. Grigolini and B. J. West, "Linear response and fluctuation–dissipation theorem for non-Poissonian renewal processes," *Europhys. Lett.* **80** 10002 (1–6) (2007).

[4] S. J. Gould, *Full House*, New York: Three Rivers Press, Random House (1996).

[5] B. B. Mandelbrot, *Fractals, Form and Chance*, San Francisco, CA: W. F. Freeman (1977); *The Fractal Geometry of Nature*, San Francisco, CA: W. J. Freeman (1983).

[6] E. W. Montroll and M. F. Shlesinger, "On $1/f$ noise and other distributions with long tails," *PNAS* **79**, 337 (1982).

[7] M. Turalska, M. Luković, B. J. West and P. Grigolini, "Complexity and synchronization," *Phys. Rev. E* **80**, 021110-1 (2009).

[8] B. J. West, V. Bhargava and A. L. Goldberger, "Beyond the principle of similtude: renormalization in the bronchial tree," *J. Appl. Physiol.* **60**, 1089–1098 (1986).

[9] B. J. West, *Where Medicine Went Wrong*, Singapore: World Scientific (2006).

[10] B. J. West, M. Bologna and P. Grigolini, *Physics of Fractal Operators*, New York: Springer (2004).

[11] B. J. West, "Fractals, physiology, complexity and the fractional calculus," in *Fractals, Diffusion, and Relaxation in Disordered Complex Systems*, eds. W. T. Coffey and Y. P. Kalmykov, New York: Wiley-Interscience (2006), pp. 1–92.

Index

bone
strength, 28
booster, 208
microscopic, 211
prototype, 213
boundary conditions, 112
brain, 304
human, 61
brain dynamics, 331
branches
dendrite, 61
breath
variability, 98
Brenner, 288
Brown, 118, 166
Brownian
motion, 118, 167
deterministic, 209
particle, 115, 121, 126, 145
BRV, 93, 96
Bury, 237
Busemeyer, 142

C. elegans, 289
calculus
fractional, 62, 228
variational, 14
Caldarelli, 271
Calder, 28
Cannon, 72
canonical representation, 112
canonically conjugate variables,
204
Cantor, 68
Cantor set, 68
Cauchy, 230
cells
Aplysai
ganglion, 50
ganglion, 49
neurons, 49
central moment, 157
channel
information, 34
chaos, 167, 168
control, 85
map, 198
network, 209
characteristic
time scale, 209
characteristic function, 138, 178, 256
Chen, 302
climate, 327
CLT, 219, 225
cluster, 1, 263
coefficient, 283

giant, 281
clusters of clusters, 178
coarse-grained, 149
cognition
noise, 346
Cohen, 302
collective variable
synchronization, 314
Collins, 96
communicate, 357
communication
efficiency, 286
complex
phenomena, 224
web, 1, 134, 190
geophysical, 247
scale-free, 268
connectedness, 264
connections, 266
connectivity, 294
Conrad, 196, 219
consensus, 303
imperfect, 312
perfect, 312
conservative
dynamics, 190
network, 107
constants of motion, 203
control, 152
convergent expansion, 184
convolution, 321
theorem, 321
cooperation, 312
correlation
velocity, 228
correlation coefficient, 148
Costa, 32
counting process, 234
coupling
all-to-all, 263
crack, 155
initiation, 237
cross-covariance, 334
cross-level effects, 220
crucial events, 157, 319
Crutchfield, 202
CTRW, 250, 254, 320,
321, 362
cumulative advantage, 265
customer lines, 274
cycles, 114

da Vinci, 51, 95, 358
Darwin, 269
data
income, 21